WIND TURBINE TECHNOLOGY

Ahmad Hemami

Associate of
McGill University, Montreal, Canada
and
Iowa Lakes Community College, IA, USA

CENGAGE
Learning™

Australia • Brazil • Japan • Korea • Mexico • Singapore • Spain • United Kingdom • United States

Wind Turbine Technology

Ahmad Hemami

Vice President, Career and Professional Editorial: Dave Garza

Director of Learning Solutions: Sandy Clark

Executive Editor: David Boelio

Managing Editor: Larry Main

Senior Product Manager: Sharon Chambliss

Editorial Assistant: Jillian Borden

Vice President, Career and Professional Marketing: Jennifer Baker

Marketing Director: Deborah Yarnell

Marketing Manager: Kathryn Hall

Associate Marketing Manager: Mark Pierro

Production Director: Wendy Troeger

Production Manager: Mark Bernard

Content Project Manager: Cheri Plasse

Senior Art Director: Benjamin Gleeksman

Technology Project Manager: Christopher Catalina

For product information and technology assistance, contact us at
Professional Group Cengage Learning Customer & Sales Support, 1-800-354-9706
For permission to use material from this text or product,
submit all requests online at **cengage.com/permissions.**
Further permissions questions can be e-mailed to
permissionrequest@cengage.com.

Library of Congress Control Number: 2010934431

ISBN-13: 978-1-4354-8646-1

ISBN-10: 1-4354-8646-3

Cengage Learning
5 Maxwell Drive
Clifton Park, NY 12065-2919
USA

Cengage Learning is a leading provider of customized learning solutions with office locations around the globe, including Singapore, the United Kingdom, Australia, Mexico, Brazil and Japan. Locate your local office at:
international.cengage.com/region

Cengage Learning products are represented in Canada by Nelson Education, Ltd.

For your lifelong learning solutions, visit **delmar.cengage.com**
Visit our corporate website at **cengage.com.**

NOTICE TO THE READER

Publisher does not warrant or guarantee any of the products described herein or perform any independent analysis in connection with any of the product information contained herein. Publisher does not assume, and expressly disclaims, any obligation to obtain and include information other than that provided to it by the manufacturer. The reader is expressly warned to consider and adopt all safety precautions that might be indicated by the activities described herein and to avoid all potential hazards. By following the instructions contained herein, the reader willingly assumes all risks in connection with such instructions. The publisher makes no representations or warranties of any kind, including but not limited to, the warranties of fitness for particular purpose or merchantability, nor are any such representations implied with respect to the material set forth herein, and the publisher takes no responsibility with respect to such material. The publisher shall not be liable for any special, consequential, or exemplary damages resulting, in whole or part, from the readers' use of, or reliance upon, this material.

Printed in the United States of America
1 2 3 4 5 XX 11 10

Dedication

*This book is dedicated to my wife Shahla,
and to our daughter Mana and our son Pooya.*

Table of Contents

Preface

This book presents an introduction to wind energy and wind turbines. Its primary purpose is to serve as a comprehensive textbook for a course taken by first-year students of a 2-year or 4-year program, who normally do not have a prior introduction to the relevant electrical and mechanical subjects, essential to follow the new subject. It also intends to be a basic self-content resource for a person interested in enhancing his or her knowledge on the subject. The principles of wind energy technology and how wind turbines work are presented assuming no previous exposure of the reader to a number of pertinent subjects.

A wind turbine is a complex electromechanical system and in order to understand its fundamentals of operation the reader has to know certain matters in mechanics and in electricity. As for a high school graduate without such knowledge, or for a wind energy enthusiast, the associated materials are presented separately in each chapter. In this way, the mainstream is dedicated more to the newer subjects on the turbine technology, and a reader with prior knowledge of the pertinent material does not need to go through everything that is presented in a chapter. On the other hand, a reader who needs an enhancement or review of a preliminary topic has the opportunity to do that without seeking other resources. This makes it convenient for a student to gain sufficient background on the relevant material in order to comprehend the main topic.

Although a good knowledge of wind turbine technology calls for proficiency in a number of subjects, such as electricity and electrical devices, mechanics and dynamics, hydraulics, and control systems, this book avoids attempting to include all of these subjects and becoming a collection of too many topics.

Whereas in the beginning the subject is treated in a simpler manner with the least necessary mathematics involved, the relevant mathematical formulas are inevitable. With gradual advancement through the book more competence is expected from readers, especially from students, as this course will be taken in parallel with other technical courses. At the end of most chapters, a separate section titled Advanced Learning is included for those who would like a deeper insight to the subject or a selected topic of the chapter.

Although the metric system of measurements is the language of choice for wind turbine technology, the U.S. customary system of measurement is always used alongside the metric system.

Supplement

An Instructor Resources CD is available with the textbook. Components of the CD include an answer key to the end of chapter Review Questions, PowerPoint™ lecture slides that present the highlights of each chapter, an ExamView® computerized test bank, and an Image Library that includes an electronic copy of the images in the book.

Acknowledgments

The author and publisher wish to thank the following educators for their helpful and constructive remarks and recommendations on ways to improve the text.

Craig Evert, Iowa Lakes Community College
Donnin Custer, Western Iowa Tech Community College
George Lister, Texas State Technical College West Texas
Chad Kurdi, Dunwoody College of Technology
Lucas J. Chavey, Cloud County Community College

From the author:

Special thanks are due to Craig Evert and Lucas Chavey who provided photographs as well as gratitude to the individuals, companies, and institutions that contributed to the pictures used in the book. Also, I wish to thank the editors of the manuscript for their high-quality professional work. Special thanks goes to my wife for her continuous and true support.

Measurement units conversion

Length

1 ft = 0.3048 m
1 in. = 0.0254 m
1 m = 3.28084 ft
1 m = 39.37 in.
1 mile = 5820 ft
1 mile = 1609.34 m
1 m = 1000 mm
1 ft = 304.80 mm

Area

1 ft^2 = 0.0929 m^2
1 m^2 = 10.764 ft^2

Volume

1 ft^3 = 0.02832 m^3
1 m^3 = 35.3147 ft^3

Mass

1 lb = 0.453 kg (kilogram)
1 kg = 2.208 lb (pound)

Force

1 lbf = 4.444 N (Newton)
1 N = 0.2250 lbf (pound-force)

Speed (Velocity)

1 mile/hr (mph) = 1.4667 ft/sec
1 mile/hr (mph) = 0.4470 m/sec
1 ft/sec = 0.682 mile/hr (mph)
1 m/sec = 2.237 mile/hr (mph)
1 km/hr = 0.27778 m/sec
1 km/hr = 0.6214 mile/hr (mph)

Density

1 lb/ft^3 = 16.0185 kg/m^3
1 kg/m^3 = 0.06243 lb/ft^3

Pressure (or stress)

1 lbf/in^2 (psi) = 6888 Pa (Pascal)
    = 6.888 kPa
1 Pa = 0.000145 psi

Torque

1 lb-ft = 1.3545 N.m (Newton-meter)*
1 N.m = 0.73827 lb-ft*

Power

1 ft-lb/sec = 1.3545 W (watt)
1 W = 0.73827 ft-lb/sec
1 W = 1 N.m
1 hp (horsepower) = 550 ft-lb/sec
1 hp = 746 W

*Note that Newton-meter and lb-ft (also ft-lb) are the units for measurement of both *torque* and *work*.

Introduction

© Cengage Learning 2012

KEY TERMS

AWEA
dispatch system
distributed generation
  system
EWEA
horizontal axis
peak hours
power plant
smart grid
vertical axis
wind farm

OBJECTIVES

After studying this chapter, you will be able to:

- Talk more professionally about wind energy.
- Use a number of technical terms with confidence.
- Explain why renewable energy is very important.
- Determine how much greenhouse gas is generated for lighting our homes.
- Realize the progress on wind turbine technology in the past 30 years.
- Talk about wind farms and what they are.
- Explain why one cannot judge the quality of wind based on one single observation.
- Describe why good wind is not sufficient for developing a wind farm.
- Realize that stormy weather is not good for a wind turbine, and it does not mean more energy.

1.0   Introduction

Today's wind turbines are huge compared to those of even a decade ago, and the trend is toward manufacturing still larger machines. Although it may not appear so, wind turbines are complicated machines to control, particularly if high performance and good efficiency are needed. The safe and high performance of these machines is possible only through technological progress in control systems, electronics, communications, and the like, and their integration with the laws of mechanics that govern the behavior of such machines.

Understanding the rules of nature and the behavior of a wind turbine, and the ways its operation can be regulated as desired, is called "wind turbine technology." This is a subject that requires a number of types of specialized knowledge, which one needs to know in order to understand how wind turbines operate, to work on them, and to carry out further research and development on their functionality.

The material related to wind turbines is plentiful. During the past 30 years a lot of work has been carried out on the associated topics, all of which cannot be included in a monograph. There are a number of good books published on the subject, some of which are listed at the end of this chapter [References 1–8]. As expected, however, each book is written with a specific goal in mind and addresses a certain category of readers. This book addresses first-year college students with little or no previous background on a number of the necessary associated subjects. It is also meant to serve readers in other disciplines with little engineering background.

1.1   Historical background

Wind and water have long been two main sources of energy for humanity. Water mills and windmills have been used for applications such as wheat grinding and irrigation, where animals would otherwise have been used as the power source. Before the invention of steam engines and conversion of energy from burning coal and later oil, energy from wind was probably the only way long-distance sea transportation could be made possible.

vertical axis
Having a vertical axis of rotation.

horizontal axis
Having its axis of rotation almost horizontal.

Early reports and ruined remains of **vertical-axis** (its axis of rotation is vertical) Persian windmills can be traced back 1000 years. The famous Dutch windmill, with a **horizontal axis** (its axis of rotation is horizontal), existed since the twelfth century.

A very good description of the history and evolution of wind turbines from the ancient time machines to the modern wind turbines is given in Reference 8 and will not be pursued further here.

The new era of interest in wind turbines for generating electricity started in the 1970s. The oil crisis in 1972 was a good stimulation. European countries were more affected than other countries and placed full attention on this renewable source of energy. Up until 1992, commercial wind turbines were much smaller (225 kilowatts), but as a result of technological progress, by the end of 2002 turbine size had grown by a factor of 10. More and more inland and offshore wind farms were developed in European countries, and in North America attention to renewable energy was brought back.

With the installation of a record 6183 MW (megawatt) wind farm (a wind power generation facility where numerous wind turbines are installed) in 2005, wind energy achieved the European Commission's 40,000 MW target for 2010, 5 years ahead of time. In 2005 Denmark provided 20% of its total electric energy requirement from wind. Between 2000 and 2008 the wind industry in the United States showed a growth of about 29% per year, with a 5249-MW installation in 2007 and an 8500-MW installation in 2008. Although the industry, like other market segments, suffered in 2009, it is envisioned that wind can supply 20% of U.S. electricity by 2030.[1]

According to the European Wind Energy Association (**EWEA**), in 2009 a total of eight new wind farms consisting of 199 offshore wind turbines, with a combined power generating capacity of 577 MW, were connected to the grid in Europe.

AWEA
American Wind Energy Association

EWEA
European Wind Energy Association.

Importance of renewable energy   1.2

Earth is a marvelous marble with countless beauties in its nature, from plants, flowers, trees, birds, and so on to underwater wonders. Despite the abundance of its marvels, the Earth is limited in size and none of its contents can be unlimited. Except for sunshine, energy from the sun, that comes from outside, everything else is from the Earth itself. If we do not take care to use resources wisely and maintain the balance, all the Earth's marvels will come to an end. It is our responsibility as the inhabitants of this planet to keep it clean and in shape.

In ancient days when the number of people was small and they did not move around so much, pollution was limited and burning wood and coal did not lead to noticeable changes. Despite that factor, the forests in the more populated areas gradually vanished and today more forests are destroyed every year. Added to that, the pollution from burning coal, oil, and gas for heating, transportation, and manufacturing is slowly destroying the environment, making the air and water dirty, raising the temperature of the planet, and disturbing our health. Eventually humans will be suffocated by this pollution, if it continues to grow in the way that it has over the past 50 years. "Why is man doing this to himself?" is a big question. Now alarms have started sounding a wake-up call for action before it is too late.

Plants absorb CO_2 and deliver oxygen. This is how nature balanced all the CO_2 produced by people and animals, and by people's burning of wood, coal, and later oil and gas. But, gradually we have fewer and fewer plants and more

---

[1] Source: American Wind Energy Association, **AWEA.**

and more CO_2 and other pollutants. We have to work with nature to keep the balance. CO_2 is called greenhouse gas.

It is everybody's responsibility to care for the environment and minimize pollution.

Man needs energy for living: for heating and for cooling as well as for transportation, and manufacturing. But what kind of living do we have if it leads to suffocation? We may not suffocate today, but down the road our descendants will. Lavishly exploiting the environment and the Earth's resources and not caring for the outcome will lead to that eventual suffocation. It is everybody's responsibility to care for the environment and minimize pollution.

Energy can come from coal, oil, and other sources. But we also have an abundance of wind, sunshine, and other natural sources of energy that do not result in pollution. We measure our energy consumption by the kilowatt-hour (kW-hr). One kilowatt-hour is the amount of energy that a 100-W lightbulb consumes in 10 hours. Pollution, in the form of generating CO_2, from coal is 0.712 kg (1.57 lb) per kW-hr. This means that *the energy required to light each 100-W lightbulb for only 10 hr* if it comes from a coal plant *generates 1.57 lb of CO_2*. At home we have TV sets, refrigerators, stoves, and heating, some of which use much more energy than a lightbulb.

The importance of renewable energy is obvious. It is clean, it is in consonance with nature, and it comes from nature without polluting it. Moreover, renewable energy such as wind and solar energy are abundant and free. It is a wise choice to use them.

If coming from a coal plant, the energy to power each 100-W lightbulb for only 10 hr generates 1.57 lb of CO_2.

1.3   Technical issues

Although wind is an abundant source of energy, and the same is true for solar energy, one has to bear in mind that it is not a continuous source with a constant supply. In this sense, we should not expect, at least not at the present time, that we can solely rely on wind for our energy needs. Any renewable energy, although plentiful, suffers from two major drawbacks: it is a low-level energy and it is not continuously available. Being a low-level energy implies that we cannot expect to have a wind turbine with the same capacity as a thermal plant. A thermal plant (steam and gas turbines) can have a capacity of 500 MW (1 megawatt = 1,000,000 watts) or more with only one or a few turbines, whereas for that capacity we may need at least 200 wind turbines in an onshore wind farm.

Moreover, a 500 MW thermal power plant is normally capable of delivering that much power on a continuous basis, whereas the output of a wind turbine depends on the wind and fluctuates with the time of the day and the month of the year.

When a wind turbine is rated for 2 MW, that is the maximum power that it can deliver when sufficient wind is available. Because of the nature of wind, and depending on region, we may not expect an average delivery of more than about 33% of that during a 24-hour period. And 33% is the maximum for many regions; it can be lower than that. On average, each megawatt of installed wind energy can power 300 American houses.[2]

Despite these shortcomings, the progress that has been made and is still being made in the use of wind turbines is very encouraging. At the present time, wind turbines with 2.5-MW capacity are made for onshore use; and with 4-MW capacity for offshore use. Comparing the technology in North America when wind turbines were used for pumping water (which disappeared in the 1930s), significant progress has been made in both size and the way the machine can be controlled. **Figures 1.1** and **1.2** show the increase in scale from a typical American windmill with 12 blades to a 1.65-MW three-blade modern wind turbine.

Although the two aforementioned shortcomings of renewable energy are real, they do not prohibit the use of wind as a 100% source of electrical energy in the future. The fact is that at the present time the technology and the facilities

© Cengage Learning 2012

FIGURE 1.1 *American windmill (40 ft, 12-m tower).*

---

[2] This is a crude number, since the power consumption depends on the region and the season, as well as other factors. But it is still a useful reference.

FIGURE 1.2 *A modern 1.65-MW wind turbine (270 ft, 81-m tower).*

are not yet in place. In a vast country, while wind does not blow in one region it blows in another region. Although more efficient, energy does not have to be generated regionally in a country. As a matter of fact, countries may have interconnected electrical energy power lines, as exist between the United States and Canada.

The question to ask is whether there is sufficient wind energy available to satisfy the demand. In other words, is there enough wind to fulfill a country's total electrical energy need? For many countries the answer is yes. For example, windy regions in the north of Quebec (Nunavik) have enough wind to produce 40% of the electrical energy consumption of Canada. Canada has many other windy regions; it has wind capacities to provide a good part of U.S. electricity needs, as well. The problem is, therefore, not in the wind but in the lack of transmission lines and the technology to supervise and control the flow of energy in a large distributed system with variable parameters (see section 1.5).

1.4   Wind farm development

A **wind farm,** sometimes called wind park, is a wind energy equivalent of a **power plant** for conventional power generation. A power plant is where electricity is generated. Compared to a fossil fuel or hydropower plant, however, it consists of a relatively large number of turbines. The number of turbines defines the capacity of the wind farm and can be augmented in future expansions of the farm.

The very first condition for a wind farm is that it has good wind and is appropriate for installation of wind turbines. To verify this, the first thing to do is to collect wind data. A piece of land may look windy and seem to be a potential place for developing a wind farm. A single observation, however, is not sufficient for a project that can cost hundreds of millions of dollars. Wind data in crude form can be found from historical data and wind resource maps that show the average wind speeds for various regions in a country. Further and more specific wind monitoring should be done on a piece of land that has been identified as a potential site for a wind farm.

Meteorological studies are necessary for wind monitoring. A meteorological tower with measurement instruments that record wind speed and direction is installed and wind data are collected and recorded. A meteorological tower (abbreviated as met tower) is a single tower, similar to a communication tower, supported by guy wires, at the top of which measurement instruments are installed. These towers can vary in height from 10 m (33 ft) to 70 m (230 ft). For better and more reliable results, if no data from a region are available, this study can continue for 3 years. Depending on the history of the region, this may be reduced to 2 years or 1 year. Based on the analysis of data and economical analysis of the cost of a project (as described in chapter 14) a go-ahead decision can be made.

A second condition to be verified is the availability of a transmission line. In fact, this is more a line capacity verification and access reservation, rather than a question of whether or not a transmission line exists. A transmission line that has reached its maximum capacity cannot be used for new projects, even though it physically exists.

Among other preconstruction activities for a wind farm is formal agreements with authorities for matters such as aviation, broadcasting, distance from public roads, distance from residential areas, and so on, as well as agreements with landowners if the land is not owned by the developer (which is most often the case). Since a wind farm developer is normally different from the utility company, negotiating a power purchase agreement is also among the preconstruction work.

After a site has been regarded as having the potential for a wind farm, it is necessary to perform soil studies in order to design the appropriate foundation for the turbines. Site design, turbine selection, and interconnection to the electrical grid are part of the engineering work for a project.

As you will see in chapter 14 in more detail, the initial cost of developing wind farms is high. For this reason, in order to facilitate it for companies and persuade them to invest in wind projects, governments introduce incentives in the form of "tax breaks" and "accelerated cost recovery" for companies. These mechanisms stimulate investments in renewable energy, without which the cost per kilowatt-hour of generated energy can be higher than that from nonrenewable power plants.

wind farm
A region where a number of wind turbines are installed for generating electric power. Also called a wind park.

power plant
An industrial unit for generation of electric power.

1.5   Wind turbine siting

Wind turbine siting implies evaluation of a site for installation of wind turbines and development of a wind farm based on the wind characteristics, wind capacity, and other criteria. A primary decision to be further supported by other criteria is the wind quality. Wind resource maps, when they exist, give a general indication of wind quality for various regions in a country. Wind quality is defined by a classification, shown in **Table 1.1,** based on wind speed. Normally in a map different colors are used to represent different classes.

As mentioned earlier, a wind resource map is not sufficient for decision making. The quality of wind in a particular location depends on topography of the terrain and the altitude. As you will learn in chapter 2, the speed of wind increases with altitude; that is, as we move farther from the ground, the wind speed becomes higher. Measurement of wind speed is done at certain standard heights, such as 10 m, 30 m, 50 m, and 70 m above the ground. Any wind map must indicate the height at which the measurements were made.

The class of wind in a specific area can be used to determine how much wind capacity is there, how much energy a turbine can generate in a year, and how long it will take for an investment to break even. It also helps to design a turbine or match a turbine speed for a wind regime. In chapter 14 we see that the economics of wind turbines depend on a number of factors; thus, a site even with fair wind may not be appropriate for a wind farm project. Also, a particular turbine may be not a good choice for a class of wind, but an acceptable one for another class.

Among other data that must be studied for a wind farm is the frequency of severe wind storms and the possibility of a hurricane. Severe weather and strong gusts of wind can break and destroy a turbine. Storms do not necessarily mean more energy. A wind turbine is stopped and put on brakes during a storm.

After a site has been identified as appropriate from a wind quality viewpoint, other matters must be investigated and the site must meet the criteria set by regulations and authorities. Most of these are environmental concerns (see chapter 15), for example, that a wind farm must not have adverse effects and disturb people, surrounding areas, ecosystems, aviation, communication systems, and so on or introduce hazards for the neighboring people, buildings, and wildlife. Other criteria or conditions to be met are as described in section 1.4.

TABLE 1.1

Definition of wind classes for wind measured at 50-m (164-ft) height		
Wind Class	**Wind Speed (m/sec)**	**Wind Speed (mph)**
Marginal	5.6–6.4	12.4–14.3
Fair	6.4–7.0	14.3–15.7
Good	7.0–7.5	15.7–16.8
Excellent	7.5–8.0	16.8–17.9
Outstanding	8.0–8.8	17.9–19.7
Superb	8.8–11.1	19.7–24.8

New and future technologies   1.6

As already mentioned, energy must be transmitted from the point of generation to the point of consumption. Transmission lines are like highways for electricity. Similarly, they have a maximum capacity that once reached, no further
charge can be added without overload, inefficiency, and breakage.

In industrialized countries, depending on the size, one or more electric
networks exist that interconnect all the electric generators together and all
the consumers are connected to this network. A **dispatch system** (communication by radio, fax, and the like from a center to all participants in an
activity) informs each generating node (power plant) about the level and
schedule of production. Many small plants either increase their production at **peak hours,** or they may work only a few hours altogether per day
during the peak hours, when the electricity demand is high. Peak hours is
the period during 24 hours when the electricity demand is at or around its
maximum.

With the growth of the wind industry and installation of new transmission
lines, it is possible to generate electricity wherever wind blows and transfer it to
the location of instantaneous demand. So, the network always has its required
supply of energy at the generating points, although the demand could be at
regions where the wind has stopped. This requires that the branches of the network have room and capacity for all the necessary electric traffic through them.
Such a capability also calls for an automated intelligent way of monitoring,
supervising, and controlling all the participating power plants. The dispatch
system is yet not fully automated and the technology has not reached the level
that it can reliably handle many more power stations, particularly to include
wind farms.

Wind farms have certain problems specific to wind energy, all stemming
from the fact that wind at one region varies with time. That is, neither its
speed nor the time of its availability is within our control. However, if many
wind farms are considered that are far from each other, then little fluctuation can be found in the total energy from them. When one goes lower in
energy, there will be another one somewhere else that goes higher. These wind
farms, interconnected to each other in a network, are the generating nodes
of a **distributed generation system** (many smaller power plants at different
locations, rather than a centralized larger power plant). Their proper control
when working together requires a new technological development that some
call a **smart grid.**

dispatch system
A communication
system in which all
the participants are
informed about the
operation. A common
example is that
used by the city cab
companies.

peak hours
The (normally few)
hours during a
24-hr period when
the electricity
consumption is the
highest.

**distributed
generation system**
An electricity
generation system
in which more than
one power plant is
involved.

smart grid
A grid equipped with
added intelligent
capabilities (measure-
ment devices, control
devices, switching
devices, and so on).

Outline of the text   1.7

This book is organized in such a way that a reader with previous knowledge
on a relevant subject can skip the corresponding chapter(s). Chapters 2, 3,
and 4 are more general, covering the energy in the wind, the fundamentals of
how the wind energy is captured, and a discussion of various wind turbines;

whereas from chapter 5 the subjects are more concerned with the propeller type turbines, which to many people represent what is meant by wind turbine.

Chapter 5 describes the various components of propeller-type wind turbines, which are further explained in subsequent chapters. A wind turbine is an electromechanical system, dealing with mechanical and electrical devices and the conversion of mechanical energy to electrical energy. Chapter 6 concerns electrical generation and how electrical sources can be connected together. This is what happens especially with wind turbines such that many units must be connected together and to the grid.

Chapter 7 discusses the operating characteristics of (propeller) wind turbines. One learns about the way the output power of a turbine can be determined and how the annual energy from a turbine can be found. The fundamentals of connection to the grid is the subject of chapter 8, where the emphasis is on the understanding of power, power relations, and the grid conditions to be satisfied. Further material on grid connection, how a turbine is physically connected to the grid, and how turbines are controlled to match the grid requirements are the subjects of chapter 11. Chapter 11 also explains the various ways that different types of turbine generators can be integrated with the grid.

Chapter 9 is devoted to mechanical systems of a wind turbine. Most notably, the gearbox is the most important mechanical system in a wind turbine, after the rotor and the tower. The gearbox has been the troublemaker in many wind turbines, and its replacement or repair costs are prohibitive. The chapter brings into the picture the importance of proper operation of gearboxes, including their lubrication, cooling, and efficiency and the general problems of gears and gear failure.

Chapter 10 concentrates on how wind turbines are controlled. Only a few control actions are required for the control of wind turbines. However, since wind turbines are complicated machines, their proper control for best performance and highest efficiency is essential and must be based on precise control schemes. This chapter focuses on the mechanical side of the power balance between turbine power input and output, whereas the electrical concerns are the subject of chapter 11.

Wind turbines include many ancillary devices. Chapter 12 describes the main wind turbine electrical ancillary devices. A reader becomes familiar with the function of a transformer, and the fundamental principles and terms used for a transformer, a device without which electricity cannot be put to work. In addition, the principle structure of the electric converters (rectifiers and inverters) used in many modern turbines are explained.

Chapter 13 covers some necessary material pertinent to the mechanical structure and mechanical strength of major wind turbine components and the problems associated with their functions. Loading on the blades, nacelle, and tower and problems such as vibration and fatigue are discussed.

The economics of a wind turbine is the driving force in any single installation or large size wind farm. Without a for-profit concept, no wind turbine project will fly. Chapter 14 brings this fact to the reader. Sufficient knowledge of economics and the analysis of turbine costs are introduced, so that the reader can make a relatively good analysis of a wind turbine project and its economic performance, if sufficient data are available.

Since commercial wind farms have became more common, particularly in Europe, many concerns have been demonstrated regarding the effect of wind turbines on the environment. More notable issues are noise, destroying the natural views, destruction of the landscape, killing birds (and bats), and distortion of broadcast and communication signals. There have been many studies, as a result, to investigate whether real adverse effects are associated with wind turbines. Chapter 15 describes all of these environmental concerns. It is necessary that a person working in the wind energy industry has sufficient knowledge about the truth of the matter.

Other aspects of wind turbines and their operations that are indispensable to know are covered in chapter 16. At the top of the list is the safety of wind turbines. Among other issues are the effect of temperature change and formation of ice in cold regions, the effect of lightning, and the possible use of wind turbines in residential and urban areas.

The final chapter is assigned to safety while working with wind turbines. Safety and compliance to safety rules is of paramount importance for people who climb and/or work on wind turbines. Normally all the people who are supposed to work on wind turbines receive a training session on safety and rescue matters by their employers. Chapter 17 gives a summary of the most important matters associated with safety at work, particularly for working on wind turbines.

Chapter summary   1.8

- Wind has been a source of energy for people since ancient days.
- The Persian windmill has vertical axis and can be traced back 1000 years.
- The Dutch windmill has horizontal axis. It can be traced back to the twelfth century.
- During the past 30 years, much research and development work has been carried out on wind turbines.
- Early wind turbines were much smaller than today's turbines, reaching a power of 4 MW.
- Earth, although large to us, is limited in size and none of its contents can be unlimited; only the sun's energy comes from outside.
- If we do not take care to use the Earth's resources wisely and maintain the balance, they will eventually run out.

- There is sufficient wind energy in the world to provide for all of our electric needs.
- Electric energy is normally measured in kilowatt-hour (kW-hr), which is 1000 W in 1 hr, or 10 hr of 100 W.
- The amount of CO_2 generated from burning coal is 0.712 kg (1.57 lb) for 1 kW-hr of energy.
- Compared to conventional power plants, wind turbines are small in capacity and their production depends on the existing wind and cannot be constant in magnitude.
- The development of wind farms is dependent on a number of criteria. Having good wind is only one of them.
- Wind monitoring is a necessary step for a piece of land having potential for wind turbine installation.
- Wind speed increases with height. Measurement at 50 m from the ground shows a higher value of the same wind measured at 10 m from the ground.
- In the future, a smart grid will allow the electric network at any time to receive energy from many wind farms at various points where wind blows with good speed.

Review questions

1. Where does the Earth's energy come from?
2. What are the advantages of wind energy compared to burning fossil fuels?
3. What are the disadvantages of wind energy with respect to burning fossil fuels?
4. Why is renewable energy very important for the future of Earth's inhabitants?
5. What is greenhouse gas?
6. How much greenhouse gas is generated by the production of 1 kW-hr of energy?
7. From the conversion table in the book find how much greenhouse gas is produced by a 200-horsepower vehicle in each hour.
8. What is the source of energy for lighting in our homes?
9. Why can't we currently use wind energy everywhere for all our energy needs?
10. What is meant by siting?
11. What is a wind farm?
12. Have you seen a wind farm?
13. Can we have a wind farm at any place we want? Why or why not?
14. What is meant by good wind?
15. If we have a wind turbine, do we get more power from wind during a storm? Why or why not?
16. Why is greenhouse gas bad?
17. What do the plants do to greenhouse gas?
18. Why should we try to generate less greenhouse gas?
19. What is "grid"?
20. What is a smart grid?

Problems

1. The amount of CO_2 corresponding to a 100 W lightbulb if it is turned on for 1 hr is 0.16 lb. Calculate how many pounds of CO_2 are generated in 1 year.
2. If in your home you have a light that is kept on for 4 hr at night, whereas nobody needs it to be on and it could be turned off, how much is the total energy (in kW-hr) used in 1 year if the lightbulb is 100 W?
3. If the price of electricity is 9 cents per kW-hr, find how much money you waste in a year under the conditions of question 2.
4. Calculate how much CO_2 is generated per year by the lightbulb of question 2.
5. Suppose that there are 1 million homes doing the same thing as in question 2. Find how much money is wasted and how much CO_2 is generated.

PROJECT

In this project, you find the approximate electricity consumption of your home. At the top of a sheet of paper write the type of home (for example, house, apartment, etc.), the number of bedrooms, and number of people regularly living there.

Make a table with five columns. In the first column, list all the electrical devices that you have at home. In the second column put the number of each. For example, you may enter 35 lightbulbs, two TV's, and so on. In the third column, put the power (wattage) of the device. This is usually mentioned on a plate at the back of the device, or printed on the lightbulbs. You must put the average value for devices where column 2 shows more than one. If you cannot find or reach the power value, ask your teacher or someone else for an approximate value. In column 4 you add the approximate number of hours that the device is on in 24 hours. For this you may have to do some calculation. For example, for an electric heater that you use only in the winter you should figure out approximately how many months in a year you use it. Then adjust your number by multiplying by the number of months in use and dividing by 12.

In column 5 write the number found by multiplying column 3 by column 4, divided by 1000.

Add all the numbers in column 5. This is the average electric consumption of your home. You may use it for calculating your cost of electricity. Also, you may find out how much CO_2 is generated if your electricity comes from a fossil power plant (for oil use half of the number given for coal, and for gas use ¼ of the number used for oil).

References

[1] Burton, T., D. Sharpe, N. Jenkins, and E. Bossanyi: *Wind Energy Handbook*, Wiley, 2001.

[2] Gipe, Paul: *Wind Power, Renewable Energy for Home, Farm, and Business,* Chelsea Green Publishing Co., 2004.

[3] Hansen, Martin O. L.: *Aerodynamics of Wind Turbines,* Springer Verlag, 2008.

[4] Harrison, R., E. Hau, and H. Snel: *Large Wind Turbines: Design and Economics,* McGraw-Hill, 2000.

[5] Hau, Erich: *Wind Turbines: Fundamentals, Technologies, Application, Economics,* Springer, 2006.

[6] Manwell, J. F., J. G. McGowan, and A. L. Rogers: *Wind Energy Explained: Theory, Design and Application,* Wiley, 2009.

[7] Sathyajith, M: *Wind Energy: Fundamentals, Resource Analysis and Economics,* Springer Verlag, 2006.

[8] Spera, D. A. (Editor): *Wind Turbine Technology,* ASME, 1998.

Energy in the Wind

© Cengage Learning 2012

KEY TERMS

Terms that you should already know: *mass, speed, flow rate, density*

New Terms
Betz limit
energy
kinetic energy
meteorological tower
"met" tower
power
power coefficient
wind data
wind harnessing
wind speed
work

OBJECTIVES

After studying this chapter, you should be able to:

- Know how much energy exists in wind.
- Know how much of that energy is usable.
- Explain why not all the energy can be used.
- Explain the effect of wind speed on wind energy.
- Describe what other parameters determine the energy in the wind.
- Understand why height is important for a turbine.
- Explain why the same turbine generates different amounts of power in winter or summer and at different times of day.
- Know the effect of buildings on the wind.
- Explain why a wind turbine cannot generate the same power all the time.

2.0 Introduction

When wind blows it can move things, depending how strong the wind is. It can rotate a turbine, for instance. So, there is *energy* in the wind (see the text on *Energy and Power* below). We need to know how much energy is in the wind and on what factors this energy depends. This chapter describes how we can determine the wind energy and how much of that can be extracted by a wind turbine. The obvious answers are the *wind speed* and the size of a turbine. But there is more to it than that.

2.1 Energy in a moving object

kinetic energy
A type of mechanical energy associated with a moving object; its magnitude depends on the mass and speed of the object.

energy
Potential to do work.

work
The resultant of spending energy. Work can be mechanical or electrical, or it can be converted to heat. Mechanical work is more tangible; when a weight is lifted, *work* is performed. Lifting a weight of 1 lb by 1 ft results in 1 ft-lb of work.

Any moving object has energy. This type of energy is called **kinetic energy.** For example, a car, a bicycle, or a ball, when moving, all have kinetic energy. If they stop, that energy is gone. The same is true for moving air, that is, the wind. The

ENERGY AND POWER

Energy is the potential to do *work* and it occurs in different forms. **Work** here implies mechanical or other type of work. Mechanical work takes place for instance when a weight is lifted, or when an engine drives a car. Mechanical work is more tangible than other types of work. For instance, a motor can do mechanical work, but a battery does not directly do mechanical work. However, it can run a motor, which does mechanical work; thus, a battery has the potential to perform work. It has electrical energy. Heating water is another form of work, although it is not mechanical; but, it can be converted to mechanical work (consider a steam turbine that can do mechanical work, where the steam is generated from heat). In this sense, heat (that can raise the temperature of a material) is a form of energy. It is "thermal energy." Examples of other types of energy are chemical energy, solar energy, wind energy, and nuclear energy.

Now, suppose that a machine or device has energy. How much energy does it have? Or, how much work can it do? At the same time, another question that can be asked is if a machine can do a certain amount of work, how long does it take to do that work? Is it capable of doing that work in 1 min or in 25 min, for example? To answer this question or understand a comparison between various energy sources, *power* is determined. **Power** is the amount of work done in 1 sec by a device that can do work. This is the measure of strength of energy sources. It can also be used to compare two or more devices. For instance, a smaller motor has less power than a larger motor; that is, it can do less work than the larger motor in the same amount of time. For example, if the smaller motor does less work in 1 sec compared to a larger motor, it will do less work in any number of seconds, say in 1 min.

power
The amount of *work* in 1 sec by a machine, a motor, an engine, and so on.

amount of energy of a moving object depends on two factors, its *mass* and its *speed*. Using the proper units for measuring mass and speed, the relationship to determine the energy of a moving object is as follows:

$$\text{Energy} = \tfrac{1}{2}\,(\text{mass})(\text{speed})^2 \qquad (2.1)$$

This equation implies that, for example, if mass is doubled (that is, if you have two objects the mass of one of which is two times the mass of the other, both having the same speed), the energy doubles. But, if the speed is doubled (if you have two objects of the same mass, but one has a speed two times that of the other), then the energy is four times more. It is very important to understand the relationship given by an equation similar to that in (2.1).

This energy can be converted to *work*. In other words, it can be used to do work (see the text on energy and power in this chapter for the technical meaning of work).

Energy in a moving fluid   2.2

Energy in a moving fluid (a gas or a liquid) can be found in the same way as described for a solid body. There are two differences, however, that must be taken into account. The first difference is that for a fluid in motion, we define the moving mass by its volume. This is so because there is no solid object with a definite mass to be considered. The second difference is that for a moving fluid, the whole volume moves.

We first consider a simpler case where the fluid is confined in a boundary. This is the case of flow in a pipe, for example. Then we consider the case where fluid is not limited by a boundary.

Fluid moving in a pipe   2.2.1

In such a case, the moving mass is represented by the size of the pipe and by flow rate. The flow rate is the amount of the volume of the liquid that moves in a minute or a second. Depending on the size of the pipe, the speed of the flowing fluid can be found. The following example can better illustrate this idea.

EXAMPLE 1:

A circular pipe has a *cross section area* of 3 square feet (ft²) (this number corresponds to a 24-inch-diameter pipe). If water flows in this pipe with a flow rate of 2 cubic feet per second (ft³/sec), how much is the speed of the water in the pipe?

Solution: In this example, the size of the pipe is defined by its cross-sectional area, which is 3 ft². The flow rate of water is 2 ft³/sec. The speed at which water is flowing in the pipe is $2 \div 3 = 0.667$ ft/sec, which is equal to 8 inches per second. **Figure 2.1** shows this pipe.

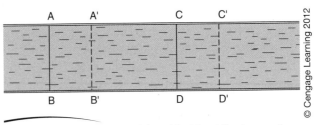

© Cengage Learning 2012

FIGURE 2.1 *In 1 sec, all particles of liquid residing in a section AB of the pipe move to section A'B'.*

One can imagine that a portion of water represented by *ABCD* in figure 2.1 moves along the pipe. After a short period it has moved from the previous position to a new position denoted by *A'B'C'D'*. (In reality, this is not what happens inside a pipe, but it shows the motion of water.) The same is true for any other liquid or gas. If it takes 1 sec for the fluid to travel from *ABCD* to *A'B'C'D'*, then the speed of the fluid is the length of the line from *A* to *A'* (which is the same for *BB'*, *CC'*, and *DD'*).

In order to see how much energy is in the water moving in a pipe it is very important to realize that, as mentioned earlier, the whole volume of water in the pipe moves. However, it is not helpful at all to consider this energy, since it depends on the full length of the pipe and other factors. Instead the power of the moving fluid is determined.

> **Important:** In the case of a moving fluid, it is not helpful to consider the *energy*, since it is the energy of the whole volume moving and depends on the full length of the pipe. Instead of energy, determining the *power* of the moving fluid is more useful.

2.2.2   Fluid moving with no boundary

When a fluid is moving with no boundary, which particularly is the case for wind, then an imaginary boundary is practically generated by the objects that lie along its passage. This is better understood from **Figure 2.2.**

A wind stream is considered as shown in figure 2.2. At some place within the stream denoted here by a plane *BB'* an object is obstructing the stream. It can be a turbine. For simplicity, assume it is round. Along the wind direction, we can assume an imaginary cylinder, the diameter of which is the size of the object. This is shown in figure 2.2 by the two parallel lines along the wind direction. You can assume that this is the boundary that defines a pipe. But in the case of no boundary flow it is *not* this imaginary cylinder that defines the boundary. Instead the nearly conical shape shown in the figure is the boundary of the flowing fluid. The reason for this is that, because of the obstruction in the flow, the particles of the fluid start moving outward when approaching the obstacle. Thus, instead of moving ahead in a straight line, they follow the curve

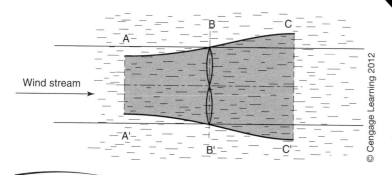

FIGURE 2.2 *An object in the path of a wind stream at section BB'.*

shown in the figure. They start moving outward at a point denoted by plane *AA'*. They resume their parallel flow at some point after the obstacle. This point is represented by plane *CC'* in figure 2.2.

Power in the wind   2.3

As was mentioned earlier, for wind or any other moving fluid it is more practical to find the power rather than the energy. In this sense, in equation 2.1 we can substitute for the mass of the air that flows in 1 sec, and that gives the power in the wind. Note that

$$Mass = (Density)(Volume) \qquad (2.2)$$

Also, note that for a fluid moving inside a pipe one can always say that

$$Volume\ moved\ in\ 1\ sec = (Cross\text{-}sectional\ area)(Speed) \qquad (2.3)$$

Substituting from equations (2.2) and (2.3) in equation (2.1) gives the power in a moving fluid, or the amount of energy in 1 sec for the area given by a pipe cross section or the size of an object in the stream of an open flow with no boundary.

$$Power = \tfrac{1}{2}\ (Density)(Cross\text{-}sectional\ area)(Speed) \times (Speed)^2$$

or

$$Power = \tfrac{1}{2}\ (Density)(Cross\text{-}sectional\ area)(Speed)^3 \qquad (2.4)$$

Equation (2.4) shows that the power in a tunnel of wind is proportional to the air density, the cross-sectional area of the tunnel of wind and the cubic power of the wind speed.

> The power in a tunnel of wind is proportional to the air density, the cross-sectional area of the tunnel of wind, and the cubic power of the wind speed.

2.4) is used to calculate the power that exists in wind, a part of
:tracted by a wind turbine. The cross-sectional area in this equa-
e size of a turbine.

ing conclusions can be derived from equation (2.4):

ne turbine and at the same time, if the wind speed doubles, the
ie wind increases by a factor of 8.
ne turbine and the same wind speed, if the weather is cold
isity), more power exists in the wind available for the turbine.
weather conditions and for the same wind speed, a turbine that
s larger in cross-sectional area than a smaller one has twice as
power available to it.

From item 2 above it can also be concluded that a turbine at a given wind
speed can produce more power in the winter than in the summer because the
weather is colder. The same conclusion can be extended from day to night if the
temperature difference is significant. This difference due to temperature change
is, nevertheless, not very much.

2.4   Power absorption by a turbine

In the previous section the available power in a stream of wind was defined in terms of the wind speed, turbine size, and the air density. This
power, however, is the power that exists in the wind when it is blowing.
A turbine cannot necessarily capture all of this power; it can only absorb
a portion of it. This depends on the type of turbine, the efficiency, and
other conditions in the operation of a turbine, as we see later in upcoming
chapters.

In order to show the fraction of power in the wind that a particular wind
turbine can harness from wind (i.e., **wind harnessing**), a coefficient is used in
equation (2.4). Thus, we can say that

Power of a wind turbine = (A coefficient)(Power in the wind)

This coefficient must, obviously, be smaller than 1. It is called **power
coefficient.**

Wind turbine power = ½ (Power coefficient)(Density)(Cross–sectional area)(Speed)³ (2.5)

The power coefficient depends on how good a turbine is in design and
how well it can grasp the wind energy. Thus, its value can be small or large.
Nevertheless, there is a maximum value that no turbine in its best performance
can exceed. It can be theoretically determined and is called the **Betz limit.** The
value for Betz limit is 16/27 = 0.59.

wind harnessing
Capturing of wind
energy by wind
turbines.

power coefficient
A number always
less than 0.59 that
determines how much
percent of the power
in the wind stream can
be harnessed by a
wind turbine.

Betz limit
The maximum value
for a wind turbine
power coefficient. It is
16/27 or 0.59.

Betz limit = 16/27 = 0.59

The power coefficients for certain turbines can reach values near the Betz limit. A value of 0.50 for a good design is acceptable. For others, this number can be smaller, say 0.25 or even 0.20. Any claim for coefficients greater than the Betz limit are groundless and can be rejected.

Wind speed variation   2.5

We learned that the wind speed is the most influential factor to how much power a wind turbine can generate. In fact, for any wind energy project the wind speed is the first decision-making factor. The nature of wind is that it is variable. This implies its direction, speed, and temperature can change. This change can be from minute to minute or sometimes from second to second. So, it is interesting and necessary to investigate the variation of wind speed.

There are four categories of wind speed variation that we are going to study. These are:

Wind speed variation with time
Wind speed variation with height
Wind speed variation with terrain
Wind speed variation with geographic zone

Wind speed variation with time   2.5.1

The variation of wind speed with time, by itself, falls into three categories: momentary changes, daily changes, and seasonal changes.

Wind is generated by a difference in temperature of air between two locations. So, it is subject to the heat from sunshine and the temperature of the surroundings in an area of the Earth. We can sometimes clearly see this variation of wind and notice that wind does not blow with a constant speed. This momentary variation can be clearly seen when the wind speed is recorded. **Figure 2.3** shows an example of recorded wind speed for some short period of time.

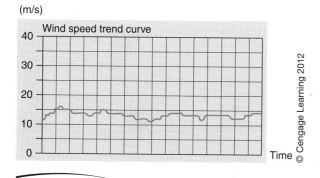

FIGURE 2.3 *A typical variation of wind in 2 min.*

In figure 2.3 each division on the horizontal axis represents 8 sec. The numbers on the vertical axis are in meters per second. The values are around 10 m/sec, which is about 33 ft/sec or 22.4 miles per hour (mph). The figure shows the variation of wind speed in a time interval of 2 min.

The daily changes of wind speed, as the name implies, is the variation of wind speed in 24 hours. For instance, in one region the highest wind speed can occur at a certain time of the day, and at other hours there is less wind or even no wind. In other words, we may be able to identify a specific pattern for the wind in a region for various times of the day. Note that this could also be true for the direction of wind.

The seasonal change in wind refers to variation of wind during a year. Again, for this case one may find a specific pattern for the wind speed for various months of the year. The best example is that in many regions we have more wind in the spring or winter compared to the summer. This information must be found from statistical data of wind history, called **wind data.**

This information shows, also, that one cannot expect the same production of power from a turbine at different seasons in a year.

wind data
The data containing the statistics of wind (speed, frequency, duration, direction, etc.) in a region.

Wind speed
The speed at which wind blows at each instant.

> **Wind speed** can change from moment to moment. This is momentary variation. During a day, the wind speed can change also. It can have the highest speed at the same time every day. This all depends on the region. A long-term variation in the wind speed is on a yearly basis. This implies that certain seasons are windier than others.

The daily changes of wind speed can be studied for any particular region. Such data are very important in evaluating the amount of energy that a turbine can absorb in 24 hours. The pattern of changes in wind speed can be represented by a curve, similar to that shown in **Figure 2.4.** In this figure two curves are shown, denoted by *A* and *B*, for two different regions. As you see, curve *B* is narrower than curve *A*. In both of these regions the average (or mean) wind speed is 16.8 mph (about 7.5 m/sec); that is where the tip is. This speed has the highest likelihood. This means that whenever it is windy, the chance that wind speed is 16.8 mph is more than any other speed. The speed, however, can be both higher and lower than that. For the region represented by curve *A*, when wind starts, it can slowly speed up or slow down; that is, its variation is small. In the other region, the wind speed reaches 16.8 mph much faster, and when it goes higher it does not last long at the higher speed.

Some other information also can be derived from these curves. For instance, we can see that when it is windy in region *A*, the chance that wind speed is 16.8 mph is 10% (corresponding to 0.1 on the vertical axis) and the chance for 11.2 mph is about 7.8%, that is, slightly lower than 10%. On the other hand, in region *B* the chance for a 16.8 mph wind speed is about 18.5%, whereas for 11.2 mph the chance is only about 5%, that is, much lower. This information can be found by reading the values on the vertical axis for the points of intersection corresponding to each wind speed with the curves. For any region, the corresponding curve can be generated from the history of records of the wind pattern information. If such a record is not available, then it can be approximated from the knowledge of the average wind speed in that region. The data from figure 2.4 can be presented in a tabular form. An example of these data is shown in **Table 2.1.**

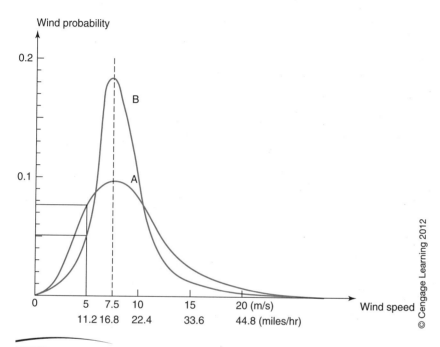

FIGURE 2.4 *Wind speed pattern for two different regions.*

TABLE 2.1

Example of wind data for region *A* in figure 2.4										
Wind speed in m/sec	1.00	2.00	3.00	4.00	5.00	6.00	7.00	7.50	8.00	9.00	10.00
Wind speed in mph	2.24	4.47	6.71	8.95	11.19	13.42	15.66	16.78	17.90	20.13	22.37
Percent of time at this speed	0.42	1.25	3.33	5.83	8.33	8.75	9.17	9.58	9.17	8.75	8.33
Duration in hour per 24 hours	0.10	0.30	0.80	1.40	2.00	2.10	2.20	2.30	2.20	2.10	2.00

Wind speed in m/sec	11.00	12.00	13.00	14.00	15.00	16.00	17.00	18.00	19.00	20.00	higher
Wind speed in mph	24.61	26.84	29.08	31.32	33.56	35.79	38.03	40.27	42.50	44.74	higher
Percent of time at this speed	6.67	5.00	3.96	2.75	2.29	2.00	1.50	1.04	0.83	0.63	0.42
Duration in hour per 24 hours	1.60	1.20	0.95	0.66	0.55	0.48	0.36	0.25	0.20	0.15	0.10

In table 2.1, the first row shows the wind speed in meters per second and the second row shows that in miles per hour. The relationship between meters per second and miles per hour is

$$1 \text{ m/sec} = 2.237 \text{ mph}$$

In table 2.1, however, all the values are rounded to one decimal point. The third row shows the percent chance that wind blows at any particular speed. This is the same information graphically shown in figure 2.4 for region *A*. Finally, the fourth row indicates the duration that in a 24-hour time period, wind blows with that speed. For example, within 24 hours for a period of 2.4 hours, that is 2 hours and 24 min, wind blows at a speed of 16.8 mph. Or, wind blows at a speed of 35.8 mph only for half an hour during a 24-hour period. Also, it is very rare that the wind speed exceeds 44.7 mph. If it does, it will last only for 6 min (0.1 hour).

2.5.2   Wind speed variation with height

In addition to the change in wind speed with time, the speed of wind is not the same at different heights. As we move away from the Earth's surface, the speed of wind gets higher. For example, if we measure the wind speed near the ground, say 10 ft from ground, as 2 mph, if we measure it at 200 ft above ground, it can be 10 mph, that is, five times more. (These figures are just an example.) This variation by height is not the same for all places. That is to say, in another location if the wind speed at a height of 10 ft from ground is 2 mph, its speed at 200 ft from ground can be more than 10 mph, or it can be less than that. This all depends on the nature of the ground in that location. That is, it depends on the roughness of the terrain, as will be discussed, next.

The relationship between the wind speed and height is not linear. This means that if, for instance, the wind speed at 10 ft from ground is 2 mph and at 100 ft above ground is 10 mph, we cannot conclude that at the height of 55 ft (the midpoint between 10 and 100 ft) the speed of wind is 6 mph. What we can say is that, the wind speed at the Earth's surface is lower than that at a relatively distant point from ground, say at 600 ft, and it increase as the height increases. However, after a certain height, the wind stream has a uniform speed and does not have any more noticeable changes.

The wind speed at the Earth's surface is lower than that at a relatively distant point above ground, say at 600 ft. The wind speed increases as the height increases. However, after a certain height, the wind stream has a uniform speed that no longer changes with height.

Wind speed variation with terrain

2.5.3

When wind blows, its direction and speed are influenced by all the obstructions that lie on its path. At a height sufficiently away from the ground surface, however, wind direction and speed are not disturbed and there is a steadier wind stream. All the obstacles to the flow of wind can slow it down. Thus, inside cities where there are many buildings, there is more obstruction to slow down the wind. On the contrary, in farmlands with short vegetations or along the surface of a lake there is not much obstruction to slow down the wind.

The following example helps you better understand this matter. The numbers in the example are based on calculations, but the formulas are beyond the scope of this book and are not mentioned here.

EXAMPLE 2:

Supposing that on a windy day the wind speed in the free (undisturbed) stream of wind at a height of 660 ft is 44 mph. Inside a city in the same region, the wind speed is 1 mph, when measured at 10 ft from ground. In a flat land sufficiently far from the city and covered with grass, however, the wind speed measured at a height of 10 ft is 26.2 mph.

You should notice also from this example that when measuring the wind speed, we always mention the height at which the measurement has been done. Otherwise, the numbers do not match. Wind measurements are normally carried out at standard heights in *meteorological stations*. The measurement device is mounted on the top of a tower. These towers are called **meteorological towers,** or just abbreviated to "met" towers.

meteorological tower
A tower with instruments to study wind and weather.

The effect of a building on the flow of wind is shown in **Figure 2.5**. Note that the whole wind speed drops because of buildings or other obstructions, but locally, the wind speed is higher on the roof, since more air must move

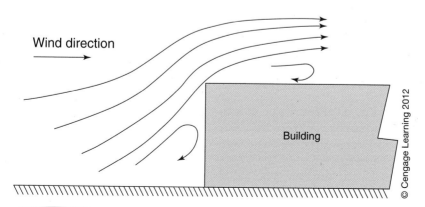

© Cengage Learning 2012

FIGURE 2.5 *Any building is an obstacle to the wind flow; the effect slows the wind.*

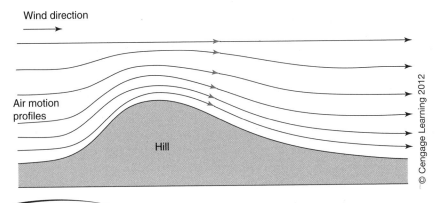

Wind direction

Air motion
profiles

Hill

© Cengage Learning 2012

FIGURE 2.6 *Profiles of wind passing over a hill.*

from a narrower area. This is shown by the profile lines being closer to each other above the roof.

The scenario on the larger scale of a hill is shown in **Figure 2.6.** After wind hits the hill from the left and passes the hilltop it speeds up on the right-hand side. In such a situation, the right-hand side of the hill is the best place to install wind turbines.

2.5.4   Wind speed variation with geographic zone

The most obvious observation that one can make is that climate is not the same everywhere. The formation of wind is affected by the change in the temperature (and thus, pressure) of air. The cooling down or warming up of a region depends on a number of factors, all related to the geographic location. For example, if a region is near the sea or surrounded by mountains, whether or not it is affected by large storms, and so on, determine the wind specifications. One region is windier than others. The average wind speed in a region is 24 mph, for example, whereas in another region it is only 10 mph. Later we will see that knowledge about this average wind speed is a key issue in deciding on a site for installing wind turbines.

2.6   Chapter summary

- Energy is the potential to do work. It can be in different forms.
- Any moving object contains some energy that is carried with it.
- This energy depends on the mass of an object and the speed at which it is moving.
- A flowing fluid, similarly, contains energy. The energy depends on the volume of the fluid in motion.
- For a fluid, we define a cross section to restrict the area, for instance the area of a pipe.

- For free-moving air (wind), we must define a cross section for the area of interest, such as the area swept by a wind turbine (i.e., the area that the blades make when rotating).
- For wind and fluids, power is more meaningful than energy, since there is no definite moving mass.
- For wind and fluids, the mass moved in 1 sec is considered. This gives the energy in 1 sec.
- Energy in 1 sec is called "power," or power is the amount of energy in 1 sec.
- Energy is equal to power multiplied by time.
- Power absorption by a turbine depends on the power in the wind and the *power coefficient* of the turbine.
- The maximum value for power coefficient for any turbine is the *Betz limit*.
- The *Betz limit* is equal to 16/27, which is equal to 0.59.
- Power absorbed by a turbine depends on air density.
- Power absorbed by a turbine is proportional to the cubic power of wind speed.
- Power absorbed by a turbine is proportional to the area of the circle made by the blades.
- Wind speed is not constant.
- Wind speed varies from hour to hour and from season to season.
- Wind speed varies with height. The higher we go from the Earth's surface, the faster are the winds.
- For any region, the wind pattern can be formed by investigating the wind speed during various hours in a day.
- The wind speed pattern can be represented by a graph, showing whether a region is windy or has less wind.
- Wind speed changes with terrain. In a built-up area, the wind speed drops because of the resistance from buildings as obstacles to the wind.
- At the top of a hill, the wind speed is greater than in a valley.

ADVANCED LEARNING

Energy in a moving object

The *energy in a moving object* is calculated from

$$E = \tfrac{1}{2}\, mv^2$$

where E is the energy, m is the mass, and v is the speed (or velocity).

If m is in pounds and v is in feet per second, then E will be in foot-pounds (ft-lb), which is one of the units for measuring energy and work.

Note that in Imperial units of measurements, pounds can be used to measure force (lb_f: pound-force) or mass (lb_m: pound-mass). If instead

(continued)

of mass, the weight of an object (in pounds) is used, then the value obtained from the equation must be divided by a factor of 32.17.

EXAMPLE:

A steel ball weighs 2 lb. If it is moving at a speed of 3 ft/sec, how much is its energy?

Solution:

$$E = \frac{1}{2}\,(2)(3)^2/32.17 = 0.28 \text{ ft-lb}$$

The power in the wind depends on the wind speed and the area of a tunnel of wind (if we do not define the area, the answer is useless), and is determined from

$$P = \frac{1}{2}\,\rho A v^3$$

where ρ (called "ro") denotes the air density, A defines the area that wind flows, and v is the wind speed. In the Imperial system of measurements ρ is in lb_m per cubic ft (pound-mass per cubic ft). Air density changes with temperature (since it expands when heated); so, at each temperature, it has a slightly different value. At the atmospheric pressure, air density at 70°F is about 0.075 lb/ft³. This refers to the weight of one cubic foot of air. (When the number of pounds is specified, but no further reference is made as to whether it represents mass or force, weight, which is force, must be considered). Thus, in the formula, a division by 32.17 becomes necessary. If ρ is given in pounds per cubic ft, A is specified in square feet, and v is expressed in feet per second, then the value of P is determined in foot-pounds per second.

Foot-pounds per second is one of the units for measurement of power. Other units of power are watts and horsepower. One horsepower is 550 ft-lb/sec.

EXAMPLE:

If wind blows with a speed of 25 ft/sec, and the ambient temperature is 70°F, how much is the power in the wind in an area of 100 ft²?

Solution: An area of 100 ft² is the area of a square 10 ft wide on each side. If the area is circular, diameter of the circle is 11.3 ft. (that is, more than the side of a square.) The air density at this temperature is 0.075 lb/cubic ft. As in the previous example, for calculation of the power, a division by 32.17, is necessary. Then

$$\text{Power} = \frac{1}{2}\,(0.075)(100)(25)^3/32.17 = 3642.76 \text{ ft-lb/sec}$$

Review questions

1. What is energy?
2. Define the relationship between work and energy.
3. Explain the difference between energy and power.
4. What is the type of energy in a moving object?
5. If a moving object comes to a stop, where does the energy go?
6. Why does the energy in the wind depend on the area of an air passage?
7. Why can not all of the energy in the air be used in a turbine?
8. What is the Betz limit?
9. Why is Betz limit important?
10. What is the value of Betz limit?
11. Explain what is meant by "power coefficient" in a turbine?
12. Can the power coefficient be 1.2?
13. What is the usual value for the power coefficient of a well-designed turbine?
14. Explain what factors determine wind speed.
15. Is wind always available all the time?
16. Does wind blow faster or slower at a higher height?
17. How do buildings influence wind speed?
18. Does wind always blow with the same speed?
19. If the average wind speed in a region is 10 ft/sec, can wind blow at a higher speed than that?
20. Does wind speed have any effect on energy from wind?
21. Can you say for the wind speed, that a turbine generates more energy at night or during the day?

Problems

1. If wind speed is 32 mph, how much is it in feet per second?
2. Find the power in the wind for an area of a circle whose diameter is 100 ft, if the air density is 0.08 lb/ft^3 and wind speed is 28 ft/sec.

PROJECT

In table 2.1 add all the values in the fourth row together. How much do you get? Is this what you expected?

Now, add all the values in row 3 together and see what you get. Discuss the result with your teacher.

Fundamentals

© Cengage Learning 2012

KEY TERMS

Terms that you should already know: *force, pressure*

New Terms
aerodynamic force
airfoil
angle of attack
cambered airfoil
chord
chord line
drag
drag coefficient
leading edge
lift
lift coefficient
mean camber line
pound per square inch (PSI)
torque
trailing edge
vector

OBJECTIVES

After studying this chapter, you should be able to:

* Realize the forces from wind on an object.
* Understand how wind generates force on an object in the wind stream.
* Know what an airfoil is.
* Explain the geometry of an airfoil.
* Explain what aerodynamic forces on an airfoil are and how they are generated.
* Understand what the moment of a force is.
* Understand how forces lead to rotation.
* Know the meanings of drag and lift.
* Understand and realize the difference between drag and lift forces.
* Understand the meaning of sum of forces.
* Learn about lift coefficient and drag coefficient.

3.0   Introduction

When wind blows, it exerts some forces on all the objects that are touched by the wind. The size and direction of these forces are evident in some cases, and you can easily observe the effect(s). For example, wind bends the trees and carries light objects. But, you cannot see the effect of wind on a solid wall. In this chapter you will learn about the nature of the two main forces from wind on a stationary object. These are *lift* and *drag*. Lift and drag are the two *components* of the total force from wind. They determine whether an airplane can fly or a wind turbine can turn. Knowledge about these force components is fundamental to understanding wind turbines. Also, there are two coefficients associated with these force components. You will learn all of these in this chapter.

3.1   Force from wind

When an object is in the path of the wind it is subject to two forces from wind. This is true for an object in the path of any fluid that moves, like water flowing in a pipe or in a river. For the sake of concentration on wind only, here we do not consider the general case and emphasize only the force from wind. An object can be of any shape and any size. The simplest case to start with is a plate with a rectangular shape. **Figure 3.1** shows a plate that lies in the path of wind at an angle with the wind direction.

The direction of wind in figure 3.1 is from left to right. As shown at the top of the figure, when wind blows it pushes the plate. This is due to a force component parallel to the wind direction. On the other hand, as wind passes through, depending on the angle of the plate with the direction of wind, it causes the pressure at the two sides of the plate to be different. The result of this pressure difference is a force that pushes the plate to one side (in figure 3.1 it pushes the plate upward). This push is from the side with higher pressure to the side with lower pressure. This lower or higher pressure is due to the speed of wind that is disturbed because of the plate. In addition to the above mentioned two force components, one more effect of the wind is to cause a rotation of the plate. This rotation is due to a **torque.**

torque
The turning effort effect of a force about a point away from the line of action of the force, also called the *moment* of the force about that point.

In figure 3.1, three cases are shown in which the angle of the plate with the wind direction are different. In (a) the general case is shown, where all three components—two force components and a torque—exist. In (b), the plate is parallel to the direction of wind and, as a result, there is no pressure difference on the two sides of the plate. Consequently there is no side force and no torque. There is only a small force pushing the plate in the same direction as the wind blows. In the third case, (c), the plate is normal (perpendicular) to the wind direction. In this case, again, there is no side force or torque. There is only a pushing force that is in the direction of wind.

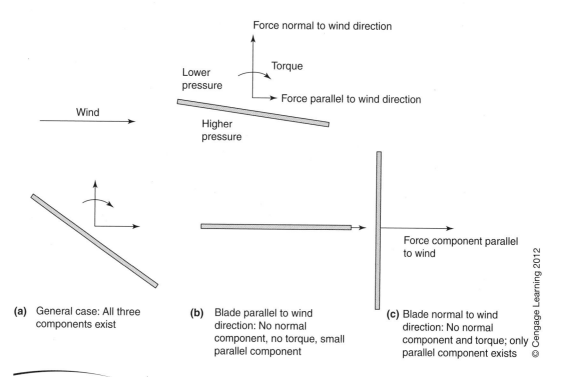

FIGURE 3.1 *Force and torque from wind on a plate.*

FORCE AND TORQUE

When a force is applied on an object it causes a motion in the same direction as the applied force. In the same way that a (linear) motion is generated by a force, a rotational motion is generated by a torque. For example, a motor generates a torque on its shaft, which causes the shaft to turn.

On the other hand, a force can make a torque, as seen in the figure below. The torque of a force depends on the magnitude of the force and the distance of the force from an axis.

A force is represented by an arrow, the length of which denotes the size of the force, and its direction shows the direction of the force, as shown in (1), (2), and (3) in the figure. The force in (2) is in the same direction as the force, in (1), but it is twice as much. The force shown in (3) is an upward vertical force. In (4) through (7) you can see a bar that is connected to a shaft at the left. In each case the bar is subject to a force.

(continued)

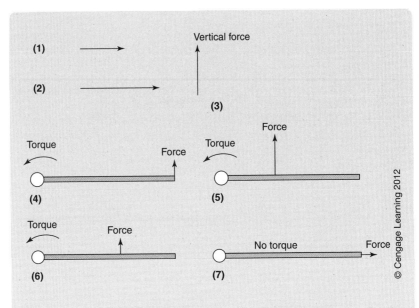

In three of these cases, the force generates a torque, as shown. This is shown by a curved arrow, which indicates the direction of the generated torque (but not its magnitude).

If force is in pounds and distance is in feet, then the torque is in pound-foot (lb-ft), also alternatively used as foot-pound (ft-lb).

angle of attack
The angle an airfoil or a thin plate makes with the direction of airflow.

drag
The component in the direction of wind of an aerodynamic force.

lift
A force perpendicular to the airflow direction caused by the difference in pressure on the two sides of a thin object in a wind stream.

The important thing is to realize the effect of the plate angle on the forces. This angle is called the **angle of attack.** In order to understand that better, refer to **Figure 3.2.** Also in figure 3.2, the proper names for the two force components are used. The force component parallel to the direction of the wind is called *aerodynamic drag* or just simply **drag,** and the force component perpendicular to the direction of the wind is called *aerodynamic lift* or just simply **lift.** It is important that you learn these two terms and their meanings. Note that the lift force is not necessarily upward. It depends on the angle of attack.

When the angle of attack is small the lift force is larger than the drag force. As the angle becomes larger the lift decreases and the drag increases. When the angle of attack is 90°, the lift is zero (for this plate) and the magnitude of drag is maximum. The drag force is the resistance of air on something that moves in the air or the force from air when wind flows over an object. On the other hand, lift is a force that moves an object to a side (it can be upward). For example, the force that keeps an airplane in the air is the lift on its wings.

VECTORS

A **vector** is used to represent an entity that has both value and direction. The best example is "force." Any force exerted on an object has a magnitude, which implies how much the force is, indicating whether it is small or large. Moreover, it has a direction, indicating at what angle the force acts. There are other entities that are better shown by a vector.

Vectors can interact with each other. For instance, they can be added together. The following figure shows a number of facts about vectors and how they are added together.

vector
A quantity that has both magnitude and direction, such as a force, as opposed to a scalar value that has only magnitude, such as temperature.

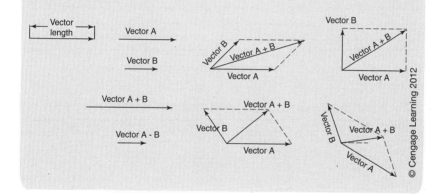

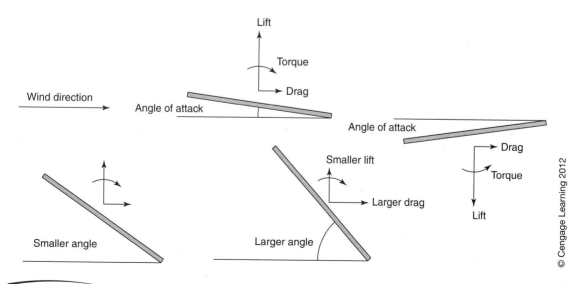

FIGURE 3.2 *Definition of lift, drag, and angle of attack.*

3.2 Aerodynamic force

aerodynamic force
Force exerted by
moving air or any gas
on an object in the air
stream.

The force from wind on a plate, just studied in the previous section, is called **aerodynamic force.** We referred to the two aerodynamic force components as lift and drag. In fact, any force can be broken into two components. Here the lift force and the drag force are the two components of the aerodynamic force on the plate under consideration. These two components are perpendicular to each other; that is, they make an angle of 90° with each other, as shown in figure 3.1.

Recall that force exerted on a surface (the plate surface, in our discussion) is always the product of the area and the pressure on the surface. That is,

$$\text{Force} = (\text{Area})\,(\text{Pressure}) \tag{3.1}$$

Thus, if the area of a plate is known (when the shape and dimensions of the plate are known), we need to find the pressure from wind and multiply that by the area of the plate in order to determine the aerodynamic force.

As you may be able to guess, the pressure from wind on a surface initially depends on the wind speed and the air density at the temperature under consideration. The exact relationship is

$$\text{Pressure from wind} = \tfrac{1}{2}\,(\text{Density})(\text{Wind speed})^2 \tag{3.2}$$

Note that in order to get correct results, density and wind speed must be expressed in their proper units.

If the magnitude of the pressure, thus obtained, is multiplied by the surface area, then the force from wind on any object can be found. In addition to the magnitude of the force, its direction is important. For the simple case of a plate, the direction of the force is perpendicular to the plate.

FORCE AND PRESSURE

In order to review the relationship between force and pressure, and their measurement units, consider the following example:

EXAMPLE:
A rectangular concrete block is 1 ft × 0.5 ft (12 in. × 6 in.) and weighs 10 lb. What is the pressure from this block to the ground when it is laid on the ground?

Solution: Remember that the weight of an object is the force of gravity to that object. When this concrete block is laid down on a surface (say ground), the force that it exerts to that surface is its weight. Thus, we have

$$\text{Force} = \text{Weight} = (\text{Area})(\text{Pressure}) = 10\text{ lb}$$

$$\text{Area} = (12)(6) = 72\text{ in}^2$$

Thus,

Pressure = 10/72 = 0.14 lb/in² = 0.14 pound per square inch = 0.14 PSI

Note the term **PSI**, standing for pound per square inch.

pound per square inch (PSI)
A unit for measuring pressure in the Imperial system of measurement.

In the following example the aerodynamic force on a rectangular flat plate is determined in order to give you an idea about the amount of force expected.

EXAMPLE 3.1:

A 3 × 1-ft rectangular plate is held in the wind stream, perpendicular to the wind direction. If the ambient temperature is 70°F (thus, the air density is 0.075 lb/ft³), and the wind blows at a speed of 25 ft/sec, find the force exerted from wind on the plate.

Solution: First, refer to figure 3.2 to observe that the values of lift and drag, the two components of the aerodynamic force, depend on the angle of attack. Here the angle of attack is 90°, and all the aerodynamic force will appear in the form of drag.

$$\text{Plate area} = (1)(3) = 3 \text{ ft}^2$$

$$\text{Aerodynamic pressure} = (\tfrac{1}{2})\,(0.075/32.17)(25)^2 = 0.73 \text{ lb/ft}^2$$

$$\text{Aerodynamic force} = (0.73)(3) = 2.19 \text{ lb}$$

Note that this force is in addition to the atmospheric pressure that is exerted on both sides of the plate. Thus, this 2.19 lb is the net force that tries to move the plate. Furthermore, it can be assumed to be concentrated at the center of mass (here, the geometric center) of the plate.

lift coefficient
The ratio of lift force to the aerodynamic force causing it. This depends on the profile of an object and not its size.

Lift and drag coefficients   3.3

In the previous sections you learned about aerodynamic force and lift and drag, which are the two components of aerodynamic force on an object in the path of wind. You also learned that the values of lift and drag are not constant and can change. In this section, two more terms are defined, which are used in determining the lift and drag force components for a given object (a flat plate, for example) and in the ratio between them (usually the lift-to-drag ratio), which is very important, as we will see later. These new terms are **lift coefficient** and **drag coefficient.**

The magnitudes of the lift and drag forces depend on the angle of attack (see figure 3.2), but at each angle their values can be defined in terms of the aerodynamic force; that is,

drag coefficient
The ratio of a drag force to the aerodynamic force causing it. This depends on the profile of an object and not its size. See appendix B for the drag coefficients of some known shapes.

$$\text{Lift force} = \text{Lift coefficient} \times \text{Aerodynamic force} \qquad (3.3)$$

$$\text{Drag force} = \text{Drag coefficient} \times \text{Aerodynamic force} \qquad (3.4)$$

where the aerodynamic force depends on the shape of an object. Based on the above formulas, the definition for lift coefficient and drag coefficient are as follows:

$$\text{Lift coefficient} = \frac{\text{Lift force}}{\text{Aerodynamic force}} \qquad (3.5)$$

$$\text{Drag coefficient} = \frac{\text{Drag force}}{\text{Aerodynamic force}} \qquad (3.6)$$

Although the "lift" and "drag" coefficients change with the angle of attack, they are highly dependent on the shape of an object. Moreover, they are the same for objects of the same shape, since these two quantities are "ratios." For example, if a flat plate of 1 ft × 6 ft is considered, its lift coefficient and drag coefficient are the same as those of a 2 ft × 12 ft flat plate. In this sense, if for the first plate the aerodynamic force in a given wind speed and angle of attack is 10 lb, the lift is 8 lb and the drag is 6 lb. Then the lift and drag coefficients are respectively 8/10 (= 0.8) and 6/10 (= 0.6). Now, if the larger plate is held with the same angle of attack in a wind stream and the dynamic force on it is 50 lb, then we can find the lift and drag as follows, using the fact that for the two plates the lift and drag coefficients are the same.

$$\text{Lift} = (50)(0.8) = 40\,\text{lb}$$

$$\text{Drag} = (50)(0.6) = 30\,\text{lb}$$

3.4 Airfoils

The importance of drag and lift forces is their contribution to the active action in the operation of a wind turbine. So far we have talked about an object, in general, and a flat plate, in particular, inside a wind stream. As we will see in the next chapter, in certain turbines it is the drag force that turns the turbine and in others it is the lift force. We learned in the previous section that the aerodynamic force depends on the shape and size of an object and that if at any angle of attack the lift and drag coefficients are known, one can determine the lift and drag forces on the object.

In a wind turbine we like to have the maximum power drawn from wind at each wind speed. For this reason, if the turbine works based on lift force, we like the lift force to be much larger than the drag force. If, on the other hand, a turbine works based on the drag force, we like the drag force to be larger than the lift. You will see later that the turbines that work based on lift force are more efficient and preferred to those based on drag force.

It may be worth mentioning here that airplanes work based on the lift force on their wings. When an airplane moves, it behaves as if it is stationary and air flows around it, like an object in the wind stream. In the case of an airplane, the drag force is opposite to the airplane motion and it is not desirable; thus, its value must be as small as possible. On the contrary, lift is the force that keeps the plane in the air; so, it is desirable that its magnitude is large.

In order to increase the lift and reduce the drag, an **airfoil** is employed instead of a flat plate. This is true for an airplanes wings as well as for the blades of the majority of the wind turbines. An airfoil has a cross section as shown in **Figure 3.3**. There is not only one airfoil, but many designs exist that are different from each other and have different characteristics.

It is quite important to note that many different sizes of airfoils can be made from one airfoil profile. In other words, if the relative sizes of various parts of an airfoil are kept the same we can have many different scales of the same airfoil. For all the different sizes, the lift and drag coefficients will be the same, but the lift and drag forces will increase with the area of the airfoil. If the area doubles, so does the aerodynamic force; and, thus, the lift and drag forces double (under the same temperature, barometric pressure, and air speed).

What is shown in figure 3.3 is the section of an airfoil. You can imagine a flat plate over which certain parts are added to shape it as having curved surfaces. These curved surfaces considerably change the aerodynamic property of the flat plate. The thin flat surface extends from the extreme point in the **leading edge** to the extreme point in the **trailing edge.** The extent between these two extreme points is called the **"chord."** The imaginary flat thin plate represents the **chord line,** a line connecting the leading edge to the trailing edge. The leading edge is a smooth curve, whereas the trailing edge is a sharp point. An airfoil can be symmetric, that is both sides have the same curve, or the upper and lower curves can be different. In the latter case the airfoil is said to be cambered **cambered airfoil**.

As you can see in figure 3.3, the upper and lower surfaces make a profile that defines the airfoil. The thickness of the airfoil at various points along the chord varies. At one point it has the thickest cross section. The midpoints between the upper and the lower curves of the airfoil section, when connected together, form the **mean camber line.** This is a curved line connecting the leading edge to the trailing edge. The mean camber line and the chord line define an area that can be thin or thick. The thinness (or thickness) of this area affects the

airfoil
A profile for the outline of an airplane wings or a turbine blades with the property of having a large lift coefficient and a very small drag coefficient.

leading edge
The edge where the moving air first strikes the airfoil.

trailing edge
The edge where the moving air leaves the airfoil.

chord
The distance between the leading edge and the trailing edge in an airfoil.

chord line
A line connecting the leading edge and the trailing edge in an airfoil.

cambered airfoil
An asymmetric airfoil (the two sides profile curves are different), as opposed to a symmetric airfoil.

mean camber line
A curve in an airfoil between the leading edge and the trailing edge, formed by connecting the midpoints of lines connecting points from upper profile curve to the lower profile curve.

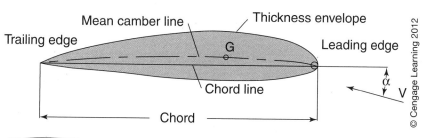

© Cengage Learning 2012

FIGURE 3.3 *Typical pattern of an airfoil and the definition of its various terms.*

aerodynamic properties of an airfoil. Figure 3.3 also shows the center of gravity of the airfoil section (point G).

The important property of an airfoil is that it possesses larger values of lift force and smaller values of the drag force for most angles of attack that arise in practice. This property is usually represented by the lift and drag coefficients and the lift-to-drag ratio, which is the ratio of the lift coefficient to the drag coefficient. Whereas with a flat plate the maximum possible value for this ratio can be a small number, with airfoils this ratio can reach values over 20.

Because of the desirable aerodynamic property of airfoils, in all the turbines for which the active force is "lift" the structure of the blades has the profile of a selected airfoil.

When a turbine blade (or an airplane wing) with an airfoil profile is inside a wind flow, various parts of it are subject to pressures of different values. The resultant of all this pressure surrounding it translates into one single force F, which has a component in the direction of wind (drag) and another component normal to the wind direction (lift). This force F acts at the center of gravity of the blade (wing). This is as schematically shown in **Figure 3.4.**

It is this lift force and the drag force that exhibit the effect of wind on the blade in a wind turbine and finally contribute to the rotation of the turbine and the amount of power it draws from the wind (see **Figures 3.5** and **3.6**).

A good airfoil is required to have the following properties:

- Gradual curves.
- Sharp trailing edge.
- Round leading edge.
- Low thickness to chord ratio.
- Smooth surfaces.
- High lift-to-drag ratio.

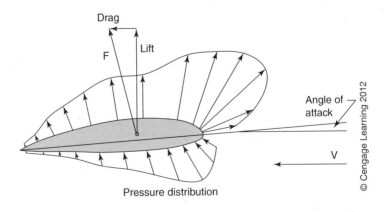

FIGURE 3.4 *Typical pattern of the forces acting on an airfoil section.*

FIGURE 3.5 *Cross section of a 41-m (134.5-ft) wind turbine blade, seen from the tip.*

FIGURE 3.6 *Picture of the 134.5-ft blade whose cross section is shown in figure 3.5.*

EXAMPLE 3.2:

In a blade with an airfoil profile the total aerodynamic force acting on the blade when wind speed is 25 ft/sec is 50 lb. The lift coefficient for this airfoil at the given angle of attack is 0.95. For the same angle of attack, if the wind speed changes to 40 ft/sec, find the lift force on the blade.

Solution: Since the angle of attack does not change, the lift coefficient remains the same at 0.95. The aerodynamic force, however, is proportional to the square of the wind speed and will increase accordingly (see equations 3.1 and 3.2). Note that since the blade dimensions do not change, the aerodynamic force for the wind speed of 40 ft/sec can be found directly without finding the aerodynamic pressure and blade surface area.

$$\text{Aerodynamic force at 40 ft/sec} = \text{aerodynamic force at 25 ft/sec} \times (40/25)^2$$

$$\text{Aerodynamic force at 40 ft/sec} = (50)(40/25)^2 = (50)(2.56) = 128 \text{ lb}$$

$$\text{Lift force at 40 ft/sec wind speed} = (0.95)(128) = 121.6 \text{ lb}$$

EXAMPLE 3.3:

A turbine has three similar blades, each with the profile of the airfoil as in example 3.2. Under the operating conditions the lift coefficient is 0.95 and the driving torque in the shaft is only due to the lift force. If the wind speed is 25 ft/sec and each blade is 22 ft long, find the torque generated in the shaft.

Solution: Based on the information in example 3.2 the aerodynamic force on each blade for the 25 ft/sec wind speed is 50 lb. Since only the lift force contributes to the turbine rotation, we first find the lift force for each blade:

$$\text{Lift force for each blade} = (50)(0.95) = 47.5 \text{ lb}$$

Assuming that the center of gravity of each blade is in its midpoint, the distance of the center of gravity from the turbine shaft is 22/2 = 11 ft.

The torque of each blade, contributing to the turbine rotation, thus can be found as

$$\text{Torque of each blade} = (47.5)(11) = 522.5 \text{ ft-lb}$$

Since there are three blades, each blade contributes to the total torque that rotates the turbine. The total torque is thus

$$\text{Total torque generated in the turbine shaft} = (3)(522.5) = 1567.5 \text{ ft-lb}$$

Chapter summary   3.5

- Any object in the wind stream is subject to a force from wind on it.
- The force on an object depends on the form and size of the object.
- The force can result in a torque, as well.
- Force tries to push an object; torque tries to rotate an object.
- In the English system of measurements force is measured in pounds and torque is measured in pound-foot (or foot-pound).
- A vector is a quantity with direction (compare to a number that does not have direction).
- Vectors are used to represent forces.
- Vectors can be added together, like numbers are added together.
- For a plate lying in the wind stream, the force from wind depends on the angle between the plate and the wind direction.
- For something thin, such as a plate, the angle between the wind direction and the plane of the object is called *angle of attack*.
- The force from wind on an object is called *aerodynamic force*.
- Aerodynamic force on a thin object has two effects: *drag*, which is a force in the direction of wind, and *lift*, which is a force at 90° from the wind direction.
- The force from wind on a thin plate-like object is proportional to the object's area and the air density, as well as the square of the wind speed (so, if wind speed doubles, the force will be quadrupled).
- *Lift coefficient* is the ratio of lift force to aerodynamic force, for an object.
- *Drag coefficient* is the ratio of drag force to aerodynamic force, for an object.
- An airfoil is a long and thin object, whose cross section has a special form. The form of the cross section gives a specific property to the airfoil; that is, a large lift coefficient and a small drag coefficient.
- An airfoil is used for airplane wings, airplane propellers, and wind turbine blades, among other things.
- An airfoil has a *leading edge*, which in operation confronts air, and a *trailing edge*, where air leaves it.
- The width of the airfoil (between the leading edge and the trailing edge) which is much larger than the airfoil thickness is called the *chord*.
- A straight line connecting the leading edge to the trailing edge is the *chord line*.
- A good airfoil has gradual curves, a round leading edge, a sharp trailing edge, a low thickness-to-chord ratio, and a high lift-to-drag ratio.

ADVANCED LEARNING

Aerodynamic forces

Referring to figure 3.4 and in connection with the definition of lift and drag, there is another effect of the wind on an object in the wind stream. In addition to the lift force and the drag force, wind causes a torque (aerodynamic torque) in the object that tries to turn it around its center of gravity. For a turbine blade, it tries to turn the blade, not the turbine. In this sense, this moment is not contributing to the energy extracted from the wind.

The following figure shows the lift, drag, and the aforementioned torque in a blade, assumed concentrated in the blade center of gravity. Note that the blade center of gravity and the airfoil section center of gravity are two separate things. The center of gravity of a blade is the point that you can assume all the mass of the blade is concentrated, whereas for each slice of the blade its center of gravity is the point at which the mass of that particular slice, having the form of the airfoil, can be assumed to be concentrated.

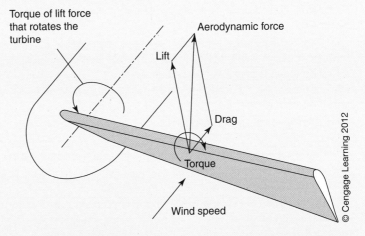

© Cengage Learning 2012

Note that the only useful contribution to the turbine rotation is from the torque of the lift force. The drag force pushes and bends the blade in the wind direction and the (aerodynamic) torque twists the blade (counter to the direction of the wind).

Review questions

1. What is aerodynamic force?
2. If an object is in the steady air (not wind) is it subject to an aerodynamic force?
3. What is the most important parameter in aerodynamic force?

4. If an object moves in the air, is there any aerodynamic force on it?
5. What is lift? Does it have a direction?
6. What is drag? When do you feel a drag?
7. Lift and drag forces are components of what?
8. Which one is more important in an airplane wing, lift or drag?
9. What is a vector used for?
10. What is the definition of a vector?
11. How does one represent a force?
12. What is the difference between force and pressure?
13. What is a torque and how is it represented?
14. Does a force always generate a torque?
15. What does a torque do?
16. Have you ever felt a torque?
17. What is an airfoil?
18. Why is an airfoil preferred to a flat plate?
19. Where in a wind turbine is an airfoil used?
20. What property is most significant in an airfoil?
21. In addition to a wind turbine, can you recall another application of an airfoil?
22. Define the lift and drag coefficients
23. In an airfoil, is the lift coefficient larger or the drag coefficient?
24. What generates lift in an airfoil?
25. If there is no wind, can you say how much the lift force on a turbine blade is?
26. What is the chord in an airfoil? Is it a force, a length, a line, or motion?
27. What is a chord line?
28. What are the requirements for a good airfoil?
29. What is the mean camber line? Is it a straight line? Can it be a straight line?
30. What is an angle of attack?

Problems

1. The blade of a wind turbine experiences a force of 100 lb when the wind speed is 10 ft/sec. To what value does the force change if under the same conditions the wind speed increases to 15 ft/sec?
2. If the lift and drag coefficients in an airfoil are 0.998 and 0.05, respectively, determine by how many times the lift force is larger than the drag force in a blade with that airfoil profile.
3. If a small 3200-lb airplane has two wings with the airfoil profile in problem 2, and the drag force on each wing at a given wind speed is 100 lb, what is the net lift force on the plane?
4. If a turbine has three blades and the center of pressure (where the lift and drag forces apply) of each blade is 10 ft from the axis of the turbine and the lift force on each blade is 100 lb, how much is the torque that rotates the turbine?

PROJECT

For this project you need an electric fan. If you do not have an electric fan handy, then generate an air flow by smoothly blowing with your mouth.

Cut a 3-in. by 6-in. rectangle out of cardboard. Hold that slightly slanted from horizontal in front of the air flow of the electric fan. Note the effect of the air flow that tries to push the cardboard in the same direction of the air flow, as well as pushing it upward.

Increase the angle and see if you can notice the difference.

Now make the cardboard have a curved form (by pushing the edges and keeping their distance shorter, using masking tape). Hold the cardboard in front of the fan. You should notice a larger upward (lift) force, compared to when the cardboard was flat.

Wind Turbine Basic Types: Analysis and Characteristics

© Cengage Learning 2012

KEY TERMS

Terms that you should already know: *lift, drag*

New Terms
Darrieus turbine
drag-type turbine
horizontal-axis wind turbine (HAWT)
lift-type turbine
propeller turbine
Savonius rotor
solidity
vertical-axis wind turbine (VAWT)

OBJECTIVES

After studying this chapter, you should be able to

- Categorize various types of wind turbines.
- Know what is meant by HAWT and VAWT.
- Know the advantage and disadvantage of HWATs and VWATs.
- Describe how wind forces a wind turbine to rotate.
- Know what a propeller turbine is and how it works.
- Understand why in some turbines the blades are twisted instead of being flat.
- Explain what a Darrieus machine and an H-rotor are.
- Know what a Savonius rotor is.
- Explain the main difference between various turbines.

4.0   Introduction

While many people think that wind turbine means a three-blade rotor on top of a tower, there are other types of wind turbines that can catch energy from wind. In this chapter, these other versions of wind turbines are introduced. You learn about the Darrieus machine, the Savonius rotor, and the terms HAWT and VAWT. But more weight is given to the three-blade turbine, which receives a thorough description. This type of turbine is the most popular, most economical, and most practical one today; this is why you see them or hear about them more often than the other turbines.

For any turbine type, certain properties are desirable, while some other properties are not and count against it in comparison with other turbines. Having a high power coefficient and a large starting torque are among the desirable characteristics, whereas having a high **solidity** is a negative point. In a propeller turbine, solidity is the ratio of the area of the blades to the area of a circle swept by the blades. For instance, a three-blade turbine has less solidity compared to a four-blade turbine with the same size blades. To compare the solidity of various types of turbines, one can use the weight ratio; for instance, out of two turbines having the same power, the one with heavier blades has more solidity.

solidity
The measure of percentage of solid area in a circle traced by a turbine rotor; for example, a four-blade turbine has more solidity than a three-blade turbine of the same size.

4.1   Turbine classification

As we learned in the previous chapter, when air flows around an object, two forces act on the object, drag and lift. Accordingly, we have turbines that work based on either of these forces. Thus, in general, we have lift-based (or lift-type) turbines and drag-based (or drag-type) turbines.

This categorization is based on the type of active force that makes the turbines turn. Turbines can also be classified based on their axis, whether it is horizontal or vertical. Axis here refers to their main shaft about which the rotating parts revolve. As we'll see in this chapter, certain turbine types can work only with a horizontal axis, while others can work with a horizontal axis or a vertical axis, and even they can be installed with their axis at an angle. In this sense, a wind turbine can be classified as a **horizontal-axis wind turbine (HAWT)** or a **vertical-axis wind turbine (VAWT).**

Even without more details about any particular turbine, one can see a major difference between a horizontal-axis wind turbine and a vertical-axis wind turbine. Since in most cases wind blows horizontally, a wind turbine whose axis is horizontal (HAWT) is sensitive to the direction of wind. This is not true for a turbine with vertical axis (VAWT), because no matter what the direction of wind, such a turbine can catch the wind.

Another advantage of a vertical-axis turbine is the fact that all the other equipment such as generator and gearbox do not need to be on the top of the tower, as is usually the case for a HAWT. So, they are easier to access when necessary.

horizontal-axis wind turbine (HAWT)
A wind turbine whose axis of rotation is horizontal or nearly horizontal.

vertical-axis wind turbine (VAWT)
A turbine whose axis of rotation is vertical. A vertical-axis turbine is not sensitive to wind direction changes.

> A horizontal-axis wind turbine is sensitive to wind direction.

Propeller wind turbine   4.2

The popular wind turbine that you know and may have seen, with three blades, on top of a mast or tower is called a **propeller turbine.** This is because it looks like the propeller of an airplane. A propeller turbine is a **lift-type turbine** since it works based on the lift force on the blades. Although usually it comes with three blades, it can have a smaller or larger number of blades. It can work with two or even one blade. It can have four, five, or more blades. Research, however, has shown that three blades are the best combination; that is, from balance, efficiency, and other viewpoints such as how it looks and the impact it has on an observer.

In a propeller turbine, wind flow is along the turbine shaft; that is, wind blows perpendicular to the blade plane (an imaginary plane that contains the blades). Since in the open air, wind normally blows horizontally, the propeller turbine shaft has to be horizontal. Thus, a propeller turbine is a horizontal-axis wind turbine. If a propeller turbine is mounted vertically on its mast, so that its blades can move horizontally, it is less likely that they will rotate at all.

A propeller turbine can be mounted in two ways as far as the wind direction is concerned: upwind and downwind, as shown in **Figure 4.1.** In the upwind configuration, blades are in front of the tower, whereas in the downwind configuration wind hits the tower before it reaches the turbine blades. In practice, the two configurations are not the same, because of the effect of the tower on the flow of wind.

propeller turbine
A turbine with blades similar to a propeller; the most common wind turbine.

lift-type turbine
A wind turbine whose power grasp from wind is based on the lift force on the blades, as contrary to a drag-type turbine.

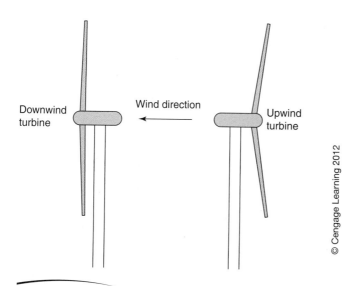

© Cengage Learning 2012

FIGURE 4.1 *Two different installation designs for propeller wind turbine.*

4.2.1 Propeller wind turbine blades and the wind force

It is interesting to know how wind turns a propeller wind turbine, while blowing toward it. This is the opposite of what happens with an electric fan. In an electric fan, the blades rotate and as a result air is pushed ahead. In a wind turbine, the flow of the air forces the blades to rotate. **Figure 4.2** shows the picture of a 131-ft (40-m) blade.

The tip of a blade is the side farther from the turbine shaft, and the root of a blade is the side near the shaft. A blade is not flat but is twisted between its root and its tip. The reason for this is explained later. Also, a blade is narrower at the tip than at the root. **Figure 4.3a** shows a schematic of a typical blade of a wind turbine. A blade of a wind turbine can be divided into different segments, each one, say, 1 ft wide, as shown in figure 4.3a. The effect of the wind on each segment is a force, pushing the segment back as well as lifting it. We refer to these forces as force components, since in combination they represent the force from wind on each segment. The sum of all the force components pushing the segments results in pushing and bending a blade. The sum of all the forces lifting the segments causes the rotation of the blade. This is shown in **Figure 4.3b.** As we can see from figure 4.3a, the effect of each segment on the blade rotation also depends on how far the segment is from the shaft axis. So, the segments that are near the blade tip have a greater share in blade rotation and the segments near the root have a lesser contribution. Note that the drag and lift forces are shown in the figure when the blades are rotating, not when they are stationary.

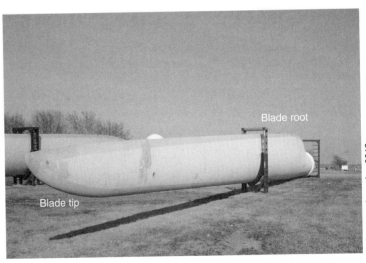

© Cengage Learning 2012

FIGURE 4.2 *A typical blade of a modern wind turbine.*

SPEED OF A POINT ON A ROTATING OBJECT

In all machinery, when a part is attached to a shaft and rotates with it, the shaft and all that rotate with it have a common "angular speed." However, various points on a rotating part do not have the same "linear speed." The linear speed for each point depends on the angular speed and the distance from the shaft axis (the axis of rotation). This is shown below

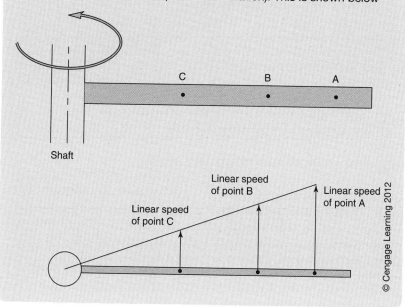

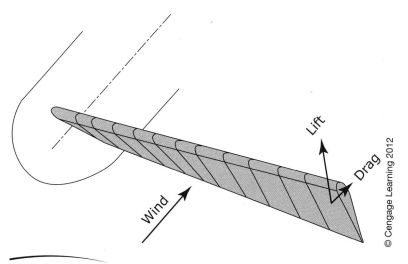

FIGURE 4.3a *For analysis a blade can be divided into a number of segments.*

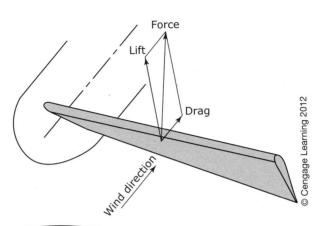

FIGURE 4.3b *Force from the wind on a blade can lead to a drag component and a lift component.*

RELATIVE MOTION AND RELATIVE SPEED

Wind direction and speed are always "relative" entities. The direction of air flow relative to a blade depends on the blade motion and the speed of rotation. Also, the relative speed of air flowing over a blade depends on the rotational speed of the blade and the distance of the segment of the blade from the axis of rotation, as well as the wind speed.

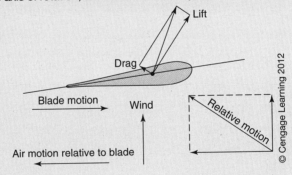

TORQUE FROM WIND ON A TURBINE SHAFT

A rotational motion is created by a torque. The torque of a force depends on the magnitude of the force and the distance of the force from an axis. In a propeller wind turbine, each blade can be divided into a number of segments. The lift force on each segment causes a torque about the turbine shaft, trying to rotate it. The turbine rotates as a result of the sum of all the torques of the lift forces on all the blade segments. The concept of dividing a blade to a number of segments is just to help our understanding of how wind causes the turbine to turn.

Blade twist

Comparing a propeller turbine blade with an airplane wing, you will see that the airplane wing is flat relative to a turbine blade, which has a twist; that is, the tip of a blade is not parallel to the blade root. This twist is not much, and can be only a few degrees, depending on the blade length. The reason why a blade must be twisted by an angle and not flat can be seen from the fact that in order to have a good lift force on a blade the air flow must hit the blade at a proper angle. This is the angle of attack, as we studied in chapter 2. When a blade rotates, the points at the tip side go faster than the points near the root. Since the wind speed is approximately the same for all the points on a blade, the relative speed of air flow with respect to the blade is different for points along the length of a blade. **Figure 4.4** illustrates the relative speed due to the combination of wind speed and the blade motion. The angle shown by φ (phi) in the figure is the angle of attack. This angle, as shown in **Figure 4.5,** is not the same for segments of the blade in the tip area, in the middle or at the blade root.

In order to have more or less the same angle of attack for all the segments of a blade, these segments must encounter the wind at different angles, as shown in figure 4.4. Otherwise some segments have very inappropriate angles of attack. In fact, for the three segments shown in figure 4.4, the direction of the relative speed determines the twist angle. In figures 4.4 and 4.5 you are looking at a blade in a direction along the blade and toward the shaft.

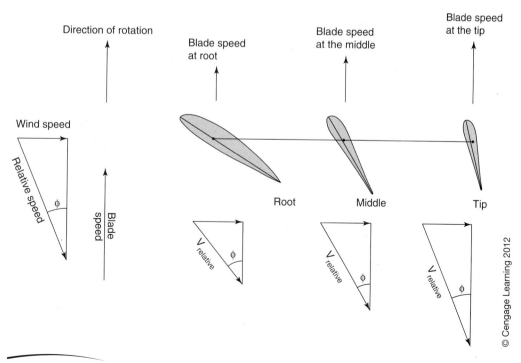

FIGURE 4.4 *Relative speed of air flow over different parts of a blade is the reason for blade twist.*

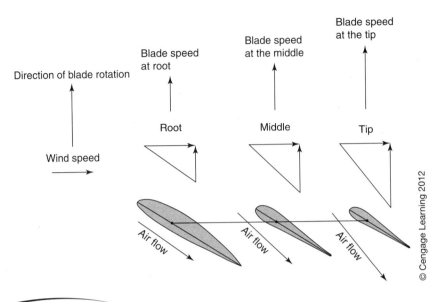

FIGURE 4.5 *If a blade is not twisted in design, a correct desired angle of attack cannot be maintained.*

Figure 4.5 shows a case where a blade has no twist. In such a case, the chord lines of all the segments of a blade are parallel. As shown, the angle of attack can be correct for certain parts of the blade, but is not correct for the other parts.

4.2.3　Yawing into wind or out of wind

Since a propeller turbine is a horizontal-axis wind turbine, its blades must be facing the wind in order to catch the maximum wind. If the body of the turbine is rotated by 90°, so that the wind direction is parallel to the blades, no energy can be caught from wind. For this reason, all the propeller turbines must be able to turn about their towers. This is called a "yaw" motion and a mechanism must be present in each turbine that detects the wind direction and turns the turbine accordingly when in operation. Conversely, when a turbine has to stop, the same mechanism yaws it out of wind, which means the turbine rotates 90° from the wind direction, so that the force trying to rotate the blades is at a minimum. In addition to yawing a turbine out of wind, other actions must be involved when a turbine is stopped, like braking. This is discussed in chapter 10.

4.3　The H-rotor

An H-rotor is a vertical-axis wind turbine in the shape of H. The two vertical segments of the letter H are the active blades, which are connected to the shaft by the middle segment. The two blades have the form of an airfoil and

SPEEDS OF DIFFERENT POINTS ON A TURBINE BLADE

Although the angular speed is the same for all segments in the blade, their linear speeds are different. **Figure 4.6** indicates the speeds of three segments at various parts of a blade.

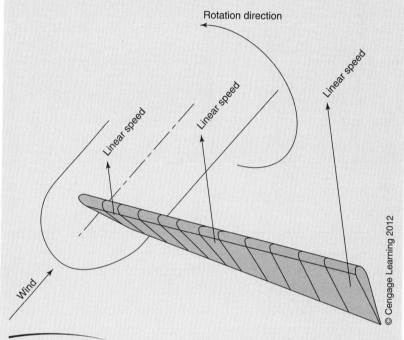

FIGURE 4.6 *Speed of different parts of a blade depend on the distance from the axis of rotation.*

the turbine works based on lift force. The schematic of this type of turbine is shown in **Figure 4.7.** Since this turbine has a vertical axis, it is not sensitive to wind direction.

The two lift forces on the two blades of an H-rotor generate a torque about the turbine shaft that rotates the turbine. This is shown in **Figure 4.8.** Note that during operation, one blade is upwind and one blade is downwind. The aerodynamic angle of attack varies constantly for each blade during rotation. The downwind blade moves in the 180° to 360° wake of the upwind blade and, thus, captures less energy than the other blade (the wind speed in this area is reduced due to the energy extracted by the upwind blade). Having two blades causes the operation of an H-rotor to pulsate. For this reason, it can have three (or more) blades in order to make its operation smoother, as shown in **Figure 4.9.**

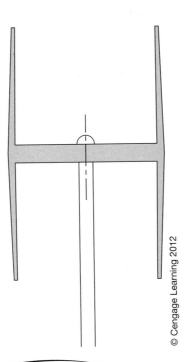

© Cengage Learning 2012

FIGURE 4.7 *Schematic of an H-rotor.*

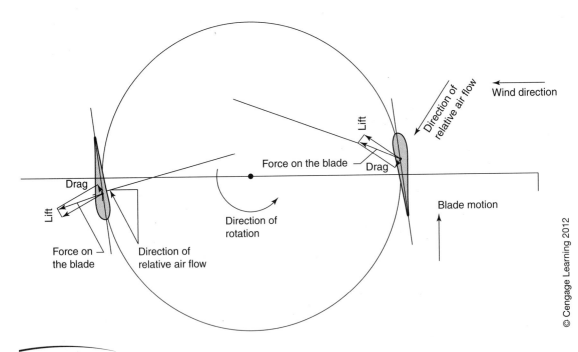

© Cengage Learning 2012

FIGURE 4.8 *Active forces on the blades of a an H-rotor.*

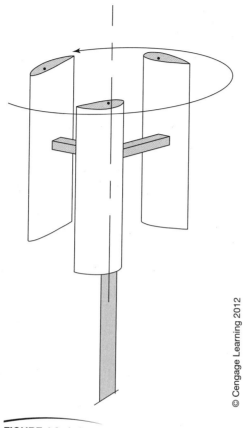

© Cengage Learning 2012

FIGURE 4.9 *A three-blade H-rotor.*

Darrieus turbine   4.4

A **Darrieus turbine** or Darrieus machine is more or less similar to an H-rotor in terms of having a vertical axis and working based on lift forces on the blades. The difference is in the way the blades are attached to the shaft. A Darrieus turbine has the shape of an egg beater. Instead of the blades being attached to the shaft in the center, they are continued from both up and down and are curved. The blades are attached to the shaft at both ends. **Figure 4.10** shows a picture of a Darrieus machine. The curvature of the blades allows the vertical shaft to be supported by a number of guy wires. In this sense, the main tower supporting the turbine shaft does not need to be as solid and strong as it must be in an H-rotor.

A disadvantage of a Darrieus machine is that it does not have a good starting torque. This means that at low wind speeds it cannot easily start to rotate. After it starts rotating, however, it has a good torque and can continue to generate electricity.

Darrieus turbine
A vertical-axis wind turbine that looks like an eggbeater.

FIGURE 4.10 *A Darrieus machine.*

Despite the two main advantages of vertical-axis turbines (not being sensitive to wind direction change and most of the components being accessed from the ground), not many Darrieus machines have been built in the past, and none is made today. From practicality, preference is given to propeller turbines.

4.5   Savonius rotor

A **Savonius rotor** is a **drag-type turbine,** named after its inventor, Sigurd J. Savonius. Its construction is relatively simple compared to the previously described wind turbines. In the simplest form a Savonius rotor consists of two half-cylinder sections fixed to a shaft in the form that their cross section makes a letter "S," as shown in **Figure 4.11.** At each moment in time one blade captures the wind while the other moves against the wind, thus opposing the wind. The net torque to rotate the turbine is the torque from the blade capturing the wind energy minus the resistive torque that the other blade receives against moving. This is the case for all the drag-type turbines.

One can add more half cylinders on the shaft in order to increase the capacity of wind capture. This is equivalent to increasing the length of the cylindrical sections, if all the sections are aligned. Alternatively a second set can be installed at 90° from the first half cylinders. This adds to the uniformity of rotational torque on the shaft, since with only two half cylinders the absorbed power pulsates (this is not uniform as the rotor turns).

A Savonius rotor has about half of the power capture capability of the other (lift-based) turbines. That is to say, its power coefficient in the best situation is

Savonius rotor
A type of drag-type wind turbine that basically consists of two half cylinders put together in the form of letter S.

drag-type turbine
A turbine that works based on the drag force on its blade(s).

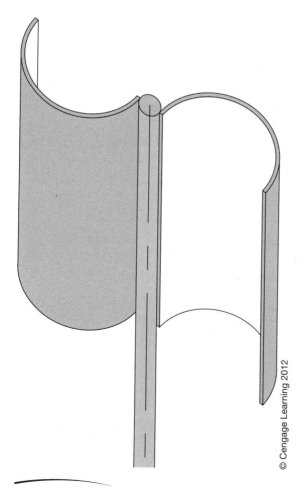

© Cengage Learning 2012

FIGURE 4.11 *Construction of a simple Savonius rotor.*

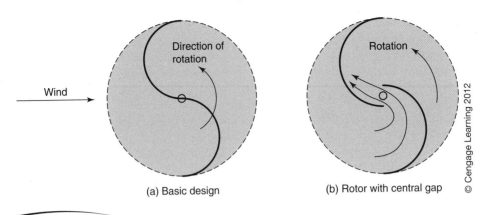

FIGURE 4.12 *Alternative design for Savonius rotor.*

about a half of the magnitude that can be reached by, say, a propeller turbine. Also, its construction inevitably involves a large mass (high solidity), which makes it very bulky.

Among the advantages of a Savonius rotor is its ability to capture low-speed winds and to have a good starting torque. Also, it can be installed either vertically or horizontally; that is, its axis can be horizontal or vertical. If a Savonius rotor is installed horizontally, then the direction of wind matters for its operation. Small units of Savonius turbines can be used for roof-top mounting if desired, provided that their operational speed is low and the building has sufficient structural strength. More on this subject is discussed later (see chapter 16).

In order to increase the power capture capability of a Savonius turbine, certain modifications to the basic design have been proposed. The first one is to have a gap in the structure where the two half cylinders are joined. This allows the air to pass through this gap from the segment capturing wind to the segment opposing wind, as depicted in **Figure 4.12.** The advantage is twofold: wind is not trapped in the capturing blade and has a more steady flow, and the opposing blade has an extra force to help it push the air.

The other design arranges the two blades to be able to swing about an axis at their outer edges. This is shown in **Figure 4.13.** This arrangement can greatly decrease the resistance to the air flow of the opposing blade. Both of these designs enhance the power coefficient, and thus the capacity, of a turbine for the same size, but introduce their own complexities.

A third design for a Savonius rotor alteration is to twist each of the half cylinders. This helps the captured power to be more uniform rather than pulsating.

Although the Savonius rotor has a simple structure and has a number of advantages over the propeller turbine, its application at the industrial level for power generation is very limited, because of two main disadvantages: one is the low power coefficient and the second is its bulkiness.

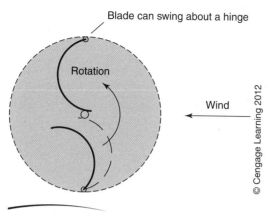

FIGURE 4.13 *Savonius rotor with swinging blades.*

Chapter summary 4.6

- There are other wind turbine types than what we usually know as a wind turbine.
- Wind turbines can have a horizontal axis or a vertical axis.
- HAWT stands for horizontal-axis wind turbine.
- VAWT stands for vertical-axis wind turbine.
- Vertical-axis wind turbines are not sensitive to wind direction. This is a major advantage of a VAWT.
- Another advantage of a VAWT over an HAWT is the easier accessibility of most of the components, since they are at ground level rather than up in the air.
- Horizontal-axis wind turbines must be yawed into wind for maximum wind capture. They must continuously follow the wind direction.
- The most common wind turbine is the propeller wind turbine. It is an HAWT. It works based on lift force on its blades.
- The Darrieus machine and H-rotor are very similar and they are VAWTs. Both are lift-type turbines and have good power coefficients, but they have weak starting torques.
- A Savonius rotor is a simple-structure drag-type turbine. It can be installed as a VAWT or as an HAWT. It is made up of at least two half cylinders mounted back to back on a shaft, but in opposite directions.
- Drag-type wind turbines have a smaller power coefficient and, thus, are less efficient than lift-type wind turbines.

ADVANCED LEARNING

Angular (rotational) speed

The angular speed or angular velocity of a rotating body is usually shown by ω (omega) and can be measured in rpm (revolutions per minute) or other units. For example, the angular speed of a shaft can be 720 rpm. If a pulley is attached to this shaft, different points along a radius on the pulley have the same angular speed, but they do not have the same (linear) speed. The points at the outer edge move faster than the points nearer to the shaft. If the word "angular" is not specified before the word "speed," then linear speed is addressed. The speed of any point on the pulley can be found from

$$v = kr\omega$$

where v denotes the speed, r is the radius or distance of that point from the axis of rotation, and ω is the angular speed. The multiplier k depends on the units used for measurement. For example, if ω is expressed in rpm, then the above formula changes to

$$v = 0.105r\omega$$

then, if r is in inches, v will be in inches per second. So, for a shaft rotating at 720 rpm the speed of the outer edge of a 10-in.-diameter pulley is

$$(0.105)(5)(720) = 378 \text{ in./sec}$$

If there is a belt on the pulley, 378 in./sec is the speed with which the belt moves.

Review questions

1. What is a horizontal-axis wind turbine?
2. What is the difference between a horizontal-axis wind turbine and a vertical-axis wind turbine in terms of how they are installed?
3. What are the advantages of a horizontal-axis wind turbine over a vertical-axis wind turbine?
4. What is a propeller wind turbine?
5. Can a propeller wind turbine have no mast?
6. What is a lift-type wind turbine?
7. What is a drag-type turbine?
8. Is a lift-type turbine generally better or worse than a drag–type turbine? Why?
9. Are the blades of a propeller turbine flat?
10. What is meant by "yawing into wind"?
11. Explain an H-rotor.

12. What is the difference between an H-rotor and a Darrieus machine?
13. Name three different types of wind turbine.
14. What is meant by HAWT and VAWT?
15. Describe a Savonius rotor.
16. Why is the Savonius rotor not popular?
17. Why are other types of wind turbines not as popular as propeller turbines?
18. Is a Savonius rotor a horizontal-axis or a vertical-axis turbine?
19. Why are the blades of a propeller turbine twisted?
20. What does a twist of 12° mean to you?

PROJECT

Make a pin wind mill with a piece of paper, a pin, and a stick (you can use a drinking straw as the stick). Observe when it turns when wind blows. Does this turbine work based on lift force or drag force? What are the problems if large wind turbines are made this way?

Three-Blade Propeller-Type Wind Turbine and Its Components

© Cengage Learning 2012

OBJECTIVES

After studying this chapter, you should be able to:

- Know what the principal components of a common wind turbine are.
- Know what a nacelle is and what components are inside the nacelle.
- Know that a turbine has a hub and three blades.
- Know what a rotor is and that the blades are moveable with respect to the hub.
- Explain what the blades are made of.
- Explain the reason why the blades are not solid.
- Describe the main difference between various towers.
- Know the advantages of each type of tower.
- Explain what makes a turbine capable of standing upright despite all the force from wind that pushes it.
- Know what other components are installed inside or near a turbine.
- Learn the function of a gearbox.
- Explain why a gearbox is necessary for a wind turbine, and know its type.
- Learn that a wind turbine needs a brake to prevent its rotation when parked.
- Learn various types of electric generators and know the differences between them.
- Know that any turbine must have a generator, which converts mechanical energy grasped from wind by the rotor to electric energy.

(continued)

- Understand the role of a transformer and the meaning of step-up or step-down transformer.
- Know that each turbine has a transformer.
- Know what an anemometer is.

5.0   Introduction

In the previous chapter, we learned that there are various types of wind turbines. However, the efficiency and practicality of horizontal-axis, three-blade, propeller-type turbines has resulted in domination of the market and, in almost all the new installations of wind farms, this is the only type that is employed for energy generation.

In this chapter, the major components of this type of turbine are discussed. Some of these components, nevertheless, are essential for all types of wind turbines, but our focus here is describing the components that a technical person working on the most common wind turbines must know about. For simplicity, however, in this chapter, and in the rest of the book, the terms "propeller type" and "three-blade" are omitted, and when "wind turbine" is used, reference is made to the three-blade propeller-type wind turbine, which is always installed with a horizontal axis and is the most common commercial wind turbine.

A wind turbine must grasp the mechanical energy from wind and convert it to electrical energy. So, it has both mechanical components and electrical components. In this chapter, these components are only introduced without entering into details. More details can be found in later chapters, but a thorough study of all this equipment is beyond the scope of this book. We first describe the mechanical components and then the electrical components. Both categories then can be divided into primary components (those without which a turbine cannot function) and the secondary components that either are much smaller, or could be deleted based on the design of each particular turbine.

Table 5.1 helps to categorize the components. In addition to the components/equipment shown in the table, a wind turbine is equipped with control systems

yaw system
The entire gears and motors involved in the yaw motion of a wind turbine.

rotor
The rotating part of a turbine

hub
The part of a HAWT wind turbine to which the blades are connected.

Wind turbine major components			TABLE 5.1
Mechanical	Primary	Tower, nacelle, rotor, foundation	
	Secondary	Gearbox, brake	
Electrical	Primary	Generator, transformer	
	Secondary	Anemometer, vane, rectifier, inverter	

© Cengage Learning 2012

without which a modern turbine cannot function. The major control systems that comprise mechanical, electrical, and/or hydraulic components are blade pitch system and turbine **yaw system.**[1] Certain modifications can be expected in the future generations of wind turbines.

Mechanical components

A wind turbine primarily consists of a tower, a nacelle, a **rotor** (a **hub** and three blades), and a **foundation,** which cannot be seen. The blades are connected to a central hub, which rotates with them. The whole assembly is called a rotor. The rotor is mechanically isolated from the rest of the turbine that does not rotate with wind. The blades and hub rotate the **main shaft,** which goes inside an enclosed space on the top of the tower. This enclosed space is called the **nacelle.** The nacelle houses the gearbox, generator, and all the other necessary components such as heat exchangers, coolers and heaters, other motors and gears and so on, as will be described later.

Figure 5.1 shows three of the major components of a modern turbine with a tubular tower. The fourth component, the foundation, is in the ground and cannot be seen.

Tower

The **tower** supports the other parts and holds them in the air (off the ground). Thus, the tower must be structurally strong to withstand the weights of the components it supports and the forces from wind that can easily bend or break the tower if it is not strong enough. In the earlier turbines **lattice towers** were used. These towers are like the ones used for overhead transmission lines, made up of a number of metallic bars that are bolted or welded together. These towers can have various designs based on the height. **Figure 5.2** shows a row of wind turbines with lattice towers.

Newer turbines have **tubular towers.** These are made up of rolled steel in the form of a cylinder or sometimes slightly tapered in the form of a conic

foundation
A massive block of concrete at the bottom of a turbine, covered by soil, containing the bolts that hold the tower.

5.1

main shaft
The low-speed shaft in a wind turbine gearbox, which is connected to the rotor.

nacelle
A room at the top of a turbine tower that houses the gearbox, the generator, and other equipment.

5.1.1
tower
A necessary part of a propeller-type wind turbine that functions to hold, and support the forces from, the rotor and nacelle.

lattice tower
A tower made from many small metal bars joined together by welding or fasteners. See figure 5.2.

---

[1] This categorization is just to help make a distinction between various components. There is not always a sharp line between primary and secondary components. For example, a wind turbine cannot function if its anemometer is broken. Nevertheless, an anemometer is an instrument that is much smaller compared to other equipment.

tubular tower
A tower that has a circular structure in the form of a cylinder or a conic shape with a slight slant.

FIGURE 5.1 *Picture of a typical modern wind turbine with tubular tower.*

FIGURE 5.2 *Turbines with lattice tower.*

section. These towers are usually made in a number of shorter segments that make both manufacturing and transportation easier. The segments are attached together by bolts. This is done on site during the turbine erection, starting from the lowest segment that is attached to the tower foundation. The bolts for attaching the bottom section are outside the tower, as depicted in **Figure 5.3,** whereas the other tower sections are connected from inside, which gives an easier accessibility and a better external look. Each segment is lifted by a crane and bolted to the segment below it. Each tower segment can be 60–70 ft (18–21 m) long. Each tower segment has a platform at its bottom. This makes it easier for climbing and descending, since each segment is isolated from others, as well as the fact that a person can rest in the middle, before continuing farther up or down. As a rule of thumb, the height of a turbine tower is the same as the diameter of the blades.

The diameter of a tubular tower can be 10–14 ft (3–4 m). If it is tapered, it has a smaller diameter on the top. Some towers can be made of a concrete lower segment and the upper segments are metallic (steel). All tubular towers have an entrance door at the bottom. From the safety point of view, this door must be kept locked and only authorized people can have access.

As a rule of thumb, the height of a turbine tower is the same as the diameter of the blades.

© Cengage Learning 2012

FIGURE 5.3 *Bolts fixing a tower to its foundation.*

The tubular towers have a number of advantages over the lattice towers. Among them are:

- Tubular towers make an enclosed space, which can be more protective for electric and communication cables and other components, such as a winch for lifting equipment, against weather conditions (snow, ice, cold, dust, sunshine, etc.).
- They are more protective for people climbing the turbine for maintenance and other purposes against wind, cold, rain, snow, and so on. This includes the periodic check on the fastening torque of the bolts attaching the tower segments together.
- They can better accommodate any personnel-lifting equipment, such as a ladder or the like. Newer turbines can be equipped with a one-person lift, which makes it easier for a maintenance person to climb the turbine.
- Birds are not tempted to make their nests on the tower elements, as frequently seen with the lattice towers.
- They have a nicer look.

On the other hand, lattice towers have the advantage of easier transportation (as the parts can be assembled on site) and certain maintenance work, such as painting them, is easier.

The most common facility to climb a turbine is a ladder that is fixed to each segment of the tower. The ladder is usually welded to the inside body of the tower. In the very latest turbines the ladder can be attached to the tower by magnets. This reduces the manpower of manufacturing the tower, as well as reducing the local stresses due to welding. **Figure 5.4** illustrates the inside of part of a turbine tower. The picture shows the ladder and the cables that carry the generator output from the top of the tower to the ground. The platform is separating the tower segments.

5.1.2   Foundation

Any turbine tower must be able to withstand all the various forces from the wind. A turbine is like a high-rise building, which is subject to heavy weight and large lateral forces from wind. For any heavy structure, the foundation must be strong enough to withstand the forces. Because soil is not strong, the foundations of large structures are mounted on a number of piles that are inserted in the ground by hammering action. The piles go deep into the ground and have the effect of being attached to the ground like nails hammered into wood.

This is not practiced for wind turbines, since the area of a tower is small (compared to a high-rise building). Instead, a large and heavy foundation is made for a turbine tower that can keep the turbine upright and can resist all the forces. Under each wind turbine, hidden in the ground, there is a relatively large concrete foundation with a sufficiently huge mass that holds the whole turbine and prevents it from turning in place, tilting to side, tipping over, or being blown away. **Figure 5.5** shows the picture of a typical foundation. After

FIGURE 5.4 *Ladder and cables inside a tubular tower.*

construction of the foundation, most of it will be covered by soil, as shown in **Figure 5.6.**

The size of a turbine foundation depends on a number of factors, including the turbine size, weather conditions in the region, the type of soil, and the terrain topography. It can be a 50 ft × 50 ft × 30 ft deep block (15 m × 15 m × 9 m deep) of concrete. The tower is bolted to the foundation by a number of bolts (see figure 5.3). These bolts are long and reach the bottom of the foundation. Only part of each bolt is out of the ground to install the tower. A plastic cover can be used for protection of bolts from rain and snow.

Rotor (blades and hub)   5.1.3

Rotor refers to all the rotating parts of a turbine. Blades are the parts of a wind turbine that catch the wind energy. As you learned in chapter 3, lift forces in the three blades give rise to a torque at the turbine shaft and turn it.

U.S. Department of Energy

FIGURE 5.5 *Concrete foundation of a turbine.*

© Cengage Learning 2012

FIGURE 5.6 *A finished foundation before installing the tower.*

In a later chapter you'll learn that turning of the shaft can be at a lower or a higher speed (rpm) and how this speed can be adjusted. Blades have the form of an airfoil. Whereas the profile of the airfoil may be the same, the size of the airfoil is not the same along the length of each blade. This is partly because of the aerodynamic property that the tip of a blade must become smaller and be rounded, and partly because of the mechanical strength that the root of a blade must be stronger, thus larger and thicker, than the other parts. You have, as well, learned (chapter 4) that in a wind turbine the blades are slightly twisted.

If you look at an electric fan, you see that all the blades are connected to a central part, the hub, which holds the blades together, so that all the blades can rotate together. In an electric fan the blades and hub are just one piece. This piece is attached to a shaft and the rotation of the shaft can rotate the blades. In a wind turbine, we have more or less the same construction, except for two fundamental differences.

The first difference is that in a wind turbine, since the size is large, the blades and hub are separate parts that can be attached together. The second difference is true for all the modern commercial turbines. In the older turbines (constructed up to 15 years ago), the blades were fixed to the hub with bolts and there was no relative motion between the hub and any of the blades. Newer turbines are equipped with **pitch control.** In a turbine with pitch control capability, a blade is not fixed to the hub and can rotate with respect to the hub about its (blade's) axis. In this way, the angle between a blade and a fixed mark on the hub can be changed. This angle, called **pitch angle,** can be changed up to 90° to 100°. A wind turbine having this capability is called a variable pitch turbine.

The effect of changing pitch angle is to modify the amount of lift force from wind on the blade, thus changing the amount of energy grasp from wind. This can be used to control a turbine. Another application of this capability is to minimize the force on blades when a turbine is supposed to be stopped.

All of the mechanisms to change the pitch angle are located inside the hub, including electric or hydraulic motors that force the blades to turn. Normally all the blades are rotated together by the same amount. When the blades are turned such that there are no aerodynamic lift forces on the blades, the blades are said to be **feathered,** or in feathered position. When the blades are not feathered they can catch the wind energy.

The hub and the blades always rotate together when a turbine is working. They drive the turbine shaft. All the energy grasped from wind is on the turbine shaft.

As we learned in chapter 2, the size of a turbine determines how much energy it can harness from wind. That translates in the blade size. In today's turbines the size of blades is much larger than those of the past, and it is still growing. Blades must be strong for all the forces that are applied on them. At the same time, they must be light. Today's blades are hollow and made of composite material. This means that their structure consists of a hollow shell, reinforced when necessary. **Figures 5.7** and **5.8** show the inside of a segment of the blade shown in figure 4.2. This is a 134.5-ft (41-m) blade. Referring to figure 4.2, note that the root of the blade where it must be attached to the hub is round, while the rest of it has the form of an airfoil.

Figure 5.7 shows the entrance of the blade. In a turbine, this entrance is inside the hub and can be opened for access when necessary. You can estimate the blade size at this end from the height of the person inside the blade. This

pitch control
The action of controlling the pitch angle of the blades in a wind turbine in order to modify the turbine performance.

pitch angle
The angle of a blade from a reference in rotation about its axis. Modifying the pitch angle of a turbine blade alters the angle of attack and, thus, the wind power capability of the blade.

feathered
A position of wind turbine blades (with respect to wind direction) with the smallest lift force and largest drag force, or the smallest aerodynamic force.

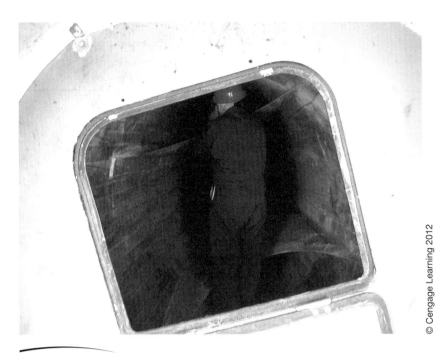

© Cengage Learning 2012

FIGURE 5.7 *Inside of a turbine blade. This picture shows the entrance of the blade and the reinforcement inside dividing it into three compartments.*

© Cengage Learning 2012

FIGURE 5.8 *The middle (of the three) compartments inside a turbine blade.*

FIGURE 5.9 *The hub and part of a nacelle.*

height is further divided into three sections by reinforcement panels shown in figure 5.7. Figure 5.8 shows the middle compartment, which is larger than the other two.

Figure 5.9 shows the hub and part of the nacelle to which it is attached. This hub is for the blades shown in figures 4.2, 5.7, and 5.8, while the openings for blade attachments are covered in this picture. In most turbines, the hub is accessible from outside; thus in order to get into the hub and the blades, when necessary, one must climb to the nacelle roof. Clipper turbines (shown in **Figure 5.14**) have a different design and one can get into the hub and the blades from inside the nacelle.

Nacelle

5.1.4

As mentioned, the rotating part of a turbine is the rotor (the blades and the hub). The intermediate part between the rotor and the tower is the nacelle. The nacelle does not rotate with the rotor, but it must rotate with respect to the tower. This rotating motion, called **yaw**, is necessary for directing the turbine to the wind stream, as the direction of wind is not fixed. This motion is provided by the yaw system, which comprises a number of yaw motors and a **yaw gear.**

The output shaft from the rotating rotor goes inside the nacelle. The shaft transfers the mechanical energy to a generator, to be converted to electrical energy. In most of today's turbines this transfer is not direct and there

yaw
The action of orienting a wind turbine to the direction of the wind.

yaw gear
The gear system to rotate the turbine for a yaw motion.

high-speed shaft
The output shaft in a wind turbine gearbox that must be aligned and coupled with the generator shaft.

is a gearbox between the main shaft (rotor output) and the **high-speed shaft** (the generator input). Thus, various equipments are housed inside the nacelle. **Figure 5.10** shows the inside of a generic nacelle, indicating the main shaft, the gearbox, the generator, and other components. An overhead crane, also shown, makes lifting and displacement of heavy objects easier during maintenance works.

The nacelle is a compartment not fixed to the tower and not fixed to the hub. "Not fixed" here implies that there are bearings between the two that allow them to move with respect to each other; that is, the assembly of hub and blades rotate with respect to the nacelle, and the nacelle rotates about the tower axis. The nacelle serves the following purposes:

1. Houses the gearbox, generator, coolers for the gearbox oil, heaters for winter time, turbine brake system, motors and gear for yaw system, the wind direction and speed measurement systems, the transformer for turbine energy supply, and other equipment based on the turbine design.
2. Allows yawing of the turbine; that is, adjusting the turbine orientation to the wind direction.
3. Provides counterweight for the hub and blades' weight.

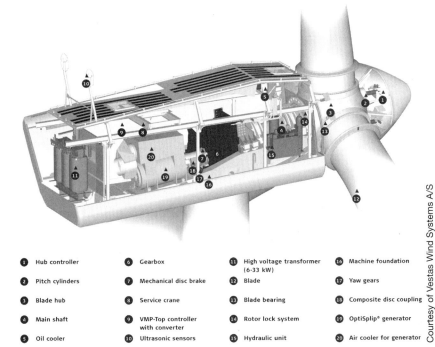

❶ Hub controller	❻ Gearbox	⓫ High voltage transformer (6-33 kW)	⓰ Machine foundation
❷ Pitch cylinders	❼ Mechanical disc brake	⓬ Blade	⓱ Yaw gears
❸ Blade hub	❽ Service crane	⓭ Blade bearing	⓲ Composite disc coupling
❹ Main shaft	❾ VMP-Top controller with converter	⓮ Rotor lock system	⓳ OptiSplip® generator
❺ Oil cooler	❿ Ultrasonic sensors	⓯ Hydraulic unit	⓴ Air cooler for generator

Courtesy of Vestas Wind Systems A/S

FIGURE 5.10 *Inside a nacelle.*

© Cengage Learning 2012

FIGURE 5.11 *The nacelle and hub of a turbine mounted on a short tower.*

All the equipment mentioned in item 1 of the previous list is among the essential components of a typical turbine, with minor differences based on the particular design of turbines by different manufacturers. We describe this equipment in this and later chapters. **Figure 5.11** shows the picture of a Vestas V90 nacelle and hub, which is mounted on a short tower and used for training purposes. The openings for attaching the blades to the hub are covered by wooden boards. This figure also shows part of the nacelle for Gamesa 87. The one-floor building at the back can be a good reference for visualizing the nacelles' size in today's turbines.

A nacelle can be a bedplate (platform) on which all the equipment is mounted, plus a cover or shell to make an enclosed room. The cover does not take any load and can be made of a light substance such as composite material. Alternatively, instead of a platform, the components themselves, particularly the gearbox, which is the largest component, can be part of the structure of the nacelle bedplate. In either way, the nacelle is usually heavy (including all the components in it), and the tower must be strong enough to hold the weight of the nacelle and the rotating parts. A bedplate type nacelle is illustrated in **Figure 5.12.**

Figure 5.13 shows the inside of a 1.65-MW wind turbine, which has a nacelle with bedplate. The nearest component in the figure is the gearbox, and the component in the far end is the generator. The brake system for this turbine has been placed between these two, on the high-speed shaft (of the gearbox). You can see only the cover of the brake in this figure. **Figure 5.14** illustrates the nacelle for a Clipper Liberty wind turbine. The structure of this

FIGURE 5.12 *A nacelle with bedplate.*

FIGURE 5.13 *Inside of a Vestas V82 nacelle.*

FIGURE 5.14 *Nacelle of a Clipper turbine.*

nacelle is different, as the body of the gearbox is a combined enclosure for the nacelle.

Gearbox

5.1.5

Gears are used when the rotational speed of a "driving element" is different from the rotational speed of a "driven element." In most cases, the speed of the driving element, which can be an electric motor, an engine, a gas turbine, a steam turbine, and so on, is higher than the speed of the part that must be driven. When one or more sets of gears are used for this purpose, all the gears are placed in an enclosed compartment, which also facilitates the lubrication and removal of the generated heat. The assembly of gears and the housing is called the **gearbox.** A typical wind turbine gearbox is depicted in **Figure 5.15.**

In a wind turbine, the rotational speed of the rotor must be matched to that of the generator. This is achieved through the turbine gearbox. Wind turbines are among the exceptional cases that the rotational speed of the driving element (here, the rotor) is smaller than the speed of the generator. A gearbox, thus, must increase the rotor speed to become appropriate for the generator. We learn in a later chapter that the typical speeds for most generators are between 900 to 1800 rpm, whereas today's wind turbines have a speed in the

gearbox
A mechanical device, consisting of various gears in an enclosure, to convert the rotational speed of an input shaft to a different speed on the output shaft.

Winergy

FIGURE 5.15 *A typical wind turbine gearbox.*

planetary gear
A set of gears
arranged in a special
form, consisting of
a sun gear in the
middle, a number
of (usually three)
plantet gears
engaging with the
sun gear and the outer
ring gear with internal
teeth. The planet
gears are mounted
on a bracket that can
turn independently,
called the arm. A
planetary gear can
accept two input
speeds. A planetary
gear is also called an
epicyclic gear.

range of 12 to 24 rpm. A wind turbine gearbox, then, must increase the speed
by a ratio of around 75 (1800 ÷ 24 = 75).

A conversion ratio of 75 is rather high and from the design viewpoint
cannot be done by only one set of gears. More details on gears and gearboxes
are covered in a later chapter in this book. At this time, it is worth mentioning that in wind turbines, the more common type of gear in use is called a
planetary gear.

Due to the friction between the gears, a large amount of heat is generated in the gearbox. This heat is inevitable and must be removed from the
gearbox. Otherwise, the gears get hotter and hotter, which eventually can
lead to overheating and damage to the parts. For this reason, and to reduce
friction, each gearbox is filled to a certain level with lubricating oil that has
the twofold function of protecting the gears from harmful rubbing together
and cooling them.

Not all turbines have the same design and similar components. The gearbox in figure 5.15 has one high-speed shaft and turns one generator. In clipper
turbines the gearbox has four outputs and simultaneously powers four generators, as shown in **Figure 5.16.**

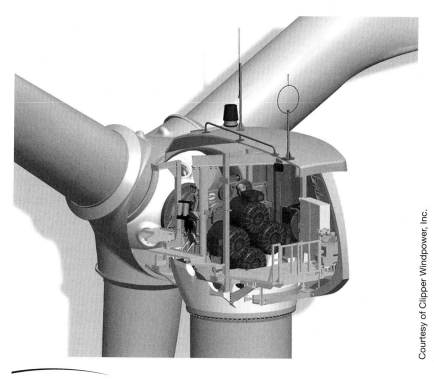

Courtesy of Clipper Windpower, Inc.

FIGURE 5.16 *Gearbox in a Clipper turbine simultaneously powers four generators.*

Turbine brake   5.1.6

Any turbine needs to be stopped on various occasions, such as for maintenance work, strong winds, a malfunction of some components, and so on.
This is irrespective of the time and weather conditions, and turbine settings.
It is necessary, therefore, that each turbine be equipped with a mechanical
braking system that prevents the rotor from turning. When a turbine is in the
shut-down condition, it is yawed out of wind and its blades are feathered, so
that there is minimum aerodynamic force to turn the rotor. This is not sufficient and the immobility must be ascertained by additional means.

Wind turbines are usually equipped with a proper brake similar to an
automobile disk brake that would be applied when not working. This brake
system is usually mounted on the high-speed shaft (before the generator). In
addition to this, for the maintenance work, or when a turbine must be stopped
for a long time, the rotor can be locked in a position by inserting a pin inside
a hole in a disk attached to the main shaft. In this way, the rotor is locked to
the body of the nacelle and cannot move. A wind turbine brake is illustrated
in **Figure 5.17.**

electric generator
A machine that
generates electricity.

single-phase
The simplest form
of AC electricity,
transmitted by two
wires, as opposed to
three-phase, which is
transmitted by three
or four wires.

three-phase
Three-phase electricity;
the sort of AC
electricity that requires
at least three wires
for transmission. The
voltage between each
pair of wires has a
sinusoidal form and has
the same frequency
and maximum
amplitude, but there is
a fixed delay between
the instants when each
of the three voltages
reaches its maximum
value.

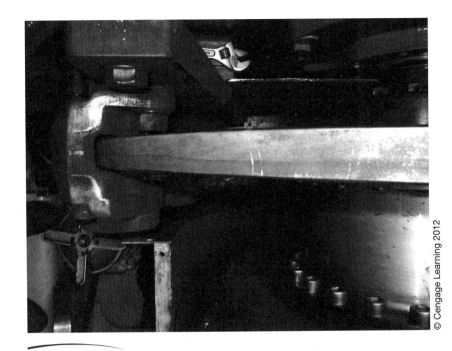

© Cengage Learning 2012

FIGURE 5.17 *A typical wind turbine brake.*

5.2   Electrical components

The major electrical components of a wind turbine are the generator and the transformer. There are many other electrical components for various functions, such as motors for yaw motion, motors for blade pitch motion, motors for oil circulation pumps, space heaters for winter time warming inside the nacelle, lighting, and so on. Also, in the control system of modern wind turbines various electrical, electromechanical, and electronic equipment is utilized. Out of these, only generators and transformers are to be explained further. Generators are discussed in more detail in chapters 6 and 11, and transformers are further discussed in chapter 12. In this chapter, these two major components are briefly introduced.

voltage
Electrical potential
difference or
electromotive force
(the active level for
electric charge, forcing
electrons to flow when
a path exists).

volt
The unit to measure
voltage, the electrical
potential difference.

5.2.1   Generators

grid
An electric network

An **electric generator,** or simply a generator, converts mechanical energy to electrical energy. It comes in different sizes, as one can see various size motors. The size is proportional to power. A generator, in fact, has the same construction as a motor. That is, it can be referred to as an electric machine; if it is fed with electricity, it behaves as a motor, and if it is turned (receives mechanical energy), it functions as a generator.

A generator can produce DC (direct current) electricity, or AC (alternating current) electricity. This depends on the design of the machine. You can learn more about DC and AC in chapter 6. At this stage, you should learn that DC electricity can be stored in a rechargeable battery, but AC electricity cannot. Of course, with proper devices AC can be changed to DC and DC can be converted to AC (see chapter 12).

AC electricity can be one-phase (or **single-phase**) or it can be three-phase. More details on this phase issue can be found in chapter 8. All the motors used in domestic products such as refrigerators and vacuum cleaners are single-phase. These motors, therefore, generate single-phase AC electricity if used as generators. Most of the motors used in industry, particularly the larger motors, are three-phase. Three-phase has the advantage of being more efficient and more economically viable.

In power generation at the commercial level (i.e., in large-size turbines) electricity is produced in **three-phase.** A three-phase electrical system is in fact three single-phase systems put together. Thus, any of the single-phase motors and other loads can be connected to one phase of a three-phase generator.

In addition to DC and AC categories, in order to use electricity, the electrical level must match. For instance, one cannot connect a flashlight lightbulb to the electricity in a home outlet. This is because of the difference in electrical level. This electrical level is called **voltage,** and is measured in **volts.** The voltage in a single battery is 1.5 volts. In a home electric outlet the voltage is 120 volts. You cannot feel a small voltage of 10–12 volts if you touch the two wires connected to electricity. A higher voltage, say 20–30 volts can give you a shock. Higher voltage, like 120 volts, kills a person. In a commercial-size wind turbine, the voltage of the electricity produced in the generator can be around 600 V, or 4000 volts.

Another term associated with AC electricity and AC generators is the "frequency." This depends on the rotational speed of a generator. But, it must match the frequency of the electrical network (**grid**). In terms of the frequency concerns, AC generators are mainly of two types: **synchronous generators** and **asynchronous generators,** also called **induction generators.**

The main operational difference between a synchronous generator and an induction generator is that a synchronous generator must have a constant speed. In all power plants that are energized by hydro, nuclear, and fossil fuels, only synchronous generators are employed. Wind turbines are the only exception. Depending on the modes of operation (discussed in chapter 11), both synchronous and induction generators are used with wind turbines. Induction generators, furthermore, are either of the type called a **squirrel-cage** or are a **doubly-fed induction generator (DFIG).**[2] More recent turbine use the latter type, which although more expensive, allows a larger portion of the wind energy to be harnessed. A wind turbine generator is shown in **Figure 5.18.** This generator has blowers for cooling. Some generators use water for cooling.

synchronous generator
An AC generator whose rotor carries one or more magnets. Electricity is generated in the stator windings when the rotor rotates. The rotor must spin at a constant speed for the generated electricity to have a fixed frequency.

asynchronous generator
Another name for induction generator; a generator that is not and does not work based on the same principle as a synchronous generator.

induction generator
An alternating current generator in which the rotor current is generated by induction rather than connection to electricity, as is the case with a synchronous generator.

squirrel-cage
A type of induction motor without windings on the rotor. Instead of windings, the rotor consists of thick copper (or aluminum) bars connected to each other at both ends by two rings, such that these metallic bars form a cage-like structure. The rotor is not connected to electricity.

---

[2] You will learn more details about these generators in chapter 11.

doubly-fed induction generator (DFIG)
An induction generator in which the rotor is made of windings and requires slip-rings to connect to electricity.

FIGURE 5.18 *Example of a generator.*

5.2.2   Transformers

transformer
An electrical device to increase or decrease the voltage in AC electricity. In the simplest form it has one primary winding where the supply electricity is connected, and one secondary winding where the output is taken.

step-up transformer
A transformer used to increase voltage.

step-down transformer
A transformer used to decrease voltage.

As you learned in the previous section, the voltage of the electricity for home use is 120 volts. This you can check with every appliance that one uses at home, including lightbulbs. On the other hand, for transmission of electricity from the point of generation (power plants, which are normally outside of cities) to the consumption points (cities), the voltage must be very high, because of technical reasons. Typical voltage for generators are 600 volts to 4000 volts.

In the same way that for rotating machinery the speeds of the driver and driven pairs must be matched by means of gears, in electrical systems the voltages at the points of connecting devices together must be matched. For this purpose a **transformer** is used. A transformer is a device that can increase or decrease a voltage, so that points to be connected together have the same voltage, or the voltage requirement is satisfied. If a transformer increases the voltage, it is called a **step-up transformer;** if it decreases the voltage it is called a **step-down transformer.**

The voltage of the electricity produced by a generator must normally be increased to a level of over 11,000 to 25,000 volts in order for connection to the rest of the electric network (the grid). Thus, a step-up transformer is used in each turbine. In most wind turbines such a transformer is at the bottom of

FIGURE 5.19 *Example of a pad transformer.*

the tower, either inside or outside. Since this transformer is mounted on a flat surface (a slab on the ground, for instance), it has a flat pad in the bottom, and, because of that, it is called a **pad-mount transformer** or a **pad transformer,** as contrary to the **pole-mount transformer,** which is mounted on the top of an electricity transmission pole. Example of a pad transformer is shown in **Figure 5.19.**

Other components   5.3

Among other components is the **anemometer,** a device without which a wind turbine cannot work.

An anemometer is used to measure the speed, and in some models also the direction, of wind. Later on we will see that for the control of a wind turbine it is necessary to know the speed of wind. Also, it is essential to yaw a turbine to the wind direction, otherwise a turbine does not get the full energy from wind, and it may even stop. If the wind speed is not measured, or the anemometer malfunctions and a zero wind speed is recorded, then the turbine controller interprets that as a no wind condition. No production, therefore, is made by the turbine.

The anemometer is usually mounted on the top of the roof of the nacelle. There are two main types of anemometer. A cap anemometer works based on rotation of a tiny turbine that generates a voltage. An ultrasonic anemometer does not have a rotating part, and its functioning is based on the speed of sound in the air. The latter is shown in **Figure 5.20.** More details on the operation of anemometers are given in chapter 12.

pad-mount transformer
A transformer suitable to be mounted on a flat slab.

pad transformer
A Pad-mount (or pad mounted) transformer

pole-mount transformer
A transformer that can be installed at the top of an electricity distribution post.

anemometer
A device to measure the speed of wind.

© Cengage Learning 2012

FIGURE 5.20 *Example of an ultrasonic anemometer.*

5.4   Chapter summary

In the following statements, wind turbine refers to a "three-blade propeller type wind turbine," which is a horizontal-axis wind turbine.

- The main components of a wind turbine are the tower, the nacelle, and the rotor. A foundation is also necessary to support a turbine.
- A rotor is the moving part, which consists of the blades and the hub.
- A nacelle is the housing on top of the tower, which holds the rotor and provides room for the gearbox, the generator, and the other components on top of the tower.
- Newer towers are tubular and can be cylindrical or slightly tapered (conic). Older turbines could have a lattice tower.
- Any turbine has a very heavy concrete foundation that makes it possible to hold the turbine despite all the force from wind and weight.
- A generator is the electricity maker in a turbine. Without a generator, a turbine is just turning, but does not produce electricity.
- Each wind turbine has a transformer at the bottom. Its role is to change (normally increase) the voltage of the electricity produced by the generator to the voltage of the electric line it must be connected to. This transformer is called a pad-mount transformer.
- Any turbine must have an anemometer that measures the direction and speed of wind. Without it a turbine cannot work. Wind direction is sometimes detected by a separate device.

Review questions

1. Name the components of a turbine that you can see from outside.
2. How many types of towers do you know?
3. What is a lattice tower?
4. Is it better to have a taller tower or a shorter tower?
5. What is the rule of thumb for tower height?
6. What are the advantages of a tubular tower with respect to a lattice tower?
7. What are the advantages of a lattice tower over a tubular tower?
8. What is meant by nacelle?
9. What is the role of a nacelle in a wind turbine?
10. Can a turbine be without a nacelle?
11. Is the nacelle fixed to the tower?
12. What are the principal components in a nacelle?
13. Why is it necessary for the turbine to have a foundation?
14. What is the name for the motion of turning a turbine in the wind direction?
15. Why is a turbine blade hollow?
16. What is meant by blade pitching?
17. What is the importance of pitch angle?
18. What is meant by feathered? Does it have a relationship with pitching a blade?
19. Should a turbine be stopped for maintenance work?
20. What makes sure that a turbine stays stationary for maintenance?
21. What is the role of a gearbox?
22. In a wind turbine, does the gearbox increase or decrease the rotational speed of the rotor shaft?
23. Name two main types of electricity.
24. Name three types of AC generators.
25. What is "voltage"?
26. What is the purpose of a transformer?
27. Is a wind turbine transformer a step-up or a step-down transformer?
28. What is an anemometer?
29. Can a wind turbine work without an anemometer? Why or why not?
30. Where is an anemometer normally installed? Can it be installed on the ground? Why or why not?

Electricity and Electric Generation

CHAPTER

6

© Cengage Learning 2012

KEY TERMS

Terms that you should already know: *energy, motor, generator, voltage, DC, AC*

New Terms
alternator
angular speed
armature
brushes
current direction
cut-in wind speed
  (cut-in speed)
cut-out wind speed
  (cut-out speed)
electric circuit
electric current
electric load
electric source
electromagnet
electromechanical system
ferrous
frequency
Hertz (Hz)
Lorentz force
magnetic field
performance curve
permanent magnet
prime mover
reluctance force
rotor
stator
synchronous speed
universal motor
wind farm
wound rotor induction
  machine (WRIM)

OBJECTIVES

After studying this chapter, you should be able to:

- Know what a generator is.
- Explain the fundamental difference between DC and AC electricity.
- Know that a generator and a motor have the same structure and that they can be referred to as an electric machine.
- Know that not all the generators are the same.
- Explain the principle of operation of all motors.
- Explain the principle of operation of all generators.
- Know that all electric machines include some sort of magnet.
- Know that the magnet in an electric machine can be a permanent magnet or an electromagnet.
- Explain the operational difference between an induction generator and a synchronous generator.
- Realize the frequency and synchronous speed.
- Explain the reason why various types of generators can be employed in wind turbines.
- Know the meanings of load and source.
- Have a better understanding of current and voltage in electricity.
- Know the "frequency" of AC.
- Know the function of a "brush" in a DC motor or generator.
- Have a better understanding of power in electricity and that it is related to current and voltage.

(continued)

89

- Understand what is meant by "electric circuit."
- Describe the conditions for connecting loads to a source for DC and for AC.
- Describe the difference when one generator is used separately, and when it must work with other generators.
- Know why generators must be connected together.
- Know what a performance curve of a wind turbine is.
- Know the difference between an isolated turbine and a grid-connected turbine.
- Understand that a wind turbine has limited power output.
- Understand that the power capacity of a wind turbine is determined by the maximum power of its generator.

6.0   Introduction

All the electricity for industrial and domestic use is produced in power plants. A wind farm is a power plant. In chapter 5 we learned that the electricity making part of a wind turbine is the electric generator (or simply, generator). Except in a battery, a fuel cell, and a solar panel, electricity is generated by an electric generator. Generators are powered by hydraulic turbines, steam turbines, diesel engines, and wind turbines. A power plant can be very large or it can be very small. On the small side, you may consider a car engine compartment as a power plant, since it has a small generator (the **alternator**) run by the car engine. This generator provides the required energy for running all the electrical devices in a car, as well as charging the battery. The battery is a reservoir of electricity that provides the essential power when the engine is not running; for example, for lights and radio as well as for starting the engine.

alternator
A machine generating alternating current electricity.

We learned in chapter 5, moreover, that there are various types of electric generators and that we have direct current and alternating current as two different types of electricity that can be utilized. These terms have been defined without further discussion of details. In this chapter, we want to learn more about generators, but because the subject is so vast, we need to first learn more about electricity and certain basic but important facts regarding practical use of electricity. In particular, power is the entity that must be understood well, since in practice it is the governing body in many applications.

The chapter does not cover materials that are normally found in detail in a book on electricity. Nevertheless, new material cannot be understood without making reference to the relevant topics in electricity. Thus, there is a need for at least a minimum description of some of the associated subjects. Also, some important facts that sometimes escape a reader's attention when studying electricity fundamentals are emphasized in this chapter. In particular, since wind turbines are connected together in order to provide more power, the subject of connecting a load to a generator and connecting generators together are highlighted.

An electric generator is a machine that has rotating and stationary parts, like a motor. That is, contrary to a battery or fuel cell, it is a mechanical system. But, since it has wire windings inside it and connects to electricity, it is in fact an **electromechanical system.** In general terms, one can say that if such a machine is connected to electricity, it turns; that is, it acts as a motor. Conversely, if it is turned (by an engine or a turbine, for instance), it produces electricity. Thus, we can define a generator as a device that converts mechanical energy to electrical energy. By the same token, a motor is a device that converts electrical energy to mechanical energy. We revisit the subject in chapter 11 to learn more about generators employed in wind turbines.

electromechanical system
Any device or equipment that has moving mechanical components and works with electricity, such as an electric motor.

> A generator has the same structure as a motor. In general, a motor receives electrical energy to turn, but if it is manually turned it generates electricity.

AC and DC   6.1

It is necessary that you have the basic knowledge about electricity and know the difference between the two terms *AC* and *DC*. As you learned in chapter 5, AC stands for *alternating current* (or *alternative current*) and DC stands for *direct current*. Both alternating and direct current are electricity that can be used in various devices. These are two forms of electricity with certain differences between them. The electricity at home that you use by plugging a device into the wall outlet is AC. The electricity that you get from a small battery or a car battery is DC.

Electricity is the flow of electrons in a device connected to an electrical supply. When you connect a device, say a lightbulb, to an electric supply you form a closed path for electrons to move. In direct current, electrons always move in one direction, whereas in alternating current they move back and forth; that is, they change their direction of motion.

For certain devices this changing of the direction of motion of electrons has no effect, because it is the motion itself that is important. For instance, in a lightbulb it does not matter in what direction the electrons move. The lightbulb lights up as far as the electrons move. On the other hand, some other devices work only with AC or with DC. A radio that works with a battery is a good example. It needs DC to be able to work.

Based on the preceding discussion, we categorize everything that deals with electricity as AC or DC. By this token, we have AC generators and DC generators. Of course, it is possible to change AC to DC and vice versa. In practice, one of the differences between DC and AC is that DC can be stored (in a rechargeable battery, like in a car), whereas AC cannot be stored, unless it is converted to DC.

The fact that DC electricity can be stored must not give the impression that DC is better or preferable to AC. Each one has advantages and drawbacks. For example, for distribution, AC has many advantages over DC. We use both AC and DC, and when necessary one is converted to the other. We need to know about both DC and AC.

The generator is one of the main components of any wind turbine. Without the generator a wind turbine cannot produce electricity. Of course, it is possible to get mechanical energy out of a wind turbine; for example, connecting a wind turbine to a pump for pumping water. This is something that has been practiced in the past with small wind turbines, but it is rarely in use today, and when we talk about a wind turbine today, it automatically implies generation of electricity.

> DC electricity can be stored (in a rechargeable battery, like in a car), whereas AC cannot be stored.

6.2 Electrical parameters

A wind turbine is a combination of mechanical and electrical systems. It is impossible to learn about wind turbines without having a fundamental knowledge of mechanics of motion, as you learned in the previous chapters, and of electrical flow. In this section, some parameters that are fundamental to electricity and electric devices are described. A good comprehension of these is very helpful in understanding how electrical devices work and how they can be repaired.

6.2.1 Electric current, current direction, and frequency

electric current
The intensity of electricity flow, measured in amps.

Before continuing, there are a few other terms that you must know. The first one is **electric current**. Electric current is a measure of how many electrons move together. In other words, it is the amount of electricity that flows. You may compare it with the current of water in a river or inside a pipe. In a river, more water is flowing, thus the current is higher. Note that current is irrespective of the speed of motion of electrons (since they always move with one speed) and irrespective of the size of a wire. You cannot say that the current in a thicker wire is more than the current in a thinner wire.

The current in a thicker wire can be larger, smaller, or the same as the current in a thinner wire. However, the current capacity of a thicker wire is more than the current capacity of a thinner wire, exactly like the capacity of a larger pipe being more than the fluid current capacity of a smaller pipe. For short, think of electric current as for comparison of how much electricity is flowing; higher current implies more electric flow, lower current implies lower electric flow. Moreover, electric current *is not* electric power. Electric power is proportional to electric current; that is, if nothing else changes except the electric current, then, if current doubles, power also doubles.

> One cannot say that the current in a thicker wire is more than the current in a thinner wire. The current in a thicker wire can be larger, smaller, or the same as the current in a thinner wire.

In conjunction with electric current, you also need to know the term **current direction.** As was mentioned in chapter 5, in DC the current is always in one direction. The current direction is a convention to make the process of understanding and analysis of the practical problems easier. **Figure 6.1** shows an example in which the direction of electric flow is shown. Here a battery energizes a few lightbulbs an other items. Each electric device has a switch. A *fuse,* which is a protective device to protect the circuit and its components, is also added. We see that the electric current direction is from the positive terminal of the battery to the switch; from the switch to the lightbulb; and from the lightbulb, back to the battery negative terminal. *The convention in this book is that the electricity flows from the positive side to the negative side.*

In many diagnostic problems, repairs, or matters that one must deal with, when working on wind turbines, the direction that the electricity flows does not really matter. On the other hand, in many cases, it makes it easier for you to tackle a problem and find a solution if you consider the current direction. You will see examples of this in your electricity course.

If the direction of the electric flow is not the same all the time, then we have alternating current (AC). In AC, the current is continuously changing direction. In other words, one can say that the polarity of the source of electricity continuously varies. This change of direction is not random; it is regular and is cyclic. In each *cycle,* the duration in which the electric flow is from right to left is exactly the same as the duration within which it is the reverse. For electricity

current direction
Direction of electron flow. Physically, outside of a battery this direction is from the negative terminal toward the positive terminal. However, conventionally, and in many books, this direction is considered to be from the positive toward the negative terminal.

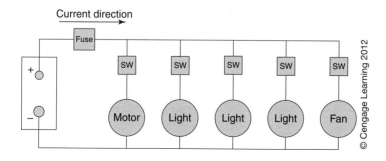

FIGURE 6.1 *Direction of electric current in an electric circuit.*

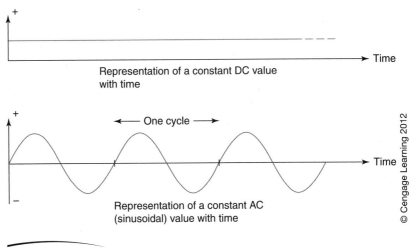

FIGURE 6.2 *Representation of DC and AC values.*

frequency
The number of repetitions, in one second, of a cyclic phenomenon.

hertz (Hz)
The unit for measuring frequency.

at home, this duration is normally a fraction of a second (like 1/100 or 1/120 of a second). **Figure 6.2** shows this variation of direction versus time, compared to a DC. A cycle is the pattern that is continuously repeated. The number of times that the cycle is repeated in one second is called **frequency** and is used in many practical problems. For example, in North America the frequency is 60. The unit for measuring frequency is **hertz (Hz).** We may say the frequency of AC electricity is 60 Hz. Hertz has the same meaning as cycles per second.

6.2.2   Voltage and power

You have already learned briefly (see chapter 5) about voltage. Voltage is a quantity representing the level of electricity. Current is a quantity associated with the electric flow. In earlier chapters, you learned about energy and power. At that point, power looked more like a mechanical quantity. But, in fact, power is as much an electrical quantity as it is mechanical. Since to follow any matter dealing with electricity one needs to have a good comprehension of current, voltage, and power, here we show an example to help you better understand the meanings of these terms. Consider four scenarios of water flow, as described below and as shown in **Figure 6.3.**

- a. A narrow stream of water, falling from a height of 60 ft
- b. A narrow stream of water, falling from a height of 10 ft
- c. A large amount of water falling from a waterfall 60 ft high
- d. The same amount of water coming down from a small fall 10 ft high

As you see, there are two variables, the height and the volume of water. Each one has two values, a small one and a large one. The first thing to understand is to realize which one of these four cases implies the highest amount of power.

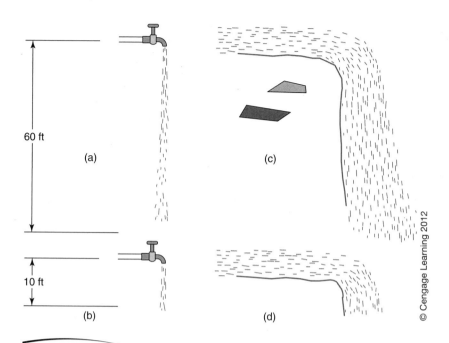

60 ft

(a)

(c)

10 ft

(b)

(d)

© Cengage Learning 2012

FIGURE 6.3 *Resemblance of water height and flow rate to electrical voltage and current.*

Your guess is right if you have picked (c), because in (c) both the height and the volume of water are the larger value.

In the above scenarios, compared to electric flow, the height is the voltage and the amount of water that flows is the current. In this sense, we may have high voltage, but low current (case a), high voltage and high current (case c), low voltage and low current (case b), and low voltage and high current (case d). This applies to anything electric. Furthermore, the power in an electric device is proportional to both voltage and current.

In an electric scenario, such as in an electric device or in the electricity at home, normally the voltage can be fixed. For instance in North America, the voltage of electricity at home is 120 volts (V). Or the voltage of the battery in your cellular phone can be 3.2 V. One more thing that you need to understand well is that while in many applications the voltage can be specified (and is considered to be fixed), power is not fixed. For example, in a motor running an elevator, the power to the motor depends on how many people are inside the elevator; the more the number of people, the more power the motor must use.

In an electric device, power is proportional to both voltage and current.

It should also be noted that, since power is proportional to voltage and current, if voltage is constant, then more power implies more current and less power denotes less current. For example, the motor in the elevator has a higher current when 10 people are moving up in the elevator compared to when only one person is in it.

Despite this power variation, the maximum power of the motor counts. No motor should be loaded for a power that is beyond its capacity. For instance, if 20 people get into an elevator the capacity of which is 12 people, this overloads the motor and is dangerous.

6.3   Electric machines

It is easier to understand how generators work if one knows how motors work. As mentioned earlier, an electric motor and an electric generator are fundamentally the same. A motor converts electrical energy to mechanical energy, and a generator does the reverse. In the study of motors and generators, thus, one can address them as electric machines. On the other hand, DC and AC machines do not work the same way and there are so many varieties of motors and generators.

This section, therefore, is confined to the very fundamental description of how electric machines work. The rest of the chapter is devoted to the special subject of connecting a motor, a generator, and the conditions that generators can be connected together. Further discussion about particular generators employed in wind turbines can be found in chapter 11.

6.3.1   Electric motors

Lorentz force
Force exerted on a wire carrying a current when placed inside a magnetic field.

ferrous
Of or related to the iron family of metals.

Most of the motors work based on **Lorentz force. Figure 6.4** shows what Lorentz force is. The right-hand side of the figure illustrates that when electric current flows through a winding that is wrapped around a **ferrous** material, a

Lorentz force

If a wire is inside a magnetic field and a current pesses through it, it is subject to a force.

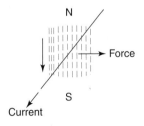

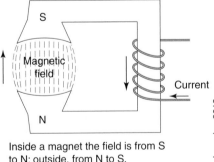

Inside a magnet the field is from S to N; outside, from N to S.

FIGURE 6.4 *Magnetic field and Lorentz force.*

magnetic field is created. A magnetic field is the area around a magnet that you can see or feel magnetism, and where a magnet works (farther from the magnet, it does not attract metal particles; that is, outside the magnetic field). Any magnet, whatever its shape is, has a magnetic field. You have seen, most likely, magnets in the form of a bar or a horseshoe. A magnetic field always has two poles, north and south. The direction of the field (outside of the magnet) is from the magnet's north pole to its south pole. The Lorentz force is shown in the left side of figure 6.4. It states that if a wire, carrying electricity, happens to be inside a magnetic field, then the wire is subject to a force; that is, a force is exerted on the wire.

In a motor, there are plenty of wires that carry electric current and at each instant are positioned inside the field of a magnet. As a result, all are subject to forces that when added together create a torque in the motor shaft and make it turn.

Some motors can work based on a different principle. In addition to Lorentz force we also have **reluctance force.** This is quite a different phenomenon, which is depicted in **Figure 6.5.** As shown, if there is a magnetic field, a force will be exerted on any ferrous material that lies in the magnetic field. This magnetic field can be that of a **permanent magnet,** or it can be generated by a winding that carries an electric current. What you know, and can buy, as a magnet is a permanent magnet that always behaves as a magnet. On the contrary, a magnet that is generated by an electric current through a wire winding loses its magnetism when there is no current in the wire. This is called an **electromagnet.**

It is further necessary to know the direction of a magnetic field when it is generated by an electric current flowing through a wound wire. **Figure 6.6** depicts this direction. The current direction in the wire is shown. Inside the winding, no matter whether or not there is a metallic bar, the magnetic field direction is as shown here. If the direction of current is changed by switching the polarity or by changing the way the wire is wound, then the field direction changes. One way

magnetic field
The area between two magnetic poles where a magnetic force exists. A magnetic field has a direction; outside a magnet, the direction of the magnetic field is from the north pole to the south pole.

reluctance force
A force by an electromagnet trying to shorten the path of a magnetic field.

permanent magnet
A magnet that has continuous and constant magnetism, as opposed to an electromagnet that can be turned on and off.

electromagnet
A magnet made of a coil (winding) and a core of ferromagnetic material inside it. This magnet can be turned on or off by connecting it to electricity.

Reluctance force

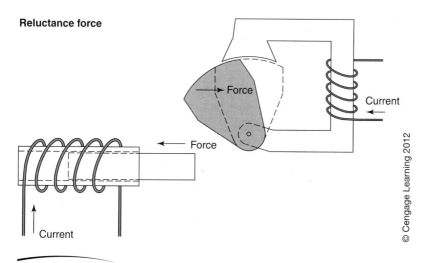

© Cengage Learning 2012

FIGURE 6.5 *Reluctance force.*

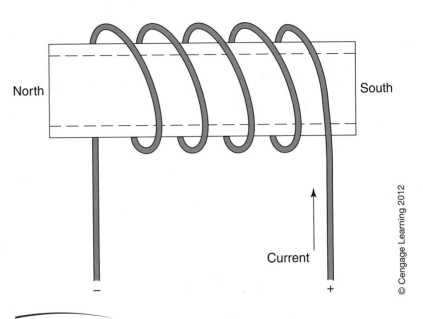

FIGURE 6.6 *Direction of magnetic field in an electromagnet.*

that you can always find the direction of the magnetic field is the *right-hand rule*. If your right-hand fingers show the direction of current, then your thumb indicates the direction of the magnetic field. This is shown in **Figure 6.7** (note that the magnetic field direction is from north to south outside of the magnet, but from south to north inside the magnet).

6.3.2   Electric generators

All generators—no matter what type, and whether DC or AC—work based on Faraday's law of induction. Faraday's law is the opposite of Lorentz force law. It states that if a wire moves in a magnetic field, then a voltage will be generated in it. As in a motor, there are plenty of wires that move together in a generator when it is turned, thus the voltages that are created in each individual wire add together and we have an electric source, a generator. All the wires in a generator are connected together and the end points come out of the generator for connection to the outside.

In a generator there is a winding that must move with respect to a magnetic field. This movement is always a rotational motion, since we need to apply the motion by a **prime mover**, like a turbine, a diesel engine, and so on, all of which have a rotational motion.

In general, the magnetic field can be provided by a permanent magnet or it can be provided by an electromagnet. In the latter case, there is a winding for the magnetic field and a winding that generates the electricity. These two windings must move with respect to each other. Thus, it is possible that the field winding remains stationary and the generating winding turns, or that the field winding turns and the generating winding remains stationary.

prime mover
A source of mechanical power that turns an electric generator.

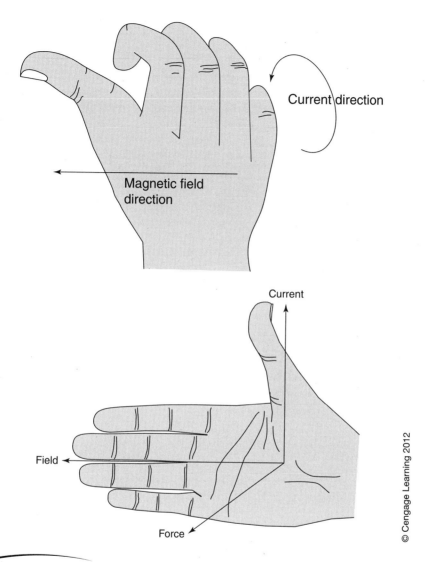

FIGURE 6.7 *Right-hand rules (a) to determine the magnetic field direction and (b) to determine the direction of the force on a wire in a magnetic field.*

Usually, smaller machines have a permanent magnet for their field and the larger machines have an electromagnet. Their magnetic field is provided by a second winding. When a winding must rotate, connection of the wire ends to the outside must be done by means of **brushes** that are stationary and slide on rings that are attached to the rotating shaft. These brushes are made of carbon, or some material containing carbon. They can be made of carbon and metal powders cemented together. Carbon is conductive of electricity and is soft, so it does not erode the other part on which it is sliding during motion. Brushes are

brush
A part made of carbon in an electric machine that slides on a conductor and allows electric transfer from a moving component to the stationary parts.

a weak point of any generator or motor due to wear. Moreover, they are prone to spark when in operation.

In a DC generator and a single-phase AC generator (see section 5.2.1), there are two wires that come out of the machine. Electricity can be obtained from these two wires. In a DC generator, one of these wires is the positive terminal and the other one is the negative terminal. In an AC generator, there are no positive or negative terminals.

In a three-phase AC generator, there are six wires coming out of the machine, one pair for each phase (see chapter 5 for single-phase and three-phase in AC).

6.4   DC machines

In this section, some general notes about DC machines are given. This book is not intended to give more in-depth detail about the operation and construction of electrical machinery. The material is just an introduction to the subject, a minimum for anyone who needs to work with electrical motors and generators, and on wind turbines.

6.4.1   General notes

armature
The rotating part in a motor, if composed of one or more windings.

In DC machines, the magnetic field is stationary and the rotating part is called **armature.** The armature winding is connected to the outside through two brushes. **Figure 6.8** illustrates the schematics of possible construction of small machines with a permanent magnet. It also shows the position of brushes, which are spring pressed on a commutator. All standard DC machines have brushes and commutators.[1] A commutator is a mechanical rectifier.

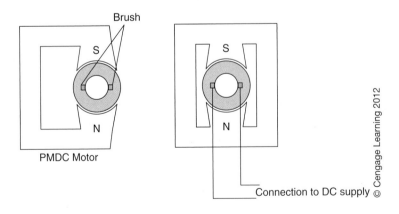

FIGURE 6.8 *Schematics of permanent magnet DC machines.*

---

[1] More recently brushless DC motors, BLDC, have become available. But these are AC machines, with an inverter as a driver. See chapter 12 for inverters.

In a DC motor, if the polarity is changed, that is, if the two connections are interchanged, then the direction of rotation changes. Similarly, in a DC generator if the direction of rotation is changed, the positive terminal becomes negative, and vice versa.

DC machines are normally larger and heavier than their equivalent (having the same power and specification) AC counterpart. A DC machine is, in fact, made up of an AC machine to which an extra component is added. If that extra part is removed then the resultant is a single-phase AC generator. This extra part is the commutator, which in a completely different way than an electronic rectifier rectifies the AC to DC.

In practice, only very small turbines can be equipped with DC generators. At the commercial level, particularly for connection to grid, no DC generator is used. It is possible, nevertheless, that AC is converted to DC for some processing (and later on converted back to AC). This is explained later on in this book in the chapter about modes of operation (see chapter 11).

Practical concerns for DC machines   6.4.2

A DC generator can be operated at any reasonable speed. Reasonable speed implies a speed at which the machine can work. For instance, speeds such as 500 rpm to 3000 rpm are acceptable, but speeds like 20 rpm or 50 rpm are very low. The power in a DC generator is proportional to its rpm. At low speeds, a generator can hardly produce any electricity. Even if it can, its efficiency is very low. Any machine must operate efficiently. Moreover, any machine is designed for particular conditions. For instance, its speed and temperature must be within certain ranges.

> Interchanging the connections in a DC motor reverses the direction of rotation. In a DC generator, if the direction of rotation is reversed the polarity of terminals switches.

Theoretically, in a DC generator if the direction of rotation is reversed, the polarity of its generated electricity reverses. Similarly, in a DC motor if the electric connections are reversed, it should rotate in the opposite direction. However, in practice, in order to do this a machine must have been designed to work in both directions. If a machine is not designed to work in both directions, then it *cannot* be run in the opposite direction. This can damage the parts, or can be dangerous. When working with any motor or generator, check the direction of rotation, as marked on the machine, or mentioned in its manual.

> If an electric machine is not designed to work in both directions, then it *cannot* be run in the opposite direction. When working with any motor or generator, check the direction of rotation (usually shown on the machine body).

6.5   AC machines

rotor
The rotating part of electric machinery.

stator
The stationary windings of an AC electric machine.

The ways DC machines and AC machines work are quite different, although both work based on Lorentz force. While there is a lot to learn about electric machinery, the scope of this book does not allow describing the details that can be regarded as unnecessary at this stage. We recall from chapter 5 that AC generators are mainly either synchronous or asynchronous (the latter also called induction generator). In AC machines the rotating part, mounted on the shaft, is called the **rotor,** and the stationary part is called the **stator.** You will learn more about AC machines in chapter 11.

6.5.1   General notes

We now make a distinction between single-phase machines and three-phase machines. As mentioned above (section 6.4), a single-phase AC generator has the construction of a DC generator. This is not true for all AC machines. On the other hand, one can turn a motor on and use it as a generator. It becomes necessary, therefore, to add a few lines about AC motors, here.

- A synchronous motor has a constant speed. It is complicated to operate and costly to run; so, it is usually made in very large sizes and in three-phase. On the other hand, synchronous machines are widely used in power generation (that is, as generators). Almost all power plants except wind turbines use synchronous generators.
- A squirrel-cage motor is the most common and less costly asynchronous type AC motor. It is made in both three-phase and single-phase. Three-phase squirrel machines are widely used in all industries as motors and in wind turbines as generators. A squirrel-cage machine does not have brushes.
- A **wound rotor induction machine (WRIM)** is another type of asynchronous machine. As a motor it is used only in special applications (it is more expensive). It is normally made in three phase. Today, it is widely used in newer wind turbines.
- There is a type of motor that works with both DC electricity and single phase AC electricity. This motor is called a **universal motor.** This machine is not made in large sizes. Examples of this machine can be found in electric drills, vacuum cleaners, and so on. If turned, this machine cannot generate AC. It generates DC.

wound rotor induction machine (WRIM)
A type of induction motor or generator whose rotor contains windings, as opposed to squirrel-cage machine that has no windings in the rotor.

universal motor
A type of motor that can work with both DC electricity and single-phase AC electricity.

Based on the preceding facts, except for the universal motor, the following is true for all AC machines: In AC machines, it is the stator winding that is connected to electricity. The stator windings do not move. In a synchronous machine, the rotor carries the magnetic field that rotates inside the stator. Depending on the size, the magnetic field can be generated by a DC machine that rotates with the shaft. Smaller machines can use a permanent magnet. In a squirrel-cage machine, the rotor rotates, but it does not need connection to electricity; so, there is no part that moves and must be connected to electricity. In the wound rotor induction machine, similarly, it is the stator that must be connected to electricity, but the rotor winding also needs connection to outside. This is done through brushes and *slip rings*.

Practical concerns for AC machines

The speed of operation of any AC machine is related to the frequency of the electricity it works with (see section 6.2). When dealing with motors and generators in AC electricity, the term **synchronous speed** is often used. Synchronous speed is a nominal speed associated with the design of any particular machine and the frequency (see equation 11.1 in chapter 11 for the corresponding formula). For 60-Hz frequency, which is the frequency of electricity in North America, any of the following numbers can represent a synchronous speed:

synchronous speed
The constant speed of a synchronous machine. This speed depends on the frequency of the supply electricity.

    3600 rpm   1800 rpm   1200 rpm   900 rpm   600 rpm   450 rpm
    225 rpm   etc.

Any other number between these is not a synchronous speed. For example, 1400 or 1500 rpm are not acceptable.

    In European countries, where the frequency is 50 Hz, the synchronous speeds are:

    3000 rpm   1500 rpm   1000 rpm   750 rpm   500 rpm   375 rpm
    187.5 rpm   etc.

Generally, a synchronous machine can have *only* one of the above speeds and it maintains that speed (some machines can accommodate two speeds by rearranging the winding connections). A synchronous generator *must* be turned at one of these speeds (depending on which one it is designed for). An induction motor runs slightly slower than these speeds (it can never run at exactly any of the synchronous speeds). For instance, for 60 Hz, an induction motor can run at 810 rpm, which is 90 rpm (10%) lower than 900 rpm. It can never run (unless additional equipment and means for this purpose are provided) at, say, 910 rpm or 1000 rpm.

    On the other hand, for 60 Hz, the same motor to work as a (induction) generator *must* turn at a speed higher than 900 rpm. How much higher it can be than 900 rpm will be discussed in chapter 11, dealing with "modes of operation."

    Contrary to the case for DC machines, the direction of rotation of a single-phase AC motor cannot be reversed by interchanging the wire connections. On the other hand, a three-phase motor has three wires to connect to electricity. By interchanging *any* two of them the direction of rotation can be reversed. Nevertheless, as mentioned earlier, for any rotating machine (including generators) to turn both directions, it must be designed that way. Otherwise it can be operated *only* in the direction shown on its frame.

Source and load in electric circuits

So far you have learned that a generator is a machine that makes electricity. You also learned the basic knowledge about DC and AC, and that AC electricity can be single-phase or three-phase. A generator, thus, can be regarded as the source of electricity, or an **electric source.**

electric source
A battery or electric generator that provides electric energy.

electric load
A consumer of electricity, that is, a device that works with electricity, when placed in an electrical circuit.

electric circuit
A setup of electrical components powered by an electric source.

wind farm
A region where a number of wind turbines are installed for generating electric power. Also called a wind park.

Every apparatus that works with electricity receives power from electricity, or we can say it consumes electricity. At home, for instance, there are lightbulbs, television and radio, refrigerator, vacuum cleaner, iron, washing machine, and so forth. You connect all of these to an electric outlet, which is ultimately connected to an electric source. The electric source is somewhere out of your house, and can be very far. You can see only the wires bringing electricity to your location. Any device that works with electricity is an **electric load** when it is connected to electricity.

From now on, for simplicity, we refer to any of the aforementioned devices as an electric load, or simply a load. *An electric source provides electricity and an electric load consumes electricity.* In this section, you learn about conditions that enable you to connect a load to a source.

An **electric circuit** was shown in figure 6.1, in which an electric source is connected to a number of loads. We could have more than one source. The simplest case for an electric circuit is one source and one load. What we want to study here is under what conditions one can connect a load to a source: also, how we can connect sources together. A good example of more sources connected together is the case of wind turbines in a **wind farm.** The generator in each turbine is a source, and all of them are connected together.

6.6.1   Connecting loads to a source

Before continuing further we want to make a distinction between different types of common loads.

1. Resistive loads can work with both DC and AC electricity. Examples of resistive loads are lightbulbs, electric heaters, and any electric item that has a heating element.
2. Electric motors. We have DC motors and AC motors. Only a universal motor can work with both DC and AC electricity.
3. Other loads, such as computers, TVs, and radios are complex inside, but from an electrical viewpoint they are considered a single load. They are like neither a resistive load nor a motor. These devices, if equipped with switching power supplies, can cause harmonics problems (see chapter 8) in power lines.
4. For AC only we have also inductive loads and capacitive loads. Inductive load is when a circuit contains an inductor (a coil) or a motor. Capacitive load is mostly due to a capacitor in the circuit (a synchronous machine can also introduce capacitive load).

Any electric device has a nameplate on which its electric specifications are mentioned. Among these are the operating voltage, the current, and the power consumption, although there is a relationship between these three entities. The values given are the *ratings* for the device; for example, power rating, voltage rating, and so on. The following list summarizes some general facts about connecting loads to a source.

- Except for resistive loads, DC loads must be connected to DC and AC loads to AC.
- The voltage rating of a load must be the same as the voltage of the source.

- The power rating of the load (its power consumption) must be lower than that of the source.
- In AC, the frequency rating of the load must match the frequency of the source.
- A single-phase AC load can be connected to a single-phase source. A three-phase source can be considered as three single-phase sources. Thus, a single-phase load can be (conditions apply) connected to one phase of a three-phase source.
- A three-phase load cannot be connected to a single-phase source.

The first thing that you can see from the preceding list is that you can never mix up DC with AC. One should not get confused if AC is first converted to DC (this is what the adaptors for cameras, cellular phones, and so on do), which is a different story, and additional devices are involved there.

Sometimes one cannot physically connect a device in a wrong way, but when there is no physical restriction if the rated values are not respected before connecting to electricity, damage or malfunctioning of the device is expected in most cases.

The preceding points are very important for larger devices, as they are more serious devices and more expensive. For example, if a lightbulb burns up because of incorrect voltage, it is not a big deal. On the contrary, if a large motor winding burns, in addition to the time for replacement, one may have to pay tens of thousands of dollars for the repairs.

One thing that is quite important to understand is the power relationship. A source is always the provider of energy. The maximum power a source can provide is limited and one must not surpass that for long time. If the power consumption of loads is larger than the power capacity of the source, in addition to some other undesirable effects, this can overheat the components in the electric source, which finally can lead to permanent damage.

> If the power demand of loads in a circuit is larger than the power capacity of the source, undesirable effects like overheating of components can result, which can lead to permanent damage.

Connecting generators together   6.6.2

In order to increase the source power, that is, to increase the electric power capacity, generators can be connected together. We refer to commercial production level, like in all wind turbines whose generators are connected in parallel and feed the same *grid* (the electrical network). This is different from when a number of 1.5-V batteries are put together. To connect sources in parallel it is required that

- The voltages of all sources are the same (AC and DC)
- All the sources have the same frequency (AC)
- All the sources are synchronized (AC)

ELECTRIC LOADS IN SERIES AND PARALLEL

In an electric circuit, loads can be combined together in order to satisfy the conditions to match the power supplies; for example, if we have two similar lightbulbs rated 120 V, but the available source voltage is 240 V. The figure below shows three resistive loads in series with each other.

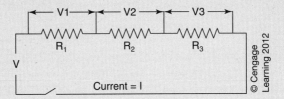

A series circuit is mostly used for electronic circuits. In a series circuit, all the loads in series with each other have the same current. If all the loads in a series are resistive loads the common current in the circuit is

$$I = \frac{V}{R_1 + R_2 + R_3}$$

The loads in series do not need to have the same resistance. The supply voltage is divided between the loads based on the value of their resistances. In the case of the lightbulb example, they must be similar (for example, both of them 120 V, 100 W), so that the 240 V is equally divided between the two. Otherwise, one light will be dim and the other brighter than normal.

Motors are almost never put together in this way. A resistor can be put, however, in series with a motor in order to drop the voltage applied to the motor.

The next figure shows a parallel circuit. This is the way almost all the lights and appliances at home are connected. The same voltage is applied to all the loads in parallel with each other. Each branch has its own current and if all the loads are resistive loads, then

$$I = I_1 + I_2 + I_3$$

where

$$I_1 = \frac{V}{R_1}; \quad I_2 = \frac{V}{R_2}; \quad I_3 = \frac{V}{R_3}$$

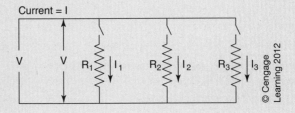

Whereas the current for a resistive load is almost constant, for a motor it depends on the motor load, which often is variable. In DC the current in the branches containing motors can be added directly to the current from resistive loads. In AC electricity, however, other type of loads and other matters come into the picture. The current relationships for other types of loads in AC are discussed further in chapter 8.

Unless a turbine works as a stand-alone unit (i.e., it is the unique source of electricity, for example in an isolated remote region), wind turbines operate in AC and are connected to a grid. What normally happens in practice is that the grid is already live (there is AC electricity in the grid), and a turbine starts to work (after maintenance or repair works or after being interrupted for any reason). Thus according to the preceding conditions, for this turbine the conditions on voltage and frequency must match those of the network. After the first two conditions are satisfied, it must be synchronized with the grid before physical connection can proceed.

Synchronization refers to timing for alternating current electricity waveforms from different sources to reach their maximum and minimum in each cycle in exactly the same instant. This can be for single-phase or three-phase AC. In wind turbines synchronization is performed automatically by the turbine controller after a turbine output has reached the required voltage and frequency. In a three-phase system, if this synchronization is performed for one of the three phases, it automatically covers the other two phases.

Connecting wind turbines together   6.6.3

When generators are connected together they share powering all the loads in their circuit. In power plants, in order to move more load to a generator its speed is slightly increased by feeding more fuel to it. In the case of wind turbines, this technique is not practicable, since the fuel is the wind, which is not in our control. Thus, we need to know what guarantees that none of the turbines feeding a grid overloads, while each one delivers its maximum power (for the wind speed at each moment). Also, how does the distribution of the loads between various wind turbines take place?

Each turbine is equipped with a control system that takes care of operation of the turbine, including this matter of how much electric load a turbine should handle at any time. Wind turbine control systems are discussed in a later chapter, since there are other variables to be controlled, as well, but in this section we explain the mechanism based on which such a system works for the above purposes.

As you have learned so far, the rotor of a wind turbine is a mechanical system that catches the wind energy and turns. The higher the speed of wind, the higher is the energy in the wind, and a rotor can grasp a higher power. Also, recall from chapter 2 that the power (that the rotor can grasp) is proportional to (equation 2.4) the cubic power of the wind speed. On the other hand, the rotor is attached (through the gearbox in many turbines) to a generator. The maximum power that a generator can deliver depends on its design and size. The maximum power that a wind turbine can grasp from wind is, thus, limited by the power capacity of its generator. This implies that, although the mechanical part of a wind turbine is capable of grasping more power, the maximum allowable power grasp is dictated by the generator. Therefore, there is a cap for maximum power grasp. This is very important to bear in mind.

> It is very important to realize that the maximum power that a wind turbine can grasp from wind is dictated by the power capacity of its generator.

angular speed
The rotational speed of an object that rotates about an axis, measured in rpm, number of degrees of rotation per second, and in radians per second.

performance curve
A curve that exhibits some relationship regarding the performance of a device, a machine, or equipment, in terms of the variation of a major parameter. *Performance curve and characteristic curve are often interchangeable terms.*

Another fact to note is that, although the power grasp increases with the wind speed, it also depends on the rotor **angular speed** (or rotational speed). As the angular speed increases, the power grasp capability of a turbine drops. This is one way that in the older turbines the angular speed was governed and there would be a limit for the speed of a turbine. Should this be not the case, a turbine would have unlimited power capacity, which is not possible. Each wind turbine has a **performance curve** that defines the variation of its power with respect to wind speed.

Figure 6.9 depicts a typical performance curve for a turbine. It shows the amount of power that the turbine could grasp from wind at each wind speed, if there was no controller. Wind speed is shown on the horizontal axis in both meter per second and miles per hour. The vertical axis shows the power in MW (megawatts). Each MW is 1000 kW. For example, at a wind speed of 8.95 mph (4 m/sec) this turbine can harness 0.1 MW (100 kW) of energy from wind.

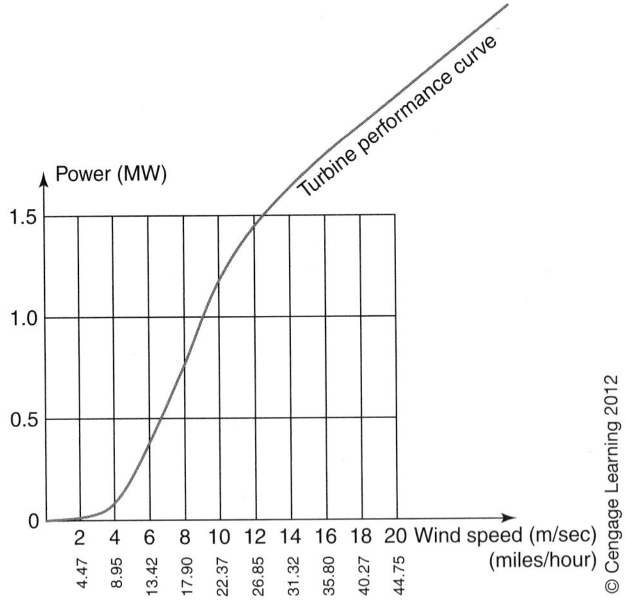

FIGURE 6.9 *Performance curve for a typical wind turbine rotor.*

Similarly, at a speed of 17.9 mph, the power that this turbine can harness is 0.8 MW. Thus from this curve you can find out for each wind speed how much energy this particular turbine can draw from wind.

The controller in a turbine has other functions too. For example, if wind speed is really low, a turbine can hardly produce any power. For this reason, when for each turbine wind speed is below certain magnitude, the turbine is turned off. That is, it is disconnected from the grid and practically it is off until wind catches up. A turbine that is turned off this way may remain stand-by until the wind speed reaches this value. This minimum wind speed is called **cut-in wind speed** (or just **cut-in speed**).

Similar to cut-in wind speed there is **cut-out wind speed** (or just **cut-out speed**). You may think that the higher the wind speed, the larger the amount of power we can get. This is a wrong perception. Based on this, in case of a storm, there is an opportunity for a turbine to generate a lot of power. You just learned that there is a cap for the power a turbine can produce, based on the size of its generator. In addition to the generator, the structural strength of a turbine also has limitation. When there is a storm, a turbine must be turned off, it must be yawed out of wind, the blades must be feathered out, and the brakes must be applied. In fact, when the speed of wind exceeds a certain magnitude any turbine must be turned off. Otherwise, the generator and various mechanical components, such as a gearbox, can be overloaded. The speed at which a turbine must be turned off is called cut-out wind speed. When a turbine is turned off at cut-out wind speed, it must be in the braked position until the wind slows down to below that speed.

Figure 6.10 shows the performance curve of the same turbine as in figure 6.9 with the exception that on the horizontal axis the lines for cut-in and

cut-in wind speed (cut-in speed) A speed of wind below which a turbine is not designed to generate electricity and is cut off from a grid. A wind turbine stays in free-wheel until the wind speed passes this speed for a certain minimum period of time.

cut-out wind speed (cut-out speed) A speed of wind beyond which a turbine is not designed to work, or is not safe to work. A turbine is shut down and disconnected from a grid; the blades turn to feather position and the brake is applied.

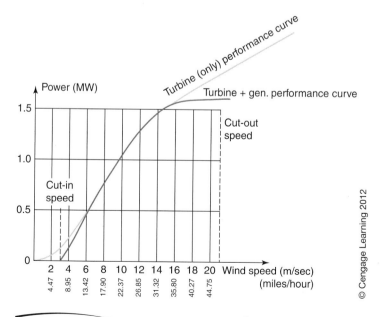

© Cengage Learning 2012

FIGURE 6.10 *Performance curve for a 1.6-MW wind turbine.*

cut-out wind speeds are shown. For this turbine, it is assumed that these two speeds are 3.2 m/sec (approximately 7.0 mph) and 21 m/sec (approximately 47 mph). Also, the effect of the turbine controller limiting the power at speeds above 15 m/sec (33.6 mph) is included. In this sense, this turbine has no output power at wind speeds below 7 mph and over 47 mph. Moreover, as can be seen from the curve, for wind speeds between 38 and 47 mph the output power of the turbine is fixed at 1600 kW (1.6 MW).

Figure 6.10 is not for any particular real turbine, but it shows what the performance curve looks like and how this curve is generated. **Figure 6.11** shows such a curve for a real turbine. The curve in this figure is for a 1650-kW turbine when the air density is 1.225 kg/m³ (0.0765 lb/ft³). The data corresponding to the performance curve of a turbine is stored in the memory of the turbine controller. The data is used as a look-up table, and for each speed of wind the turbine output is controlled not to exceed the value indicated on the curve.

You recall from chapter 2 that a turbine generates more power in a colder region and less power in a warmer region (the density of cold air is larger than the density of warm air). The maximum power for this turbine is 1650 kW for wind speeds between 12 to 20 m/sec (26.85 to 44.75 mph). A good controller takes into account a number of parameters that can change, such as the effect of temperature, in the control algorithm.

Based on the preceding discussion, the wind turbine controller, in addition to taking care of the voltage and frequency at all times, and synchronization at the time of connecting to a grid, regulates the power output of the turbine not to exceed the capacity of the turbine. So, depending on the wind speed variation the power output is governed so that no overloading happens. In this way, any turbine that is connected to the grid makes its contribution to the grid power, without taking extra load.

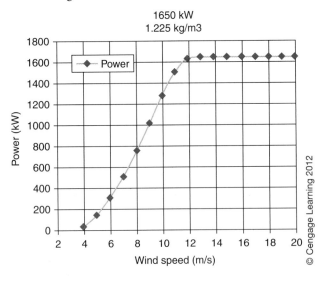

FIGURE 6.11 *Performance curve for a real 1650-kW wind turbine.*

Chapter summary

- A generator converts mechanical energy to electrical energy.
- A motor converts electrical energy to mechanical energy.
- A generator and a motor have the same structure, and they are referred to as electrical machines. Thus, if a motor is turned, it generates electricity.
- AC and DC electric machines cannot operate interchangeably. AC machines require AC electricity and DC machines require DC electricity.
- Electric voltage is a measure of level of electricity and electric current is a measure of electric flow in a circuit. In resemblance with a waterfall, the height represents the voltage and the amount of water flowing down is the current. The voltage is measured in volts and the current is measured in amperes.
- In AC, frequency is the number of times that the alternating cycle repeats in each second. It is measured in Hz (Hertz).
- In North American countries, electric power is provided at 60 Hz. In European countries it is at 50 Hz.
- In an electric circuit, electric load is any device that consumes electricity. Electric source is any device that generates or provides electricity. A battery and a generator are examples of source. A lightbulb and a motor are examples of load.
- In connecting loads to electricity, the voltage of loads and source must match.
- The power capacity of a source must be more than the powers of loads connected to it.
- In connecting loads to source for AC the frequency must also be the same. This is particularly important for motors in order to get a desired rpm.
- All electric motors work based on the Lorentz force, which states that if a current carrying wire lies inside a magnetic field (the area where the magnet is active), a force is exerted on it.
- All generators work on the principle that if a wire moves inside a magnetic field, a voltage will be generated in it.
- A magnet is considered a permanent magnet (PM) since the magnetism does not diminish. A magnet can be temporary, generated by wound wire carrying electricity. Such a magnet is called an electromagnet.
- The right-hand rule expresses that if in a winding we keep our fingers of the right hand in the direction of current, then the thumb shows the side where the north pole is.
- A DC motor can have any (acceptable) rpm. By interchanging the wires in a DC motor, its direction of motion can be reversed.
- An AC motor can have speeds at synchronous or below but not so much less than synchronous speed.
- Synchronous speed of an electric AC machine depends on the frequency of electricity.
- Generators can be connected together, the same as it is done in wind farms.

- AC and DC generators cannot be put together.
- For DC generators to be connected together, the voltages must be the same.
- For AC generators to be connected together, in addition to voltages, the frequencies must be the same, and they must be synchronized, as well. Synchronized means that the time their currents switch direction must be exactly the same.
- When connecting wind turbines together, the power capacity of each unit must remain within their limits at all times. The wind turbine controller is responsible to take care of this requirement.
- A wind turbine does not generate any power when wind speed is below cut-in wind speed. It also does not generate electricity for wind speeds over the cut-out speed.
- During strong winds at elevated speed a wind turbine must be taken out of service, that is, it must be stopped and brakes must be applied.

ADVANCED LEARNING

Power in DC circuits

In any electric device, the current can be determined based on the power and the voltage. The operating voltage for many devices is shown on them or on a nameplate, for example, on a lightbulb or a motor nameplate. One should not connect the device to a voltage that is very different from that mentioned on it (a slight difference is acceptable).

In DC electric circuits, the relationship between power, voltage, and current is:

$$\text{Power} = \text{Voltage} \times \text{Current}$$

If voltage is in volts and current is in amperes, then power is in watts. This implies, also, that if power is given in watts and voltage is given in volts, then current is in amperes.

The relationship for power in AC is more complicated and cannot be included here.

Frequency and period in AC

In alternating current electricity, the time that the flow of electrons alternates their direction depends on frequency. The duration of one cycle, shown in figure 6.2, is called the *period*, and in equations it is denoted by the letter T, measured in seconds; for example, $T = 1/60$ sec. The relationship between frequency (denoted by f) and period is

$$f = 1/T \quad \text{and} \quad T = 1/f$$

If T is measured in seconds, then f is found in Hertz. For instance in North America, the frequency of the AC power (the electricity distribution in cities and industry) is 60 Hz. Then the period of the current and voltage is 1/60 or 0.017 sec. The electrons in all wires carrying electricity continuously change direction after each half period, that is, each 0.0085 sec. Note that it is the electron movement in the filament of a lightbulb that heats it and makes it hot and bright.

Metric and Imperial unit conversion for wind speed

For converting the wind speed from meters per second to miles per hour you can multiply the value in meters per second by 2.237. For example, a wind speed of 10 m/sec is 22.37 mph. Conversely, for changing from miles per hour to meters per second, divide the number by 2.237 (or multiply the number by 0.447). For instance, a wind speed of 20 mph is equivalent to 8.94 m/sec.

Review questions

1. In a wind turbine, what is the component that makes electricity?
2. Can a wind turbine work without a generator?
3. Can a wind turbine produce electricity without a generator?
4. What does DC stand for?
5. What does AC stand for?
6. What is the main difference between DC and AC?
7. Name a device that works with both DC and AC.
8. In what sense are a motor and a generator similar?
9. Describe the Lorentz force.
10. In electricity, what does the term "load" refer to?
11. What is meant by "source" in electricity?
12. What is your understanding of current?
13. What is the power of a load?
14. How is "power" related to "current"?
15. What is the voltage of electricity at home?
16. What is the voltage of electricity of a single AA size battery?
17. Can you say what the difference is between an "A" size and a "AAA" size battery?
18. What is the voltage of your cellular phone battery?
19. Is your cellular phone battery AC or DC?
20. Can we say a battery is an electric source?
21. Where is the source of electricity at home?
22. What is the difference between DC and AC in terms of electricity storage?
23. Can we store AC?
24. What is frequency?

25. Does DC have a frequency?
26. What is an electric circuit?
27. Can an electric circuit have more than one load? Explain and provide an example.
28. Can an electric circuit have more than one source? Provide an example.
29. Can an AC generator be connected to a DC generator?
30. What is a brush in a DC machine for?
31. What are the conditions for a DC load to be connected to a DC source?
32. What are the conditions for an AC load to be connected to an AC generator?
33. What does the term "squirrel cage" bring to your mind?
34. Can squirrel-cage be a type of generator?
35. Synchronous and asynchronous are types of what?
36. What is meant by synchronous speed?
37. What is the main difference between synchronous and asynchronous?
38. What is the other name for an asynchronous generator?
39. What are the conditions to connect two DC generators together?
40. What are the conditions to connect two AC generators together?
41. What determines the power capacity of a wind turbine?
42. If two wind turbines are connected together, should they have the same power?
43. What governs the condition that a wind turbine does not become over-loaded?
44. Does a wind turbine always generate the same power?
45. Does a wind turbine generate more power in a storm?
46. What must be done to a wind turbine in case of a storm?

Problems

1. In figure 6.4 the direction of current in the winding is shown. Using the right-hand rule determine the direction of the force on a wire perpendicular to the paper and carrying a current the direction of which is coming out of the paper.
2. The synchronous speed for a motor is 1800 rpm. If this motor runs at 1750 rpm, how much percentage is it running slower than synchronous speed?
3. In North America, a motor runs at 1500 rpm; what kind of motor can it be?
4. From the two sets of numbers for rpm in section 6.5.2 of this chapter, find a relationship between each pair of numbers.
5. If a motor in North America has 1800 rpm, what is its speed if taken to Europe?
6. A motor works when it is connected to AC. By mistake you connect it to DC, but you see it also works with DC. What kind of motor is it?
7. From figure 6.10 determine the power production of the wind turbine for a wind speed of 10 m/sec.

8. From figures 6.9 and 6.10, for a wind speed of 18 mph determine how much is the power production of the turbine whose performance curve is shown in figure 6.10.

9. From figure 6.10 find the wind speed (in meters per second) at which the turbine generates 1200 kW. How much is this wind speed in miles per hour?

Projects

PROJECT 1

Find an electric device at home, such as a fan, vacuum cleaner, kitchen utensils, and so on. Find out if it has a nameplate.

From the nameplate, determine and write down any of the operating specifications, such as voltage, rpm, power, current, and so on. (voltage is specified by a letter V after the value, current by a letter A, power by W or kW). Do this for at least two devices. Compare the power values. Next find a lightbulb and see if you find any of those specifications on the lightbulb (or its box). Compare the power values.

Discuss this with your teacher.

PROJECT 2

For each set of rpm values in section 6.5.2 of this chapter, determine the pattern that each number makes from the first (the largest) number in the set.

Propeller-Type Wind Turbine Characteristics

© Cengage Learning 2012

OBJECTIVES

After studying this chapter, you should be able to:

- Know that a turbine ultimately converts mechanical power to electrical power.
- Determine the mechanical power of a rotating body, such as a turbine rotor.
- Know how torque, speed, and power are related.
- Understand what a diagram exhibits and how a mathematical relationship can be shown graphically.
- Graphically demonstrate certain relationships that you learn in this chapter.
- Determine the speed of any point on a rotating body, such as a turbine rotor.
- Describe what tip speed and tip speed ratio are.
- Calculate the tip speed ratio.
- Understand what is meant by angular speed.
- Describe how a turbine rotational speed affects its grasped power from wind.
- Realize that the power grasp capability of a turbine depends on the blade pitch angle.
- Realize that for a fixed blade pitch angle, the power grasp changes between a maximum value and zero.

- Realize that at some speed, a turbine can just turn without being able to deliver any power.
- Understand for a turbine with a fixed blade pitch angle how the power grasp changes with wind speed.
- Determine the daily production from a wind turbine.
- Determine the annual energy production of a turbine.

7.0   Introduction

In this chapter, you learn more about the characteristics and properties of propeller wind turbines. As discussed previously, the word propeller will be omitted and throughout the chapter "wind turbine" implies the common three-blade horizontal-axis wind turbine. In the course of learning the new material, it is necessary that some technical terms be known and certain techniques that are widely used in scientific and industrial application be very well understood. For this reason, some discussion at the beginning of the chapter is devoted to these terms and techniques. In particular, understanding of the term "power" in the mechanical parts of a wind turbine is quite important. You also learn about characteristic curves (or characteristic diagrams) and how the relationship between two or more variables can be represented by a curve. Understanding and interpreting the meaning of a curve and the information embedded in a curve is very important.

7.1   Mechanical power

A wind turbine consists of the rotor, which catches the energy from wind and concentrates it on the shaft, and the generator, which receives the mechanical energy from the shaft (usually through the gearbox) and converts it to electrical energy. In chapter 2, you studied the power in the wind and learned that not all of this power is usable in a turbine. You also learned some facts about generators in chapter 6. You have most likely noticed that power in the wind depends on the wind speed, the size of a turbine, and other parameters (see chapter 2).

When this power is transformed to a shaft, we should be able to express this power in terms of the specifics about motion of the shaft. Similarly, when this power moves to a generator, we should be able to express the power in terms of electric parameters, such as voltage and current. In this section, we are going to learn about mechanical power in something that moves on a straight line, or rotates (although in a turbine we have only rotational motion, it is worth learning about power in both rotation and translational motion).

> Mechanical power is "power" defined in terms of mechanical parameters, such as force, speed, torque, and the like. Compare mechanical power to electrical power, which is defined in terms of voltage, current, and so on.

Power in linear motion
7.1.1

Figure 7.1 shows an object that is under the effect of a force. It is assumed that the force passes through the object mass center and causes the object to move along a straight path with a constant velocity.[1] In such a case, the object receives power to maintain its motion. In other words, in order to maintain the motion at the same speed, some power must be provided to the object. This power can be defined as

$$\text{Power} = \text{Speed} \times \text{Force} \qquad (7.1)$$

This expression for power is in terms of the speed and the force, both of which are mechanical parameters related to motion. Note that the force has the same direction as the motion (velocity).

Here it is worth mentioning some units that one can use in order to calculate the amount of power in equation (7.1) and similar equations for rotational motion. Any entity (force, power, etc.) can be measured in various units. Here we stick with the units that you already know, or are going to use most. You are already familiar with miles per hour (mph) for measuring speed and rpm for measuring angular speed (rotational speed). Also, the units you know for force and torque are pounds and foot-pounds, respectively. One of the units for measuring power is the "watt," for which we use the symbol "W." Since a watt is a small quantity, a kilowatt (one thousand times a watt) is more often used. The symbol for kilowatts is "kW."

In equation (7.1), if the value for speed is in miles per hour and the value for force is in pounds, the answer does not appear in watts and a conversion becomes necessary; that is, a constant multiplier must be included in the equation. Equation (7.1), thus, changes to

$$\text{Power (in watts)} = 1.987 \times \text{Miles per hour} \times \text{Pounds}$$

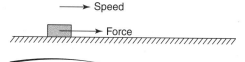

FIGURE 7.1 *An object under the effect of a force.*
© Cengage Learning 2012

---

[1] Force *F* has to overcome the friction forces so that it maintains a constant speed.

For simplicity, we can say

$$\text{Power (in watts)} = 2 \times \text{(speed in miles per hour)} \times \text{(force in pounds)} \quad (7.2)$$

In the metric system of units, if speed is measured in meters per second (m/sec) and force is in newtons (N), the value of power is determined directly in watts.

EXAMPLE 7.1:

In order to pull a small trailer, your car must exert 300 lb of force. If you run at 25 mph, how much power are you using to pull the trailer?

Solution: The answer can be found by entering the values directly in equation (7.2)

$$\text{Power} = (2)(25)(300) = 15,000\ \text{W} = 15\ \text{kW}$$

As you may know, there are other units for measuring or calculating power, such as foot-pounds per second and horsepower. But here *we intentionally want to determine the power in kilowatts*, which is also a unit for measurement of electrical power.

In order to review your understanding of the relationship between power and energy, let's see a second example. It is normally the *energy* that we pay for; for instance, we pay for electrical energy at the end of each month. When we pay for gas at a pump station, we are paying for the amount of *energy* we use.

EXAMPLE 7.2:

If you pull the trailer in example 7.1 for 3 hours (hr), how much energy have you spent?

Solution: The answer to this question can be found from the following relationship

$$\text{Energy} = \text{Power} \times \text{Time} \quad (7.3)$$

Thus, for 3 hr at 15 kW, the amount of energy is

$$15\ \text{(kilowatts)} \times 3\ \text{(hours)} = 45\ \text{(kilowatt-hours)}$$

Pay attention to the unit for energy. The kilowatt-hour is the common unit for calculating the amount of electrical energy. It can also be used for other types of energy, as in this example.

The purpose of example 7.2 was to introduce the kilowatt-hour, which is a unit for measuring energy. Note that at home we pay for the number of kilowatt-hours that we have consumed. For instance, we pay 8 cents for each kilowatt-hour of electricity in the electricity bill.

Power in rotational motion   7.1.2

The object in figure 7.1 has a translational motion. A turbine shaft has a rotational motion. Similar to speed and force in translational motion, in rotational motion one has "angular speed" and "torque." You have already heard these terms in a previous chapter.

For a shaft that, under the effect of a torque, rotates with a constant angular speed, the power is

$$\text{Power} = \text{Angular speed} \times \text{Torque} \qquad (7.4)$$

Similar to what we did for equation (7.2) to include a desired set of units, equation (7.4) can be written as

$$\text{Power (in watts)} = 0.142 \times \text{rpm} \times \text{foot-pounds} \qquad (7.5)$$

In the metric system if the speed is in rpm and the torque is in N.m, then the constant for equation is approximately 0.105 in order to find the power in watts.

EXAMPLE 7.3:

In order to turn a shaft at a constant angular speed of 1200 rpm, a motor attached to the shaft must apply a constant torque of 35 ft-lb. How much power does the shaft receive from the motor?

Solution: Since the units for the angular speed and the torque are rpm and ft-lb, respectively, we can directly use equation (7.4) in order to find the required power in watts. Thus,

$$\text{Power} = (0.142)(1200)(35) = 5964 \text{ W} = 5.96 \text{ kW}$$

EXAMPLE 7.4:

If 1 kW-hr costs 7 cents, how much is the cost of running the shaft in example (7.3) for 8 hr?

Solution: We need to calculate the amount of energy in 8 hr and multiply the amount by 7 cents.

$$5.96 \text{ (kW)} \times 8 \text{ (hr)} \times 7 \text{ (cents/kW-hr)} = 333.76 \text{ cents} = \$3.34$$

Note that if you multiply the units inside the parentheses, the terms kilowatts and hours cancel and the only term that remains is "cents."

7.2   Diagrams

A diagram can represent the mathematical relationship between two (and sometimes more) variables. The graphic representation is often very convenient in terms of demonstrating the facts that can be mathematically embedded in an equation. In addition, it sometimes serves better than the formula based on which it is drawn. When a diagram deals with a relationship defining the performance of a machine or a device, it is called the **characteristic curve** or **characteristic diagram** for that device. The concept and the application of diagrams can be better understood by considering the following examples.

7.2.1   First example

Consider first the simple case of the relationship for the power in a shaft in terms of angular speed and torque, expressed in equation (7.4). This is one relationship between three variables. Note that, since there are three variables, two more equations can readily be arranged out of this equation; that is, one equation defining angular speed in terms of power and torque, and another equation defining torque in terms of power and angular speed.

Figure 7.2 shows a diagram in which the horizontal axis depicts the values of angular speed, and the vertical axis depicts the values of power. This diagram is used when the value of torque remains constant (i.e., it does not change). Based on this diagram, instead of calculating the value of power for each value of angular speed, one can read the corresponding values for each variable in terms of the value for the other.

In this diagram you observe that (a) what you see in addition to the two axes is a "line" and (b) that it passes through the origin. The origin is where the value for both variables is zero. Although in this case you see a line, one may refer to it as a "curve," which is the general term. The reason that this curve (line in this case) passes through the origin is the fact that when angular speed is zero, the value for power is also zero. This may be not the case for other equations.

Pay attention to the fact that the value of torque was (and must be) constant for plotting this diagram. The reason is obvious; we can have only two variables, one represented on the horizontal axis and one on the vertical axis. Also, note that on each axis, in addition to the name for the variable on that axis, the unit for its measurement is specified. Without the names and the units of measurement, a diagram cannot be useful.

The diagram in figure 7.2 can represent a particular shaft whose torque is constant. This diagram thus can be regarded as the characteristic curve for the shaft. For every value of angular speed, you can directly read the corresponding magnitude of power, and vice versa. For instance, for an angular speed of 1000 rpm (point A) the corresponding power is 300 kW. For any other value, one can use the same process as depicted for point A; that is, draw a line from the specified point (point B) parallel to the axis until it hits the curve (point C). Then from the intersection point, draw a line parallel to the other

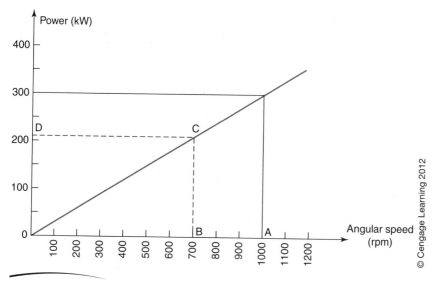

FIGURE 7.2 *Example of a simple diagram.*

axis. The intersection point shows the answer (point D). The answer, thus, is 210 kW. You may need to use approximation for the points whose values are not exactly marked.

> Without the names of variables and the units of measurement a diagram is not complete.

When the characteristic curve for a specific device is a straight line, we may say that the relationship between the two variables is *linear*. Although in this example the diagram was built based on a known formula, in some other cases it may have been generated by measurements and experiments, where there is no formula. Two more points in this example are worth mentioning:

- The values of angular speed and power are shown only for the range of their variation in the device. For instance, this shaft can have speeds up to an angular speed of 1200 rpm; so, there is no need to extend the graph for values beyond that.
- Neither the scales nor the lengths of the two axes need to be equal.

Second example   7.2.2

As a second example consider again the same relationship for a shaft torque, angular speed and power. This time we consider a case that can often arise in practice. Suppose that a shaft is driven by a motor and both torque and angular speed can be changed. However, since the motor has a limited capacity, in order

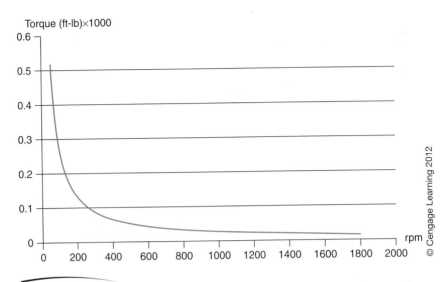

FIGURE 7.3 *Relationship between torque and angular speed in a shaft with constant power of 3.6 kW.*

not to overload the motor we want to keep the power limited to 3.6 kW. You can always use equations (7.4) or (7.5) if the units are rpm and foot-pounds, but a diagram can be more convenient if you do not have your calculator handy.

Suppose that the unit for angular speed measurement is rpm and the unit for torque measurement is foot-pounds. Thus, we use equation (7.5). You can immediately see from either of equation (7.4) or (7.5) that in order to keep the power at a given value, if rpm increases then the torque must be lowered, and if the torque is increased then rpm must decrease. The corresponding curve is shown in **Figure 7.3.**

As is clearly shown in the figure, small values of rpm correspond to larger values of torque, and vice versa. For example, if the torque is 400 ft-lb, the angular speed is around 64 rpm, whereas for a speed of 500 rpm the torque can be only 50 ft-lb. Pay attention to the fact that the values on the vertical axis (torque) are multiplied by 1000. Note also that since not all the values can be marked on a diagram, the magnitudes read are approximate. Moreover, in using a diagram, a ruler is very helpful.

7.3   Mechanical power in a turbine

Mechanical power in a turbine is the amount of power that the turbine harnesses from wind. Based on what we learned in chapter 2, we can determine the available power based on the wind speed, air density, and the turbine size (blade diameter). Out of this amount of power, the rotor of a turbine can harness up to a maximum theoretical value of 59%. In fact, a turbine rotor can harness between 0% and a maximum value, which cannot exceed 59%.

That maximum value depends on the quality of the rotor design. For instance, for a particular turbine it can be 52%. This is the energy on the shaft before the gearbox represents its **efficiency.** We learn in a later chapter that any device, such as a gearbox or a generator, has a definite efficiency. So, the final energy output from the turbine, after the gearbox and the generator, will be even less.

The efficiency determines how much of the input power is available on the device output. For example, in a generator, how many kilowatts of electrical energy can it deliver for each 100 kW of mechanical power input (certainly it is less than 100, since part of the power converts to heat during the process)?

Based on the preceding discussion, the energy grasped by the rotor of a turbine is further reduced in the succeeding components. In this section, nevertheless, we are not going to bring the effect of the efficiency of the components into picture. The intention is to study the power grasped by a turbine rotor. In particular, we want to see if there is any difference between cases when a rotor rotates at different angular speeds.

efficiency
The measure of how much of the energy given to a device or machine is consumed by it in order to function, when modifying a specification of that energy; efficiency is the ratio of the output energy to the input energy.

SPEED OF A POINT ON THE BLADE OF A TURBINE

You learned in chapter 4 that for any object rotating around a shaft, the (linear) speed of points at different distances from the shaft is not the same. Here we apply that to points on a turbine blade, and provide a formula with which the speed of any point can be found in terms of its distance from the shaft axis and the rotational speed of the shaft.

Speed (ft/sec) = 0.105 × Distance from axis (ft) × Angular speed (rpm)

In this formula if the distance is entered in feet and the angular speed is in rpm, then the speed is determined in feet per second.

Note that the two terms "angular speed" and "rotational speed" are the same, and "speed" (without any other word before it) implies "linear speed," where the word "linear" is normally dropped for simplicity.

If instead of feet per second, it is desired to find the speed in miles per hour, then the following formula is used.

Speed (mph) = 0.071 × Distance from axis (ft) × Speed (rpm)

In a turbine blade, usually the speed of the tip of the blade is of more interest. The distance of this point from the axis of the rotor shaft is the radius of the rotor, which is constant for any given turbine. This point has the highest speed in a turbine.

As an example, consider the blade of a Vestas V82 turbine. The radius of the blade circle is 41 m (134.5 ft). For an angular speed of 14 rpm, the tip of each blade moves with a speed of 197.2 ft/sec or 134.5 mph.

When a turbine is stationary (it is not working), it grasps 0% of the wind energy. This is when a turbine is yawed out of wind and its blades are feathered. In this case, one wants a turbine not to grasp any power from wind. The position when a turbine is yawed out of wind and its blades are feathered corresponds to the minimum power grasp. Any small amount of power that the rotor may grasp is canceled by the rotor brakes (in order to make sure that there is no rotor motion).

When a turbine is yawed into the wind, the blades capture the wind and a torque is created in the rotor shaft. In addition to the wind speed, the air density, the blade size, and the blade airfoil form, the magnitude of this torque depends on the pitch angle of the blades, if this angle can vary. In fact, changing the pitch angle alters the design of the blade. So, we need to study

a. The effect of the angular speed change in a turbine, and
b. The effect of changing the blade pitch angle.

7.3.1   Effect of the angular speed change and tip speed ratio in a turbine

We recall from chapter 2 [equation (2.5)] that power grasp in a turbine rotor is proportional to the air density, the cubic power of wind speed, the area swept by the blades when rotating, and the power coefficient. We furthermore understood that the magnitude of the power coefficient in a turbine can be as much as the Betz limit of 16/27. What we did not study is whether for a turbine the power coefficient is constant or changes with the variation of some parameters.

In this section, we learn that the power coefficient is not constant for a turbine and it depends on the angular speed. We see this variation in the form of a curve, which is very much illustrative of the performance (of the rotor) of any particular turbine.

> The power coefficient is not constant for a turbine; its value changes when the angular speed changes.

Before continuing further, we are going to learn another term that is frequently used in wind turbines. We recall, as well, that a wind turbine blade works based on the lift force on the blade, and that the lift force depends on (relative) air speed. We furthermore know that the relative air speed depends on the angular speed, and that one advantage of a propeller wind turbine is that its blades can have speeds higher than the wind speed. Moreover, from chapter 4, we know that various points on a turbine blade have different (linear) speeds for the same angular speed of the rotor. In particular, the tip of a blade has the highest speed, since it is the farthest point from the axis of rotation.

From all the aforementioned points, we see that the tip of a blade has the highest speed and has the highest lift force, and this lift force depends on the speed of rotation. The speed of the tip of blade is called **tip speed.** Since the wind speed is not in our control, normally the ratio of tip speed to the wind speed is of more importance, because it defines some information irrespective of the wind speed. For this reason the term **tip speed ratio** is used. Tip speed ratio is a number (without dimensions) that defines how many more times the tip speed is than the wind speed. For example, a tip speed ratio of 10 implies that the turbine rotor has such an rpm that the tip of the blade has a speed 10 times the wind speed. A tip speed ratio of 1 corresponds to the rotational speed that the tip of the blade has the same speed as the wind.

> **tip speed**
> The speed of the tip of a blade in a wind turbine.

> **tip speed ratio**
> The ratio of the speed of the tip of a turbine blade to the wind speed.

> The *tip speed ratio* is a (dimensionless) number that determines how many more times the speed of the tip of blades in a turbine is than the speed of wind.

EXAMPLE 7.5:

The blades of a turbine are 130 ft long and the hub has a diameter of 10 ft. What is the tip speed ratio for this turbine if the turbine operates at 15 rpm and the wind speed is 20 mph?

Solution: Since the hub has a diameter of 10 ft and the length of the blade is 130 ft, the distance of the blade tip from the axis of rotation is 135 ft. The speed of the tip of the blade in miles per hr, thus, is (see the text for *Speed of a point on the blade of a turbine* in this chapter)

$$0.071 \times 135 \times 15 = 143.78 \text{ mph}$$

and the tip speed ratio for a 20-mph wind speed is

$$143.78 \div 20 = 7.2$$

Note that since the wind speed was given in miles per hour the speed of the tip of the blade should also be found in miles per hour, so that the two numbers that are divided together have the same units.

Now we can utilize the knowledge we gained about characteristic diagrams for study of variation of power in a turbine with the variation of the turbine angular speed. Increasing the tip speed ratio (TSR) is equivalent to increasing the angular speed, and increasing the power coefficient implies increasing the power. **Figure 7.4** shows the typical form of the variation of the power coefficient in a turbine versus the tip speed ratio. As can readily be seen from the curve, we understand that as the TSR increases (starting from zero, corresponding to 0 rpm), the power coefficient starts from zero, reaches a maximum value, and then decays to zero again. The interpretation of the curve is that, at 0 rpm (the tip speed equal to zero), there is no power harnessed from wind by the rotor. As the angular speed increases (and so does the tip speed), the power harness increases until it reaches a maximum at TSR = 8 (this could be some other value for a different turbine).

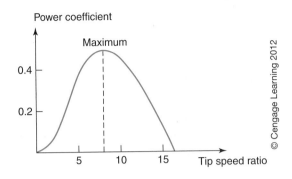

FIGURE 7.4 *Typical characteristic curve of a wind turbine.*

Then as the angular speed increases (and the TSR increases accordingly), the aerodynamic forces on the blades change in such a way that there is no more power grasped from wind (when the power coefficient is zero, so is the power). This point corresponds to when all the forces from wind produce zero torque (the generated torques cancel each other) and, thus, no power is generated.

The characteristic curve in figure 7.4 represents a typical curve. The scale on the vertical axis depicts the value of power coefficient, and the scale on the horizontal axis varies between zero and the maximum limit of the tip speed ratio. Each particular turbine has a characteristic curve that looks similar to this, passing through the origin, but the values for the maximum point (for both power coefficient and corresponding TSR) and the upper limit for TSR (where the power coefficient is zero again) can be different.

We see that increasing the rotational speed (thus having a higher TSR) is not always favorable. This is so owing to the fact that the aerodynamic forces (and therefore lift and drag forces) vary, but not by the same proportion. This curve implies another important fact. Since we want the turbine to grasp the highest power, a turbine must rotate, as far as possible, at a speed where the power coefficient is maximum.

Important: High values of TSR are not desirable or practical. This is a matter that must be considered during the design of a turbine. An upper limit for the TSR is 20. Practical values are in the range of 14–16 in older turbines (faster turbines) and lower than that, say, 7, for modern turbines.

The last statement implies that if we want a turbine to work always at its maximum power coefficient it is necessary that for each wind speed, the angular speed is altered accordingly. For example, based on the numbers in figure 7.4, for each wind speed, the tip speed must be 8 times more. So, if for a wind speed of 15 mph, the rpm is 18 (so that the speed of the blade tip is 144 mph, for the wind speed of 30 mph, it must change to 36 rpm.

The characteristic curve in figure 7.4 corresponds to a rotor with fixed blades; that is, blades whose pitch angle cannot be modified. Alternatively, it can correspond to a particular position of the blades in a rotor with an adjustable (regulated) pitch angle (almost all newer turbines have adjustable pitch angle blades). For a turbine with adjustable blades, there is one such a curve for each pitch angle.

> In order to grasp the highest power from wind, a turbine must operate with a rotational speed such that the power coefficient is maximum (or near maximum).

In order to better understand the characteristic curve in figure 7.4, assume that the wind speed does not change. In such a case, any change in the tip speed ratio reflects a change in the angular speed; that is, as one moves to the right on the horizontal axis, the angular speed increases. Thus, for any turbine, as the operating speed increases, the power coefficient changes.

Effect of changing the blade pitch angle   7.3.2

In the previous section, you learned that in order for a wind turbine to always harness the maximum possible power from wind, it must operate at an angular speed that is adjustable based on the wind speed. In practice, however, this is very difficult to manage because of the way generators work. A synchronous generator must operate at a constant rotational speed and an induction generator can only slightly deviate from synchronous speed (you learn more about modes of operation of wind turbines in a later chapter). In addition, continuously changing the rotational speed of a turbine not only requires additional control devices, which is costly, it introduces technical difficulties such as extra shocks and vibrations. Consequently, a better solution is sought.

> The aerodynamic performance of a rotor in a wind turbine is very much affected by the blade pitch angle.

Adding the capability to adjust the blade pitch angles in a wind turbine implies additional cost. The benefits introduced by variable pitch angle, nevertheless, compensate for and justify the extra cost. Today's wind turbines are equipped with devices for rotating each blade about its longitudinal axis and modifying the pitch angle; they are, thus, variable pitch angle turbines. Normally, all the blades in a turbine with a variable pitch angle are rotated simultaneously and by the same magnitude (in future designs this may change). For each pitch angle, the performance of the rotor alters and a different characteristic curve represents its aerodynamic characteristics. **Figure 7.5** depicts some of these curves plotted on the same coordinate system for a particular turbine.

> The effect of changing the pitch angle of blades is to decrease or increase the power coefficient of a turbine; that is, to modify the power capture capability of a turbine.

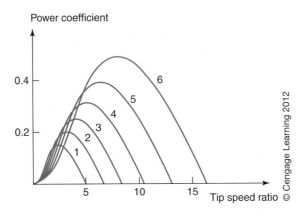

FIGURE 7.5 *A set of characteristic curves for different pitch angles of blades in a wind turbine.*

This figure shows how the power coefficient of a turbine can be modified by altering the blade pitch angles. In this figure, for clarity, only six curves are shown, numbered from 1 to 6. Each curve corresponds to a different pitch angle, from a minimum value to a maximum value that are possible in a rotor. Curve 6 corresponds to the normal operation of the turbine, whereas curve 1 (or possibly one before it, not shown) corresponds to when the blades are feathered, and not much power is captured. This is for the time when a turbine is to be yawed out of wind and stopped.

> In a turbine with variable pitch angle, all the blades are rotated simultaneously and by the same amount.

Another application of altering the pitch angle is to limit the power grasp in a turbine. For example, when the wind speed increases and power in the wind goes beyond the limit for a turbine (see figure 6.10 in chapter 6), we need to reduce the amount of power capture. This is done by adjusting the pitch angles of the blades. As a result, though the power in the wind is higher, owing to a reduced power coefficient a lesser percent of the power is captured by the rotor.

> In a variable pitch angle turbine, if the wind speed increases so that the power capture can overload the turbine components, the power coefficient is reduced by adjusting the blade pitch angle in order to limit the power capture capability of the rotor.

Turbine power variation with angular speed

From the content of the previous section we learned that the power coefficient in a turbine changes with the tip speed ratio. Since the tip speed ratio depends on the angular speed and the harnessed power depends on the tip speed ratio, we can define a different curve that shows the variation of rotor power with respect to the angular speed.

Imagine that the wind speed does not change. The power, thus, depends only on the air density and power coefficient [equation (2.5)]. For some finite time, the air density remains constant, thus the effect of air density change is negligible. Owing to the fact that the power coefficient changes with angular speed, if we find the power in a rotor for different values of angular speed and plot this variation, a curve that more or less looks similar to the characteristic curve in figure 7.4 will be obtained. Such a curve is shown in **Figure 7.6**. This curve is for a fixed value of blade pitch angle.

In figure 7.6, the amount of power grasped by a rotor when the wind speed is 25 mph (11.2 m/s) is shown if it rotates at different speeds. The difference is due to the change in power coefficient. For instance, if this turbine operates at 17 rpm, its rotor captures the maximum power of 1400 kW. At a speed of 10 rpm, the amount of power captured is only 918.4 kW. If the turbine does not turn at all, obviously it grasps 0 kW. Also, if it is allowed to turn very fast at 32 rpm, the rotor can only turn, but it cannot deliver any power, since all the power is used to turn the rotor only (no load can be connected to the turbine).

The last statement can be interpreted differently, as well. Suppose that wind blows at 25 mph and the turbine operates at its normal speed of 17 rpm. If due

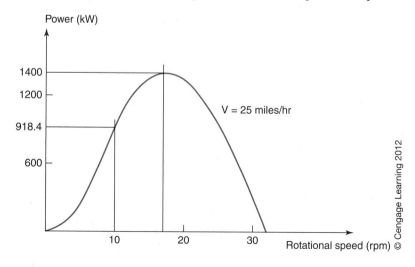

© Cengage Learning 2012

FIGURE 7.6 *Variation of power capture versus rotor angular speed (for a fixed pitch angle and a constant wind speed).*

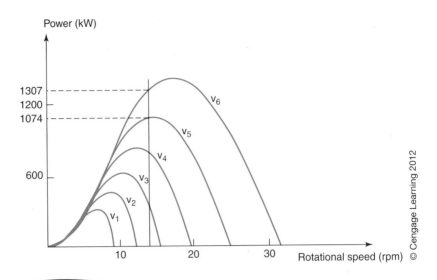

Power (kW)

1307
1200
1074

600

v_6
v_5
v_4
v_3
v_2
v_1

10   20   30   Rotational speed (rpm)

© Cengage Learning 2012

FIGURE 7.7 *Power capture curves versus rotor angular speed for different wind speeds.*

to a fault the load is disengaged from turbine (turbine is disconnected from grid), then the turbine speeds up. The turbine speed increases continuously until it reaches 32 rpm. Unless something is done about it, the turbine continues to turn at this speed. Such a scenario (overspeed) is not good and must be avoided. It can lead to a dangerous and destructive situation.

We can draw more curves for different wind speeds, since for each wind speed we can generate one curve. This is illustrated in **Figure 7.7,** which shows a set of similar looking curves, each one for a particular wind speed. Wind speeds are shown here as v_1, v_2, and so on, where $v_1 < v_2 < v_3 < v_4 < v_5 < v_6$.

From figure 7.7, we can see that if this turbine rotates at 14 rpm, then if the wind speed is v_6, its rotor captures 1307 kW from wind. If the wind speed drops to v_5, the power capture drops to 1074 kW (no change in pitch angle is assumed). Then, for a wind speed of v_3, the power delivered by the rotor is much less, and if the wind speed is v_2, no output power is expected from the turbine.

7.5   Daily energy production by a turbine

In this section, you learn how to determine the amount of energy that a turbine produces in a day. In order to better comprehend this section you may want to review the content of chapter 2 and the previous sections of this chapter first. We recall from chapter 2 that wind does not blow with a constant speed during the 24 hr of a day. We also recall that energy in the wind depends on the wind speed. Furthermore, in this chapter, we learned that the power coefficient is not always the same for a turbine and depends on the turbine rotational speed and the wind speed. Now, we are going to put all this information together and describe the way the daily production of energy from a wind turbine can be determined. This can be better explained using an illustrative example.

For this purpose we need to use equation (2.5). In addition to that, the information about the specific turbine and the pattern of wind in the region where the turbine is installed is essential. First, let us set some units for equation (2.5). If the turbine blade size is to be expressed in feet, the wind speed in miles per hour, the air density in pounds per cubic feet, and the power is to be found in watts, equation (2.5) will have the form

$$\text{Power (watts)} = 0.05 \times \text{Power coefficient} \times \text{Air density (lb/ft}^3)$$
$$\times \text{ diameter}^2 \text{ (ft)} \times \text{wind speed}^3 \text{ (mph)} \qquad (7.6)$$

Note that the swept area is always a circle. Thus, in the equation it is replaced by diameter squared and the constant (π) is included in the multiplier (0.05).

Problem statement   7.5.1

We state the problem as follows:

If the wind speed in 24 hr for a region is as shown in **Table 7.1**, and the air density in the season is 0.076 lb/ft³, find the power production of a wind turbine with a blade diameter of 262 ft (80 m), operating at 14 rpm. The cut-in and the cut-out speeds are 8.95 mph (4 m/sec) and 35.8 mph (16 m/sec), respectively. The values of the power coefficient for the turbine, between the cut-in and cut-out speeds, are shown in **Table 7.2**.

The fourth row of table 7.1 depicts the number (or fraction) of hours for each wind speed during 24 hr. Since the numbers in the first row are discrete, we may interpret the table as follows:

Wind blows with a speed between zero and 2.24 mph for 0.10 hr (6 min)
Wind blows with a speed between 2.24 and 4.47 mph for 0.30 hr (18 min)
Wind blows with a speed between 4.47 and 6.71 mph for 0.80 hr (48 min)
And so on

TABLE 7.1

Wind profile for the region

Wind speed in m/sec	1.00	2.00	3.00	4.00	5.00	6.00	7.00	7.50	8.00	9.00	10.00
Wind speed in mph	2.24	4.47	6.71	8.95	11.19	13.42	15.66	16.78	17.90	20.13	22.37
Percent of time at this speed	0.42	1.25	3.33	5.83	8.33	8.75	9.17	9.58	9.17	8.75	8.33
Duration in hour per 24 hr	0.10	0.30	0.80	1.40	2.00	2.10	2.20	2.30	2.20	2.10	2.00

Wind speed in m/sec	11.00	12.00	13.00	14.00	15.00	16.00	17.00	18.00	19.00	20.00	higher
Wind speed in mph	24.61	26.84	29.08	31.32	33.56	35.79	38.03	40.27	42.50	44.74	higher
Percent of time at this speed	6.67	5.00	3.96	2.75	2.29	2.00	1.50	1.04	0.83	0.63	0.42
Duration in hour per 24 hr	1.60	1.20	0.95	0.66	0.55	0.48	0.36	0.25	0.20	0.15	0.10

TABLE 7.2

Power coefficients values of the turbine between cut-in and cut-out speeds												
Wind speed (m/sec) 16	15	14	13	12	11	10	9	8	7	6	5	4
Wind speed (mph) 35.80	33.56	31.32	29.09	26.87	24.61	22.37	20.14	17.90	15.66	13.42	11.19	8.95
Power coefficient 0.16	0.25	0.32	0.38	0.40	0.44	0.46	0.43	0.39	0.37	0.34	0.30	0.20
Tip speed ratio 3.64	3.88	4.16	4.48	4.85	5.29	5.82	6.47	7.62	8.32	9.70	11.64	14.55

© Cengage Learning 2012

The third row shows the percentage of time for each range of wind speed. But we are using the fourth row only.

Note that in table 7.2 the power coefficients are given in terms of the wind speed. This is possible since the diameter of turbine is fixed and it operates at a constant rpm. It is more convenient to have the values of power coefficient as given in this table, rather than in terms of the tip speed ratio values. Otherwise, extra calculation is required to find the power coefficient for each wind speed. The values of TSR, corresponding to the wind speeds shown in the table and an operating speed of 14 rpm, are given in a separate row in table 7.2.

7.5.2   Solution

The procedure to solve this problem is to break 24 hr into smaller portions of time; then for each portion determine the amount of energy based on the length of time and the wind speed for that time interval. The amounts of energy, then, must be added together.

To understand this better, assume first that wind speed does not change in 24 hr. In such a case, you recall that

$$\text{Energy} = \text{Power} \times \text{Time} \qquad \text{(7.3 repeat)}$$

and for the amount of energy in 24 hr equation (7.3) leads to

$$\text{Energy (W-hr)} = 0.05 \times \text{Power coefficient} \times \text{Air density (pound/ft}^3) \times \text{Diameter}^2 \text{ (ft)} \times \text{Wind speed}^3 \text{ (mph)} \times 24 \text{ (hr)} \qquad \text{(7.7)}$$

Now, since the wind speed is not constant during the 24 hr of the day, this duration will be broken into smaller time intervals.

Before starting with calculations pay attention to the following facts:

a. The turbine production depends on the air density. Therefore, with all the data unchanged, the same turbine will produce more energy if the air becomes denser, that is, when the air is colder. This implies that, if the given air density is for the summer, then in the winter the productivity is slightly higher. Likewise, if the given air density is for the winter, the productivity will be slightly smaller during the summer.

b. In chapter 2 you learned about energy and power. Power is the work (or ability to work) during a one-second interval. If the power is multiplied by the time, then it gives the amount of work or energy (see the text box in chapter 2 about energy and power). If power is given in kilowatts and time in hours then the unit for measuring the amount of energy is kilowatt-hours (see section 7.1.1).

c. Since the numbers in table 7.1 are based on wind speed, we are going to break the 24 hr of a day into a number of segments based on the duration of wind for each wind speed. For this table, we cannot have segments for wind speed changes smaller than 1 m/sec (2.24 mph). However, we can combine the segments for higher values, such as a 2 m/sec change, but the accuracy then will be reduced.

In order to understand item c consider the following derivation from table 7.1:

> *Wind blows with a speed between zero and 4.47 mph for 0.40 hr (24 min)*
> *Wind blows with a speed between 4.47 and 8.95 mph for 2.2 hr (2 hr and 12 min)*
> *Wind blows with a speed between 8.95 and 13.42 mph for 4.10 hr (4 hr and 6 min)*
> *And so forth*

We are going to use the smaller time intervals for more precision.

> The procedure to solve this problem is to break the 24 hr into smaller intervals of time; then for each interval, determine the energy based on the length of time and the wind speed for that time interval. The sum of all the values obtained in this way, then, represents the total daily energy production.

In using equation (7.1), note that the constant 0.05, the air density, and the diameter squared terms are constant for all interval segments that we want to add together. For simplicity, these terms can be factored out. The calculation of energy production assumes the following form:

Energy =

$(0.05)(0.076)(262^2) \{(0.2)(8.95^3)(1.4) + (0.30)(11.19^3)(2.0) + (0.34)(13.42^3)(2.1)$
$+ (0.37)(15.66^3)(2.2) + (0.37)(16.78^3)(2.3) + (0.39)(17.90^3)(2.2)$
$+ (0.43)(20.13^3)(2.1) + (0.46)(22.37^3)(2.0) + (0.44)(24.61^3)(1.6)$
$+ (0.40)(26.84^3)(1.2) + (0.38)(29.08^3)(0.95) + (0.32)(31.32^3)(0.66)$
$+ (0.25)(33.56^3)(0.55) + (0.16)(35.80^3)(0.48)\} = (260.85) \{200.737 + 840.701$
$+ 1725.662 + 3126.077 + 4020.735 + 4920.921 + 7365.786 + 10298.780$
$+ 10493.190 + 9280.782 + 8877.494 + 6488.722 + 5197.190 + 3523.792\}$
$= (260.85)(76360.66) = 19,918,677$ watt-hour $= 19919$ kW-hr.

The preceding turbine as described, therefore, generates 19,919 kW-hr energy in 24 hr. If we divide this number by 24, the average power is about 830 kW. To better understand the context, the values in the calculation are arranged in **Table 7.3**.

7.5.3   Variable pitch turbine

Although nothing was said about pitch angle in the problem statement for the previous example, the fact that the cut-out speed was only 35.8 mph (16 m/sec) reveals that this is an old-type turbine that is not equipped with pitch control capability.

TABLE 7.3

Summary of calculations for a 24-hr generation

Wind speed (mph)	8.95	11.19	13.42	15.66	16.78	17.9	20.13	22.37
Power coeff.	0.2	0.3	0.34	0.37	0.37	0.39	0.43	0.46
Power (kW)	37.4	109.65	214.35	370.65	456.0	583.46	914.94	1343.2
Duration (hr)	1.4	2.0	2.1	2.2	2.3	2.2	2.1	2.0
Energy (kW-hr)	52.36	219.3	450.14	815.44	1048.8	1283.6	1921.4	2686.4

Wind speed (mph)	24.61	26.84	29.08	31.32	33.56	35.80
Power coeff.	0.44	0.40	0.38	0.32	0.25	0.16
Power (kW)	1710.7	2017.4	2437.6	2564.5	2464.9	1915.0
Duration (hr)	1.6	1.2	0.95	0.66	0.55	0.48
Energy (kW-hr)	2737.1	2420.9	2315.7	1692.6	1344.7	919.2

The turbine is cut off from production at a rather low-speed wind. Moreover, the values of power coefficients for each speed are fixed, as represented in table 7.2.

In modern wind turbines, the blade pitch angle can be adjusted and the cut-out speed is higher. The power capture capability at the higher wind speeds remains at the maximum power level by adjusting the blades' pitch angle. In this sense, the performance of the turbine looks like the curves in figures 6.9 and 6.10 in chapter 6. In order to see the difference in productivity, we are going to extend the previous example by adding a new condition to the problem statement. Thus the new problem statement is:

- The wind speed data for a region is given in table 7.1, and the air density in the season is 0.076 lb/ft³. The cut-in speed for a turbine is 8.95 mph (4 m/sec). The turbine has blade pitch control and the maximum power is maintained at higher wind speeds. If the cut-out speed is 44.75 mph (20 m/sec), find the power production of a 262-ft-diameter wind turbine with the power coefficient values as depicted in table 7.2.

Solution: The turbine in the previous example did not operate after the wind speed exceeded 35.8 mph (16 m/sec). Based on the wind data in table 2.1, nevertheless, wind can have speeds between 35.8 mph and the cut-out speed of 44.75 mph for almost 1 hr each day (0.96 hr, to be more precise, based on the numbers in the last row of the last 5 columns of table 7.1. This can be seen by adding together the time intervals in those columns). With the previous turbine, we missed 1 hr of production per day, as the turbine could not produce any energy.

For the new turbine in the current example we need to add the extra energy for this 1 hr. For this hour, the power capture of the turbine is at its maximum. Moreover, since the turbine has pitch angle control, we can assume that the power output remains at maximum for the whole range of speeds after reaching maximum power, based on figures 6.9 and 6.10 in chapter 6.

According to table 7.3, the maximum power of the turbine is 2564.5 kW and corresponds to a 31.32-mph wind speed. The turbine operates at this power for any wind speed higher than 31.32 mph, up to the cut-out wind speed; that is, for a period of 2.65 hr (2 hr and 39 min).

The energy produced by the turbine, thus, is

Energy =

$(0.05)(0.076)(262^2) \{(0.2)(8.95^3)(1.4) + (0.30)(11.19^3)(2.0) + (0.34)(13.42^3)(2.1)$

$+ (0.37)(15.66^3)(2.2) + (0.37)(16.78^3)(2.3) + (0.39)(17.90^3)(2.2)$

$+ (0.43)(20.13^3)(2.1) + (0.46)(22.37^3)(2.0) + (0.44)(24.61^3)(1.6)$

$+ (0.40)(26.84^3)(1.2) + (0.38)(29.08^3)(0.95) + (0.32)(31.32^3)(0.66 + 0.55$

$+ 0.48 + 0.36 + 0.25 + 0.20 + 0.15)\} = (260.85) \{200.737 + 840.701$

$+ 1725.662 + 3126.077 + 4020.735 + 4920.921 + 7365.786 + 10298.780$

$+ 10493.190 + 9280.782 + 8877.494 + 26053.2\}$

$= (260.85) (87204.26) = 22747231$ watt-hour $= 22747$ kW-hr

The total energy in 24 hr is 2828 kW-hr more, and the average power for the pitch controlled turbine is

$$22747 \div 24 = 947.8 \text{ kW}$$

which is considerably higher than the value of 830 kW for the previous turbine (that is, 14% higher).

Table 7.4 illustrates the details of calculation for this case. It is to be noted, however, that the numbers in the above examples are just to illustrate how the daily energy is calculated. They are for demonstration purposes and do not reflect a particular real turbine.

TABLE 7.4

Summary of calculations for a turbine with pitch control

Wind speed (mph)	8.95	11.19	13.42	15.66	16.78	17.9
Power coefficient	0.2	0.3	0.34	0.37	0.37	0.39
Power (kW)	37.4	109.65	214.35	370.65	456.0	583.46
Duration	1.4	2.0	2.1	2.2	2.3	2.2
Energy (kW-hr)	52.36	219.3	450.14	815.44	1048.8	1283.6

Wind speed (mph)	20.13	22.37	24.61	26.84	29.08	31.32
Power coefficient	0.43	0.46	0.44	0.40	0.38	0.32
Power (kW)	914.94	1343.2	1710.7	2017.4	2437.6	2564.5
Duration	2.1	2.0	1.6	1.2	0.95	0.66
Energy (kW-hr)	1921.4	2686.4	2737.1	2420.9	2315.7	1692.6

Wind speed (mph)	33.56	35.80	38.0	40.27	42.50	44.74
Power coefficient	reg'd	reg'd	reg'd	reg'd	reg'd	reg'd
Power (kW)	2564.5	2564.5	2564.5	2564.5	2564.5	2564.5
Duration	0.55	0.48	0.36	0.25	0.20	0.15
Energy (kW-hr)	1410.5	1231.0	923.2	641.1	512.9	384.7

7.6   Annual energy production

Once the way of the daily energy generation is understood, it is easy to extend it to yearly production. The simplest scenario is that we assume the same production for every day of the year. This gives a very crude answer that may include a large percentage of error. A more accurate method is to repeat the aforementioned calculation for the daily energy production for various seasons of the year, based on the data for availability and distribution of wind and the ambient temperature (air density changes). Then for each season (not necessarily four and equal number of days' seasons), the daily production is multiplied by the number of days.

The preceding method is still an estimate. If the seasonal data about the number of days with no wind are available, the estimate can be more accurate if those days are not included in the calculation. Moreover, if there are a number of days for scheduled maintenance that a turbine will be out of service, those days are also to be excluded.

All turbines today keep a record of the history of major events (such as shut-down and faults) and the production data. This data can be used to see the production of a particular turbine in the previous years, as well as using them for the approximate forecast of the following year(s).

Figure 7.8 shows the production of a wind farm at various months of a year. This wind farm has 50 MW of installed power, consisting of 25 2-MW turbines.

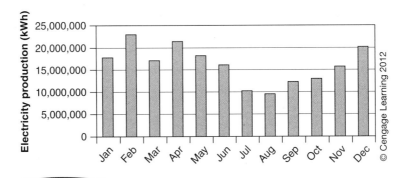

FIGURE 7.8 *Electricity production of a 50-MW installed wind power in a year.*

EXAMPLE 7.6:

As an example, we consider the turbine for which the daily production was determined in the previous section. We assume that the previous calculation was for a warm season and that in the cold season the air density is slightly higher, reaching 0.0775 lb/ft³ (1.24 kg/m³). Also, for simplicity, the pattern of wind is taken to be the same all year round. As well, we also assume that the data show that about 32 days in a year there is no wind and that in 9 days per year the turbine has been shut down because of maintenance and other reasons. We want to determine the yearly energy production of this turbine.

Solution: The daily production, based on the previous calculation, was 22,747 kW-hour for the warmer days. First, we find the daily production for a cold day. Since the wind pattern does not change, the only difference is the air density. The production in the winter time is higher by a ratio of 0.0775/0.076 (the ratio between the two values of air density), that is, 1.02 or 2% more.

Thus, we add 2% to the previous figure to obtain the wintertime daily production.

$$22,745 \times 1.02 = 23,199 \text{ kW-hr}$$

If the cut-out days are equally distributed between the two seasons, in each season the turbine operates during (365 − 41)/2, or 162 days. The yearly production, thus, is

$$162 \times (22747 + 23199) = 7,442,928 \text{ kW-hr} = 7.4 \text{ MW-hr}$$

Chapter summary   7.7

- A diagram can illustrate the relationship between two variables in an equation.
- It can be faster to use a diagram to see the variation of two related quantities with respect to each other.
- For anything that moves, there is a "power" moving it.
- Power is proportional to the force required to move an object and how fast the object moves (its speed).
- If the motion (of an object) is rotational, power is proportional to the torque required to turn the object and how fast the object rotates (its angular speed).
- Power grasped from wind by a turbine is affected by how fast the turbine rotates and the pitch angle of the blades.
- The relationship between the power grasp of a turbine and rotational (angular) speed is not linear. This relationship is shown by a curve (figure 7.6).
- TSR stands for *tip speed ratio*, which is the ratio of the speed of the tip of a blade (when a turbine operates) to the wind speed.
- The power grasp in a turbine depends on the power coefficient. The power coefficient is not constant for a turbine.
- The power coefficient of a turbine changes with tip speed ratio according to a diagram similar to figure 7.4.
- The tip speed ratio at any instant depends on wind speed and turbine angular speed.

- As far as possible, a turbine must operate at a speed where the power coefficient is around its maximum value.
- To find the daily production of a turbine one must have the data for the pattern of wind variation during 24 hr.
- Daily production of a turbine is calculated by adding together the "power multiplied by time" for various values of turbine power (which depends on the wind speed) and wind duration (at that speed), during a 24-hr period.
- Annual energy production is determined by multiplying the daily production by the number of days that a turbine works during a year. An adjustment is necessary for the effect of temperature on air density for winter days.

ADVANCED LEARNING

Design Tip Speed Ratio

Any turbine is designed for a specific *tip speed ratio*. This is determined based on the most probable wind speed in the region (the wind speed with longest duration), and the operating speed of the turbine (a turbine may have been designed to operate at 15 rpm, for instance). The preceding two variables (wind speed and turbine angular speed) determine the TSR, since the rotor size is fixed.

During the design process, the blades of the rotor are so designed that the *designed TSR* (the TSR for which the turbine is designed) gives the maximum power coefficient when the characteristic curve is drawn. This is shown in the following figure.

In this figure, the pitch angles for blades are indicated by the Greek letter φ (phi). Each curve corresponds to a constant pitch angle, and the largest curve corresponds to the normal operation of turbine. The design TSR, indicated by the vertical line, is where the maximum of this curve occurs, and is about 6.

All other curves belong to different values of pitch angle φ, and their maximum is away from design TSR. Those curves show the characteristic diagram of the turbine when the pitch angle is modified from that for the normal operation.

In the literature, the tip speed ratio is normally denoted by the Greek letter λ (Lambda: pronounced "landa"). The design TSR is denoted by λ_d. Also, the power coefficient is represented by C_p.

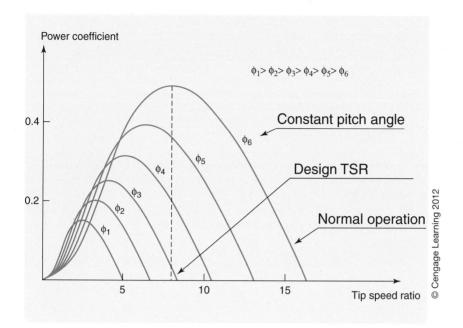

Review questions

1. What is a diagram used for?
2. Why is a diagram preferred to a mathematical relationship?
3. What is the drawback of a diagram?
4. What is meant by mechanical power?
5. In an object moving in a straight line, power is the product of which parameters?
6. In an object with a rotational motion, which two parameters define power?
7. Is the power grasp in a wind turbine constant?
8. What is angular speed?
9. If a turbine turns faster, does it grasp more power from wind?
10. If the power grasp of a turbine at a particular angular speed is known, can you find the torque on the turbine shaft?
11. How do you determine the speed of a point, such as the tip or the middle point, on the blade of a turbine?
12. What is meant by tip speed?
13. What is tip speed ratio?
14. What does a characteristic curve of a turbine represent?
15. Is tip speed ratio a constant for a turbine?
16. If a turbine turns with a constant (angular) speed, and the wind speed increases, what happens to the (tip speed ratio) TSR?
17. If the rotational speed of a turbine doubles and the wind speed doubles, too, what changes are made in the TSR value?

18. What is TSR value when a turbine is braked and does not turn?
19. How much approximately is the power coefficient if the TSR has a value of 5 in the turbine whose characteristic curve is shown in figure 7.4?
20. What will ultimately change if in a turbine the blades' pitch angles are changed?

Problems

1. A turbine operates at a constant speed of 18 rpm. If at a certain time the power generated by the rotor is 1000 kW, how much is the torque on the rotor shaft?
2. If a force of 600 lb is necessary to move a vehicle and the engine provides a power of 75,000 W to the wheels, what will be the speed of the vehicle?
3. Based on figure 7.3 (use a ruler), determine the torque when the speed of the shaft is 1000 rpm. Compare it with the value obtained from calculation [equation (7.5)].
4. A turbine has an angular speed of 18 rpm. The hub has a diameter of 10 ft and the blades are 140 ft each. Find the speed of a screw that is 40 ft from the tip of the blade.
5. For the turbine in question 4, what is the speed of the tip of the blade in miles/hour? Find the speed in feet per second.
6. If wind blows at 25 mph, how much is the tip speed ratio for the turbine in the previous question?
7. The same turbine as in question 4 is used for another site for which the wind speed is 18 mph. If the tip speed ratio is to remain the same, find the new operating rpm of the turbine in this site.
8. A wind turbine is designed to operate in a region where the average wind speed is 18 mph and the tip speed ratio is 8. The turbine has a constant rotational speed. If the cut-in and cut-out speeds are 9 and 35 mph, respectively, find the minimum and maximum values for the tip speed ratio during the turbine operation.
9. The daily wind pattern for a region is shown in **Table 7.5** and the air density is 0.076 lb/ft³. Determine the daily energy generation for a turbine with the characteristics as shown in **Table 7.6**. The cut-in and cut-out speeds are 9 and 35 mph, respectively. Suppose that the turbine is pitch regulated (has adjustable blades) and the maximum power occurs at a speed of 19 mph, after which this maximum power is maintained by pitch control. The rotor diameter is 270 ft and the rotational speed is 15 rpm.
10. For the turbine in question 9 find the yearly production if the turbine operates 6 months during winter and 5 months during summer, including the shut-down days. The air density in the winter time is 0.08 lb/ft³.

TABLE 7.5

Wind profile																		
Wind speed (mph)	3	5	7	9	11	13	15	17	19	21	23	25	27	29	31	33	35	37
Duration (hr)		0.2	0.3	0.8	1.4	1.8	2.0	2.1	2.2	2.6	2.1	2.0	1.7	1.5	1.2	1.0	0.7	0.3 0.1

© Cengage Learning 2012

TABLE 7.6

Power coefficient values (C_p)															
TSR	2	3	4	5	6	7	8	9	10	11	12	13	14	15	larger
C_p	0.1	0.2	0.25	0.32	0.39	0.46	0.45	0.4	0.38	0.33	0.3	0.25	0.2	0.15	0.12

© Cengage Learning 2012

PROJECT

On a piece of graph paper make two perpendicular lines to be used as horizontal and vertical axes. On the horizontal axis mark spaces of 1 in. each, up to 5 or 6 in. Starting from left, at each mark write values of 10, 20, 30, and so on. At the extreme right on this axis, make an arrow and write "miles per hour" along the line.

Do the same thing on the vertical axis, but with divisions a half inch apart; that is, at each half-inch-long interval write numbers 10, 20, 30, and so on. At the top of the line make an arrow and write "feet per second" above it. Also, make a mark at the point at 2.2 in. from the origin.

Now, from the third mark on the horizontal axis draw a vertical thin line up. Similarly, from the point corresponding to 2.2 in., draw a horizontal thin line. These two lines intersect at a point. Connect this point to the origin to make a line, and continue it from the right side until it covers the whole area of your graph.

In this way, you have made a diagram for converting speeds from miles per hour to feet per second and vice versa. You can use this diagram by drawing parallel lines from the point corresponding to the value you know, until it hits the sloped line in the middle. From the point of intersection, you draw another parallel line until the other axis is hit. This point shows the corresponding value you are looking for.

You can subdivide your marks for an easier reading.

Grid Connection

KEY TERMS

active power
apparent power
balanced load
capacitive load
capacitive reactance
collector
delta connection
harmonics
inductive load
inductive reactance
islanding
lagging
leading
line current
line voltage
phase angle
phase current
phase voltage
power factor
power factor correction
reactive power
resistive load
star connection
substation
three-phase system
volt-ampere (VA)
volt-ampere-reactive
  (VAR)
wye connection

OBJECTIVES

After studying this chapter, you should be able to:

- Know how in operation a turbine can be connected to an already existing grid, which provides electricity, and the newly connected turbine contributes to a grid's power generation.
- Know how to determine the power for a load connected to DC electricity.
- Know that in AC electricity there is more than one type of load.
- Talk about resistive load, capacitive load, and inductive load.
- Know that the term for loads other than resistors is reactance.
- Understand what a phase angle is in AC.
- Understand that voltage waveforms and current waveforms can be similar but do not reach their maximum and minimum values together (at the same time).
- Recognize three different powers in alternating current.
- Determine the three powers in AC based on the given data.
- Know what power factor is and where it comes from.
- Understand the role of power factor.

- Realize that in an ideal situation the power factor must be unity (1), and in practice near to unity, as much as possible.
- Understand how to improve the power factor in an AC circuit.
- Know about leading and lagging, or expressions such as "current leads the voltage."
- Understand three-phase systems and why they are used.
- Understand the relationship between the three phases and the importance of their order.
- Describe balanced loads in a three-phase system.
- Understand the various methods of connection in three-phase AC.
- Explain the difference between a delta (Δ) and a wye (Y) connection.
- Realize the nature of a star connection in three-phase AC.
- Calculate phase voltage from line voltage, and vice versa.
- Calculate phase current and line current.
- Understand new terms such as substation.
- Know what a collector is in a wind farm.
- Explain power quality and why it is important.
- Realize the conditions for power generation and connecting a turbine to grid.
- Understand what harmonics are and why they should not be present in AC current.
- Know the meaning of islanding and that it is not good if it happens.

8.0   Introduction

Except for small turbines that can power an isolated installation or community, the turbines in a wind farm, which are part of the generating force in a network or power production facility, must be connected to a grid. In this sense, a turbine is a member of the group of electricity sources that work together and must be in technical accord with each other. The role of each turbine when connected to a whole is to inject power into the grid without disturbing the system or becoming a load instead of contributing as a source.

A turbine that is already connected may become disconnected, or a turbine may start working from rest. It can be reconnected to the grid only if it is ready, by satisfying some technical requirement. In this chapter, the requirements for a turbine to be connected to a grid are discussed. Before the main discussion, however, we need to describe "power" in electrical terms in a circuit.

Electrical power   8.1

In chapter 7, we learned about mechanical power, which is the power in a mechanical system in terms of the mechanical concepts such as torque and rpm. This section describes electrical power, which is the power in an electrical system in terms of electrical entities, such as voltage and current. A turbine is an electromechanical system that receives mechanical power and turns that into electrical power. The produced electrical power is less than the power harnessed from wind (it cannot be higher or even equal to that power, because of the losses in the components). On the other hand, it often becomes necessary to reduce the power harnessed to match the power requirement, or to avoid overloading. (This is what happens when a turbine is pitch controlled to limit its output power in the higher wind speeds. See figure 6.10 compared to figure 6.9.)

Power in DC electricity   8.1.1

In direct current electrical circuits we have two independent measurable entities: current and voltage. Power in such a circuit is obtained by the product of these two entities.

$$\text{Power (watts)} = \text{Voltage (volts)} \times \text{Current (amps)} \qquad (8.1)$$

The following examples help provide a better understanding of the matter.

EXAMPLE 8.1:

A DC motor draws a current of 35 amps (A) when working. If the working voltage of the motor is 100 V, determine the value of the motor power.

Solution:

$$\text{Motor (electric) power} = 100 \times 35 = 3500 \text{ W} = 3.5 \text{ kW}$$

Note that this is the amount of power the motor consumes.

EXAMPLE 8.2:

A voltage of 90 V is applied to a load. If the resistance of the load is 18 ohms (Ω), find the power consumption of the load.

Solution:

First find the current from Ohm's law (voltage = resistance × current).

$$\text{Current} = 90 \div 18 = 5 \text{ A}$$

$$\text{Power} = 90 \times 5 = 450 \text{ W}$$

Important: For a resistive load (as in a heater with only heating elements), it does not matter if the connection is to DC or AC. You can always use the preceding formula.

EXAMPLE 8.3:

If the power consumption of a heater is 5000 W, and it its operating voltage is 230 V, find the current in the heating element.

Solution:

$$\text{Current} = 5000 \div 240 = 21.74 \text{ A}$$

It is important to note that the amount of power in the preceding three examples is what the load demands from a source (load is the device that consumes electricity; source is the device that provides electricity). If the source cannot deliver what is demanded from the load, then the values of voltage, current, or both will drop such that the product of current and voltage becomes equal to what the source can deliver.

> Always, the power capacity of a source for providing electricity must be higher than the power demand by a load that is connected to that source.

8.1.2   Power in AC electricity

resistive load
A load that exhibits only resistance to the electric current.

capacitive load
A load that consists of capacitors.

inductive load
A load that contains motors and electric windings.

Power in alternating current electricity is more involved than in direct current. This is due to the *capacitive* and *inductive* effects in AC that do not come into the picture with DC. In alternating current, we have three types of load, compared to direct current where we have only one type. In DC, all the loads can be assumed to be **resistive load.** In addition to resistive load, in AC electricity we have **capacitive load** and **inductive load.** A resistive load dissipates energy in the form of heat. An electric heater or a light bulb is a resistive load, implying that it has resistance (to the electric current). When an AC circuit contains a motor or an electromagnet, we have inductive load. Note that a load can be inductive and resistive at the same time. A capacitive load corresponds to capacitors in an AC circuit. For DC, inductive and capacitive loads do not carry a meaning; a capacitor does not conduct and a motor does not introduce inductive load.

When in a circuit, the load has only elements that have ohmic resistance, such as lightbulbs or a heater without any motor, the circuit is said to have resistive load. If there is any other type of load, for instance a motor or a device containing a motor, the load cannot be considered resistive. As mentioned in example 8.2, if the load in an AC circuit is resistive, then you can use the same relationship as in DC in order to determine the power that the load takes from a source. If the load is not resistive, then one must know the *phase angle* associated with the load.

Capacitive and inductive loads and phase angle

When a capacitor is connected to DC it behaves like an open circuit and there is no current through its circuit. In an AC circuit, however, a capacitor behaves as a conductor and there is a current associated with it. This current defines the presence of a resistance closing the circuit. The electrical resistance exhibited by the capacitor is called **capacitive reactance,** and in terms of load to the circuit, it is considered as a capacitive load. If the alteration of the voltage is sinusoidal, alteration of this current is also sinusoidal.

Nevertheless, the current is always ahead of the voltage, meaning that it reaches its maximum (minimum) value *before* the voltage reaches its maximum (minimum) value. We say the current is **leading** the voltage. Equivalently, we may say the voltage is **lagging** the current. The meaning of leading and lagging is illustrated in **Figure 8.1.** The leading (or lagging) can be defined as the fraction of the time of one cycle. Depending on a circuit and the (other) components in it, the difference in time that current leads voltage varies.

cpacitive reactance
The apparent resistance toward electric flow (thus, limiting current) that a capacitor exhibits in an alternating current circuit.

leading
A term concerning the power factor in an AC circuit, when the current is ahead of the voltage; thus, the circuit has a leading power factor as opposed to the lagging power factor (see also lagging).

lagging
A term concerning the power factor in an AC circuit, when the current is behind the voltage (voltage reaches its maximum value in its cycle before the current reaches its maximum).

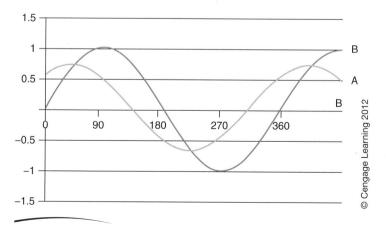

© Cengage Learning 2012

FIGURE 8.1 *Meaning of leading and lagging: A is leading B (B is lagging A).*

When a capacitor is connected in an AC circuit, the electrical resistance exhibited by the capacitor is called *capacitive reactance*, and in terms of load to the circuit it is considered as a capacitive load.

phase angle
The angle between the expressions for voltage waveform and the current waveform in alternating current electricity. Physically, it translates in a time delay between the two waveforms in an electrical circuit.

Since one cycle is 360°, the amount of leading can be described in terms of number of degrees (instead of the corresponding time); for instance 40°. The number of degrees represents an angle, called the **phase angle** of the circuit. The phase angle determines by how much (as an angle; in degrees, for instance) the current in a circuit leads the voltage. For a purely (100%) capacitive load, this angle is 90°. Because the behavior of a capacitor depends on the frequency of the alternating current, and since a circuit may have various components, including resistors, the amount of current lead may assume any angle between 0° and 90°; that is, not a negative number and no value larger than 90°.

The phase angle is of great importance in finding the power in AC electricity. When a coil (or winding) is connected to a DC electric source, the only effect is the resistance of the wire in the coil. The current in the coil circuit depends on this resistance. On the other hand, in an AC circuit, the winding of the coil has a more pronounced effect on the current. Associated with winding size, the number of turns and so on, and the frequency of the electric source, a form of resistance is exhibited by the coil, which reduces the current compared with when it is connected to DC.

In an AC circuit, a coil is considered to be an *inductor* when its ohmic resistance (the wire resistance) is ignored. If the current through a coil when connected to DC is much higher, say 10 times more, than when connected to AC, the resistance of the coil can be ignored. An inductor implies the pure effect of exhibiting a braking behavior to the flow of current in an AC circuit. This braking effect translates to limiting the current. Associated with the current one can find a resistance value, based on the voltage. This electrical resistance in a circuit with an inductor is called **inductive reactance,** and the circuit having it is said to have an inductive load.

inductive reactance
The apparent resistance toward electric flow (thus, limiting current) that an inductor exhibits in an alternating current circuit.

When an AC circuit contains inductive load, the up and down current variation in the waveform always follows (is behind) the voltage, meaning that current reaches its maximum (minimum) value *after* the voltage reaches its maximum (minimum) value. We say the *current is lagging the voltage*. Depending on the way a circuit is made and the other components in it, the amount of time that current lags voltage (delay between current and voltage) varies.

When an inductor is connected in an AC circuit, the electrical resistance exhibited by the inductor is called *inductive reactance*, and in terms of load to the circuit it is considered as an inductive load.

If the only load in an AC circuit is an inductor, the phase angle of the circuit is −90°. The negative sign is due to the fact that the reference is the voltage and since in this case the current is behind the voltage this angle is negative. We can say that for a purely inductive load the phase angle is −90° and for any other combination of resistive and inductive load, the phase angle is somewhere between 0° and −90°.

In a circuit with only resistive load, the phase angle is zero. We may say this fact differently; that is, for a resistive load the phase angle is zero.

> The phase angle is always between the voltage and the current in a circuit. For a purely resistive load, the phase angle is zero. For a purely capacitive load the phase angle is +90°. For a purely inductive load, the phase angle is −90°.

When an AC circuit contains both capacitive and resistive loads, the phase angle is between 0° and 90°. If it contains both resistive and inductive loads the phase angle is between −90 and 0. In a general circuit, containing all the three loads (resistive load, capacitive load, and inductive load), the phase angle can be anything between −90° and +90°. This all depends on the values of the loads.

At home we usually have lights, a refrigerator, a washing machine, and so on. The load is not purely resistive. In industry it is the same; motors and other devices make the loads a mixture of resistive, inductive, and capacitive. So, in an AC circuit with many branches (to various users, say) each user has a current (depending on the amount of electricity usage) at some phase angle. The total phase angle for the circuit is determined by adding all the currents together. In the following examples, use is made of the vector representation for voltage and current in AC circuits, which makes the calculations more convenient and it is widely used in practice (see the text for *use of vectors in AC electricity*, in this chapter).

EXAMPLE 8.4:

Figure 8.2 shows a circuit in which all the resistive loads are represented by $R = 4\ \Omega$, all the capacitive loads are shown as $X_C = 6\ \Omega$, and all the inductive

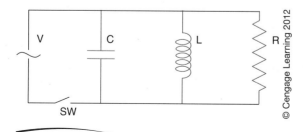

FIGURE 8.2 *Example of loads applied to an AC source.*

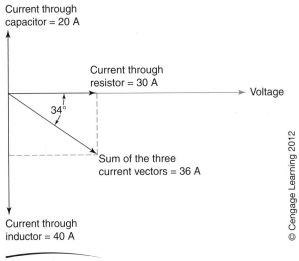

Current through
capacitor = 20 A

Current through
resistor = 30 A

Voltage

34°

Sum of the three
current vectors = 36 A

Current through
inductor = 40 A

© Cengage Learning 2012

FIGURE 8.3 *Solution to example 8.4.*

loads are put together and denoted by $X_L = 3\ \Omega$. Find the current and phase angle for this circuit, if the applied voltage is 120 V.

Solution: The numerical values for current through the resistor, the inductor and the capacitor, respectively, are:

$$\text{Current in } R = I_R = 120 \div 4 = 30\ \text{A}$$
$$\text{Current in } L = I_L = 120 \div 3 = 40\ \text{A}$$
$$\text{Current in } C = I_C = 120 \div 6 = 20\ \text{A}$$

The vectors corresponding to the three values of currents, based on the fact that for R, L, and C, the phase angles (angle between the current and the voltage) are $0°$, $-90°$, and $+90°$, and based on their values, are shown in **Figure 8.3.**

The sum of the three currents in the circuit is graphically determined to be about 36 A, and the phase angle can be measured to be $-34°$. This indicates that for this circuit the current is lagging the voltage by 34°.

Note that the current through the inductor is 40 A, but the total current of the circuit is 34 A (that is, less than 40). This is all normal and happens in AC electricity. The reason for this is the effect of the capacitor, which behaves opposite to the inductor. In fact, if the capacitor is such that the current in it is equal to that of the inductor, then it cancels the current in the inductor, and the total current will reduce to that of the resistor, only.

USE OF VECTORS IN AC ELECTRICITY

A convenient way to show the phase angle in an AC circuit is the use of vectors to show the values for voltage, current, and power. This is described here for parallel circuits, which are the usual circuits in many applications, including wind turbines and their connection to a grid.

Since in a parallel circuit voltage is the common entity between all branches, voltage is selected as the reference, which is shown as a vector along the horizontal axis in the graphical presentation, as shown in the following figure. The positive direction for measurement of angles is counterclockwise (CCW).

For each load in all branches of the circuit, based on the type of load (resistive, capacitive, and inductive load), the current is obtained and shown on the graph. In such a case, for a purely resistive load, the vector representing its current is along the same direction as the voltage vector (horizontal), the current in a purely capacitive load (any capacitor in the circuit) is shown vertical and upward ($+90°$ with respect to the voltage vector), and the current through any inductive load (any inductor in the circuit) is depicted by a downward vertical vector.

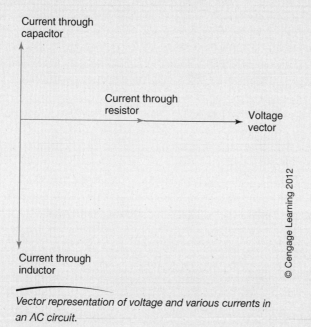

Vector representation of voltage and various currents in an AC circuit.

The phase angle of this circuit is determined by finding the sum of the currents; that is, the sum of the current vectors defines a vector whose angle with the voltage vector (positive or negative) defines the phase angle of the circuit.

(continued)

The sum of the three current vectors is shown in the following figure, which also shows the phase angle for the circuit. The phase angle is −21°.

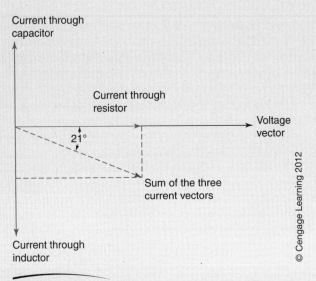

Finding the sum of various currents in an AC circuit.

The vector sum is the total current in a circuit. Note that we can measure the two components of the vector sum and define the total circuit current by its horizontal and vertical components.

EXAMPLE 8.5:

A local generator supplies electricity to four consumers. **Figure 8.4** shows these four loads in the form of vectors, representing the currents they take from the source and their phase angles. Determine the total current that the source must provide.

Solution:
As illustrated in figure 8.4 the currents are as follows:

Load 1: 20 A with +20° phase angle
Load 2: 30 A with 0° phase angle
Load 3: 36 A with −35° phase angle
Load 4: 40 A with −55° phase angle

The total current can be found graphically (by adding vectors), or numerically. The answer is as shown in the figure, which is a vector with the length of 111.4 at an angle of −25°. Thus, the total current is 111.4 A, and the phase angle is −25°.

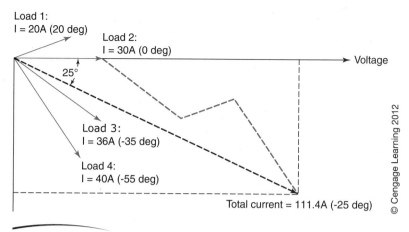

FIGURE 8.4 *Problem in example 8.5, with solution.*

Active, reactive, and apparent power

8.1.2.2

Now that you have learned about the difference that comes into effect when capacitive loads and inductive loads are used in alternating current circuits, we can consider studying the power in AC circuits. In AC circuits there are three types of power and we cannot just simply use the word "power" without specifying which one of the three is addressed. You may expect that we use the same terms as we used for load—nominally resistive, capacitive and inductive—for power, too. But this is not the case.

The three categories of power in an alternating current circuit are *active power*, *reactive power*, and *apparent power*. It is important to understand the meanings of these terms and how in practice they affect the power generation and consumption.

Active power is the true power that an electric device converts to work or other types of energy. For example, in a motor this is the mechanical power that can be determined (measured) in terms of torque and rpm. In a heater with heating elements, this is the amount of heat that is produced per second, and can be measured. Active power is also called *true power* or *real power* in some books.

Reactive power is associated with a capacitor (or capacitive load) and an inductor (or inductive load) in a circuit; this power *must* be present so that the device works. Both a capacitor and an inductor work based on *borrowing* power from a source, but they give back this power to the source. In half a cycle of the (mostly sinusoidal) variation of voltage, they take energy (from the circuit they are connected to) and save it, and in the other half cycle they inject it back to the circuit. Nevertheless, the power (or energy for each second) must be available. That is, the source must be able to supply this extra power. We say *extra* since this power is not converted to true work.

Since active power and reactive power are not of the same nature, they cannot be added together, numerically. They must be added together as vectors.

Active power
The net power that converts to heat or work.

Reactive power
The amount of power that is stored in a capacitor or an inductor in each half cycle and then is given back to the circuit in the next half cycle. This power is used for magnetization in motors, for instance.

Active power and reactive power can be represented by vectors that are apart by 90°. In this way, they can be added together in order to find out how much power a source must provide for a certain load.

Figure 8.5 depicts a load that has both an inductive component and a resistive component. There are many examples of such a load. Any motor represents such a load. In other words, the windings of all motors behave as a purely resistive load in parallel with a purely inductive load.

If we show the two components of the total current that the circuit must provide for this motor, the horizontal component is associated with the current through the resistor, and the vertical component (downward for an inductor) denotes the current in the inductor (this is the current that the winding takes to store energy). These two currents and their sum are shown in **Figure 8.6.**

Figure 8.6 also illustrates the vector representations of the power in the AC motor. Three power terms are shown in the figure, two of which have been already defined. The active power is shown along the horizontal axis, the reactive power is shown along the vertical axis; thus, at 90° from each other. The third power term, which is the vector sum of the active and reactive power, is called **apparent power,** and represents the total power that must be supplied by the source providing energy to the circuit.

Furthermore, as depicted in figure 8.6, the reactive power is associated with the current in the inductor. This power is obtained by the product of circuit

apparent power
The power that a generator must supply to cover for the consumed active power and reactive power in a circuit; but that is not equal to their algebraic sum, due to the phase difference.

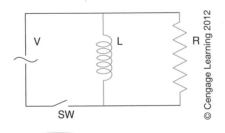

© Cengage Learning 2012

FIGURE 8.5 *Representation of winding in an electric motor.*

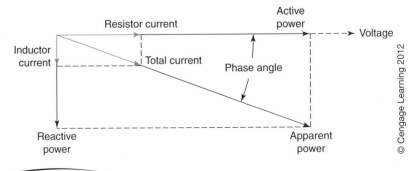

© Cengage Learning 2012

FIGURE 8.6 *Currents through the components of the circuit in figure 8.5.*

voltage and the current through the inductor. Similarly, the active power is obtained by multiplying the horizontal component of the total current by the magnitude of the circuit voltage (the voltage of the circuit at the point it is applied to the motor). The apparent power is determined from the multiplication of the total current by the circuit voltage.

The expressions for the three power terms in any (single-phase) AC circuit are as follows (the word "total" for current is unnecessary and is omitted):

$$\text{Active power} = \text{Voltage} \times \text{Current} \times \cos(\text{phase angle}) \qquad (8.2)$$

$$\text{Reactive power} = \text{Voltage} \times \text{Current} \times \sin(\text{phase angle}) \qquad (8.3)$$

$$\text{Apparent power} = \text{Voltage} \times \text{Current} \qquad (8.4)$$

In order to differentiate these three powers from each other in practice, three different units of measurement are employed. Active power, which reflects a true physical work, is measured in watts (W). The apparent power, which defines the numbers of volts and amperes, is measured in **volt-amperes (VA).** The unit for reactive power is **volt-ampere-reactive (VAR),** usually pronounced "var."

The phase angle is normally denoted by the Greek letter θ (theta) and the quantity $\cos\theta$ is called **power factor.** The power factor is always smaller than 1.

Comparing the circuit in figure 8.5 with the one in figure 8.2 demonstrates that, if in addition to a motor, there existed a capacitor in the circuit, the difference would be reflected in the magnitude of the phase angle θ. Most of the industrial AC motors have capacitors in parallel with them, partly to reduce the magnitude of the phase angle θ.

A good understanding of the meaning of each of these power terms is very important. The magnitudes found from calculation based on equations (8.2) and (8.3) all represent power, and all are associated with the same load in an AC circuit. Apparent power is the magnitude of power in terms of voltage and current that the circuit (source) must be capable of providing at the point the load is connected. Based on the power factor, only part of this power goes toward useful work, but the rest must be there as a reserve. This must not be mistaken when we are considering efficiency in a device, where we may use the same statement that part of the energy converts to useful work. In the case of efficiency, the rest of the spent energy turns into heat, and the values must be added numerically (no vector is involved).

Figure 8.7 gives an insight to the meaning of power in an AC circuit. It shows a reservoir having an outlet not at the very bottom. The water in the tank denotes the power. If the outlet was at the bottom of the tank all that water (power) would be available for consumption. Having the outlet at some distance from the bottom, however, implies that only part of the water can be consumed, although the tank is full. Thus, for the same magnitude of power generation, in DC electricity all the power in the line is available for usage, whereas in AC electricity, only a fraction of that is available, but all has to be present.

Since

$$\text{Active power} = \text{Apparent power} \times \cos\theta$$
$$= \text{Apparent power} \times \text{Power factor} \qquad (8.5)$$

volt-amperes (VA)
A unit of measurement of electric power in AC electricity. It is normally used to express apparent power. It is particularly used when voltage and current are out of phase.

volt-amperes-reactive (VAR)
A unit for measuring reactive power in AC circuits. Reactive power exists in a circuit containing inductors (coils, magnetism, and motors) and capacitors.

power factor
The ratio of active power to apparent power. This ratio is smaller than unity.

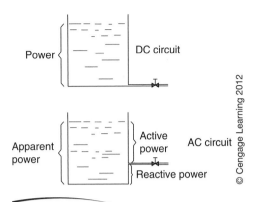

© Cengage Learning 2012

FIGURE 8.7 *Analogy to indicate active and reactive power definition.*

it is preferable that for any load the power factor is as much as possible close to unity, or in other words, the phase angle θ is close to 0. Otherwise, there is a lot of waste of energy due to the extra current in the lines, as well as maintaining the reserve power (which is the reactive power).

It is not from the load side that we are concerned about power factor; it is from the generator side, which must provide the extra power. Having a large phase angle (small power factor) is not desirable. However, the power factor is a property of any particular load and on the load side we try to reduce the phase angle and improve the power factor. Nevertheless, it is not always possible to raise the power factor at the consumer side and, therefore, provisions must be made for correcting the power factor at the generating side.

EXAMPLE 8.6:

An AC motor draws 7.5 A when connected to 120 VAC (note here the notation VAC, which simply implies AC volts). If the power factor of this motor is 0.68, how much are the apparent power, active power, and the reactive power?

Solution:

$$\text{Apparent power} = (120)(7.5) = 900 \text{ VA}$$
$$\text{Active power} = (900)(0.68) = 612 \text{ W}$$

In order to find the reactive power we can use any of the following calculations:

$$\text{Reactive power} = \sqrt{\text{App.power}^2 - \text{active power}^2} = \sqrt{900^2 - 612^2}$$

$$= 659.9 \approx 660 \text{ VAR}$$

Reactive power $= (900)(\sqrt{1 - \text{powerfactor}^2}) = (900)\sqrt{1 - 0.68^2}$

$$= (900)(0.733) \approx 660 \text{ VAR}$$

Reactive power $=$ apparent power $\times \sin \theta$

For the last relationship we need to first find the phase angle. This formula is normally used when the phase angle is already determined.

$$\theta = \cos^{-1}(0.68) = 47.2°$$

Reactive power $= (900) \times \sin(47.2°) = (900)(0.733) = 660 \text{ VAR}$

Note that a motor is an inductive load, and when connected to an AC circuit the current lags the voltage. Thus, the angle 47.2° is negative if the vectors for current or power should be drawn, similar to figure 8.6.

> When a motor is connected to AC circuit, the current lags the voltage by the value of the phase angle. This is a lagging circuit.

EXAMPLE 8.7:

If in an AC circuit the apparent power is 2000 VA and the active power is 1500 W, what is the phase angle of the circuit?

Solution: From the ratio of the active power to apparent power we find the power factor

$$\text{Power factor} = 1500 \div 2000 = 0.75$$
$$\text{Phase angle} = \theta = \cos^{-1}(0.75) = 41.4°$$

Note: In this problem the nature of the load (if capacitive or inductive) is not defined. Thus, depending on the load, the circuit can be lagging or leading.

EXAMPLE 8.8:

The active power in a circuit is 5000 W and the power factor is 0.7. How much is the current in the circuit, if the applied voltage is 220 V?

Solution:
It follows from equation (8.2) that

$$\text{Current} = \frac{\text{Active power}}{\text{Voltage} \times \text{power factor}} = \frac{5000}{(220)(0.7)} = 32.5 \text{ A}$$

EXAMPLE 8.9:

The reactive power in a 220-V circuit is 4000 VAR and the power factor is 0.65 lagging. In order to improve the power factor a capacitor is added (inserted parallel to the circuit). If the current due to the capacitor (the current that the capacitor draws from the circuit) is 15 A, what is the new power factor?

Solution: Referring to figure 8.2, if a capacitive load is in the circuit its current vector is upward (i.e., its current leads the voltage). The lagging circuit (i.e., before the capacitor is connected) is due to an inductive load whose current vector is downward. The effect of the capacitor is to partially or totally cancel the current of the inductive load.

$$\text{Current of the inductive load} = 4000 \div 220 = 18.22 \text{ A}$$

The current of the inductive load after the capacitor is added = 18.22 − 15 = 3.22 A.

Now we need to find the component of the total current that does not change (that is, the component along the voltage vector, which is not affected by capacitor). This can be found by finding the total current in the circuit, based on the reactive power, and the power factor

$$\text{Total current} = \frac{18.22}{\sqrt{1 - 0.65^2}} = 23.93 \text{ A}$$

$$\text{Current component not affected by capacitor} = (23.93)(0.65) = 15.55 \text{ A}$$

Total current after the capacitor is inserted is found from the values of the currents thus obtained:

$$\text{New total current} = \sqrt{15.55^2 + 3.22^2} = 15.88 \text{ A}$$

$$\text{New power factor} = 15.55 \div 15.88 = 0.98 \text{ A}$$

What is performed in this example is a common practice in industry. In order to improve the power factor (or correct the power factor), that is, raising its value, capacitors are added to a circuit. The ideal value for power factor is unity.

Figure 8.8 helps to show the various values of currents in the process of solving the problem in this example. As shown in the figure and calculations, the total current to be supplied by a source has decreased from 23.93 A to 15.88 A, due to the addition of the capacitor, while the active power has not changed. The higher current leads to considerable waste of energy in the transmission lines, as well.

power factor correction
Improving the power factor of alternating current circuits by introducing a bank of capacitors.

The ideal value for power factor of an AC circuit is unity. Most loads are inductive and their power factor can be improved by putting an appropriate capacitor in parallel with the load. This is a usual practice and is called **power factor correction.**

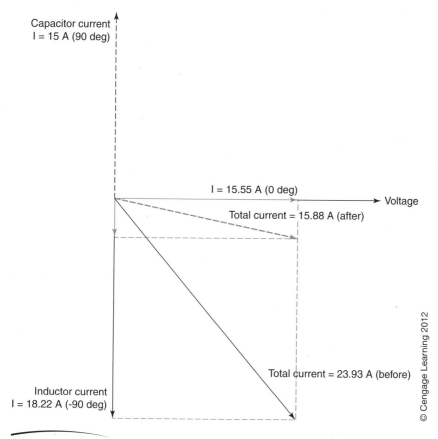

FIGURE 8.8 *Graphical solution to problem of example 8.9.*

Example 8.9 showed how to find the corrected power factor of a load or a circuit when a capacitor is inserted in the circuit in parallel with the load(s). It has to be noted that, as can be understood from example 8.9, if the capacitor current is higher than 18.22, then the current leads the voltage. In this sense, if the capacitor current is considerably higher, then the power factor still can be high, and the benefit from adding the capacitor is lost.

In practice, there is not only one load connected to a circuit. Also, loads are not usually constant and they can vary. Selection of the right capacitor, therefore, must be based on the most probable value for all loads considering the length or percentage of time any load is connected and the rate at which a particular load operates. For instance, a machine can work at 80% of its capacity, 8 hr per day, on average.

The following example shows how to select the capacitor for an AC circuit. This selection is based on the reactance of a capacitor, which is described in this chapter (see *Capacitor and Inductor Reactance*).

EXAMPLE 8.10:

An AC motor has a rating of 240 V, 3500 W. The frequency of the electric circuit is 50 Hz. If the motor mostly works at 80% of its capacity with a power factor of 0.76, find the capacitor to put in parallel with the motor in order to raise the power factor to 1.

Solution: We find the right capacitor to change the power factor to 1, knowing that the power factor may never be exactly equal to, but fluctuate around, 1. We first find the current for 80% capacity of the motor.

$$\text{Current} = \frac{(3500)(0.80)}{(240)(0.76)} = 15.35 \text{ A}$$

The portion of this current due to the induction in the motor winding is

$$\text{Inductor current} = (15.35)(\sqrt{1 - 0.76^2}) = 9.98 \text{ A} \approx 10 \text{ A}$$

We should now find a capacitor that if connected to 240 V the current is 10 A. Thus, the capacitive reactance of this capacitor is 24 Ω (240 ÷ 10). The value of the capacitor can be found by using the appropriate formula.

$$\text{Capacitance} = \frac{1}{(2)(\pi)(f)(X_c)} = \frac{1}{(2)(3.14)(50)(24)}$$

$$= 1.33 \times 10^{-4} \text{ farads} = 133 \ \mu\text{F}$$

The answer is, thus, 133 microfarads (μF). We have to find the nearest standard capacitor.

8.1.2.3   Power in three-phase AC electricity

The relationships (8.2) to (8.4) can be used for single-phase AC electricity. Power generation at the industrial and commercial level is always in three-phase. While all that has been said in the previous section for power in AC electricity holds for both single-phase and three-phase systems, the formulas for power have to be slightly modified for calculation of power in three-phase electricity. In this section, without going into the details of three-phase systems the equations to be used for apparent power, active power, and reactive power are given.

A three-phase system by all means is more efficient than single-phase AC. In fact, a **three-phase system** consists of three synchronized single-phase systems, together. The sinusoidal variations of the three phases do not take place simultaneously, though, meaning that not all the phases reach their maximum at the same time. This is shown in **Figure 8.9,** which also illustrates the sequence of the three phases. AC electricity is generated and transmitted as three-phase.

three-phase system
Three-phase electricity; the sort of AC electricity that requires at least three wires for transmission. The voltage between each pair of wires has a sinusoidal form and has the same frequency and maximum amplitude, but there is a fixed delay between the instants when each of the three voltages reaches its maximum value.

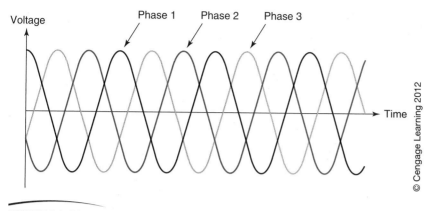

FIGURE 8.9 *Waveforms of a three-phase alternating current electricity.*

Then at consumption site it is either delivered in three-phase to consumers with large consumption, such as industry, factories, and large commercial buildings, or only one phase is delivered, for instance at homes and in small buildings. The distribution at the city level is three-phase. For example, in three streets, each gets one phase of the three phases (the single phase delivered to the houses in that street). From a technical viewpoint, this separation of consumers on three phases must be done in such a way that, as much as possible, the loads on each phase are equal. If the loads on the three phases are equal then the system has a **balanced load.** Having a balanced load is better for a generation and distribution system, and less waste is involved.

balanced load
When the three loads in a three-phase electric system have equal values.

delta connection
A method of connecting the three wires of three-phase electricity to a load in which the three individual loads form a triangle.

wye connection
Also called *star connection*. A way of connecting the terminals of a three-phase load to the supply wires. A three-phase load has six terminals. If three of them are connected together and the other three are connected to the supply voltage, this forms a Y connection. The order of which terminals to choose is important.

star connection
A method of connecting wires in a three-phase load, where the three connected wires form a letter Y, as opposed to forming a Δ.

THREE-PHASE ELECTRICITY

It is necessary that a reader know about some fundamentals of three-phase electricity in order to follow the formulations given for three-phase power. Since a three-phase system is equivalent to three single-phase systems, one can always break such a system to its equivalent three single phases. However, this may become more tedious and confusing. A three-phase system may use three wires or four wires. This is described in the following paragraphs.

A three-phase load consists of three similar elements (three resistors, for example) or three similar groups of elements (coils and capacitors, for instance in most motors). A three-phase load can be connected to a circuit in two different ways, by either a **delta connection** or a **wye connection** (also called **star connection**).

(continued)

If the elements in a three-phase load are connected together in a way that they form a triangle (Δ, a Greek letter called delta) and then each corner of the triangle is connected to three electricity lines, this is called a delta connection. If the three elements are connected together from one side, so that they form a letter Y, and then their other ends are connected to wires carrying electricity, this is called a wye connection. These arrangements are shown in the following figures.

Note that in the figures there are three lines denoting the three phases. The phase lines are numbered 1 to 3. This numbering is important and their sequence must always be respected. A fourth line, called, the null (or neutral), is added in the figure. The neutral line and one of the three phase lines are what we get at home. The neutral line is normally grounded. As shown in the figure, between each phase line and the neutral line the voltage is 120 V. This is nominal household voltage in North America. Corresponding to the 120-V voltage difference between each phase line and the null line, the voltage difference between each pair of phase lines is 208 V (i.e., $120\sqrt{3}$).

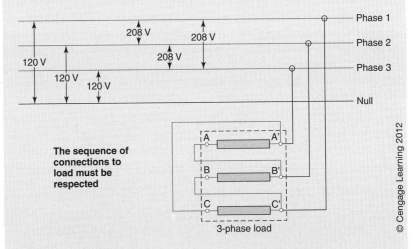

The sequence of connections to load must be respected

3-phase load

The preceding figure shows a load with a delta connection (or connected in delta). Notice that for the Δ connection there is no place to connect to the neutral line. Indeed, it is not necessary for connecting to the fourth (neutral) line.

The next figure illustrates the same load shown differently, but still with a delta connection. Sometimes, this form of representation is preferred, since it is easier to follow the type of connection. This figure shows the general terms for the two voltage differences, namely, **phase voltage** and **line voltage**. Phase voltage is the voltage difference between each phase line and the null line, and line voltage is the one between each pair of phase lines.

phase voltage
The voltage across each individual load connected to a three-phase system. The relationship between *phase voltage* and *line voltage* depends on the type of connection (delta or wye) used.

line voltage
The voltage between the supply lines of a three-phase system.

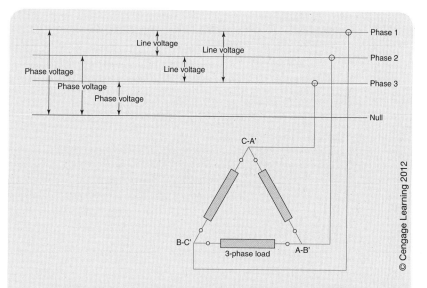

The same load can be connected using a wye connection. In such a case, the common points of the three elements in the load can be connected to the neutral line, if desired.

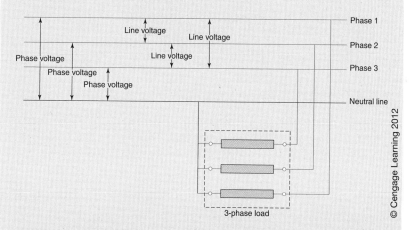

A single-phase load is normally connected between the neutral line and one of the phase lines. It is possible, however, to connect a single-phase load to two of the phase lines. In this way, a higher voltage is applied to the load.

In conjunction with phase and line voltage terms, we also have phase current and line current. Line current is the current that one can measure in each of the phase lines connected to a load. **Phase current** is the current in the element(s) connected between two phase lines. These are illustrated in the following figure.

(*continued*)

Phase current
The current passing through a load connected between each of two lines of a three-phase system. The relationship between *phase current* and *line current* depends on the type of connection (delta or wye) used.

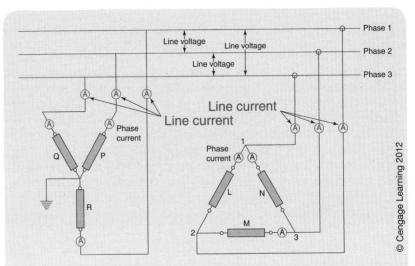

In this figure there are two three-phase loads; one consisting of elements P, Q, and R has a Y connection and the other, consisting of elements L, M, and N, has a Δ connection. The circles with a letter "A" inside them represent an ammeter. Note that there are only three wires shown. The common point in the Y connection is grounded. This is equivalent to connecting that to a neutral line, which is also grounded. These facts in three-phase systems are illustrated in this figure.

1. In a Y connection the phase current and the line current are equal.
2. In a Y connection the voltage applied to each element is less than the line voltage.
3. In a Δ connection the voltage across each element is equal to the line voltage.
4. In a Δ connection the phase current is less than the line current.

Although it is not possible to guarantee a balanced load when a variety of consumers are involved (for example, all houses do not normally have the same energy consumption at the same time), all the relevant calculations are based on a balanced load assumption. Whereas for electricity distribution in a city all the loads are single phase (like all the appliances at home), in industry there are many three-phase loads. For instance, all large motors are three-phase, since a three-phase motor is smaller and runs smoother than the equivalent single-phase motor. All three-phase devices are balanced loads when connected to a source or a three-phase circuit. Before continuing with the rest of this section you should read the basic facts about the three-phase systems and its vocabulary in this section.

In a three phase system the power equations are as follows:

$$\text{Active power} = \sqrt{3}\,\text{Line voltage} \times \text{Line current} \times \text{Power factor} \quad (8.6)$$

$$\text{Reactive power} = \sqrt{3}\,\text{Line voltage} \times \text{Line current} \times \sin\theta \quad (8.7)$$

$$\text{Apparent power} = \sqrt{3}\,\text{Line voltage} \times \text{Line current} \quad (8.8)$$

See the last figure in *Three-phase electricity* in this section for the definition of **line current**. These equations are more appropriate when line voltage
is known and the current a device draws is measured, no matter whether the
connection is delta (Δ) or wye (Y). Power factor is usually abbreviated as *pf*.
Now taking into account that in a Y connection:

- The phase current and the line currents are equal, but the voltage applied
  to each element is equal to phase voltage,
- Line voltage = $\sqrt{3}$ Phase voltage,

and that in Δ connection

- Line current = $\sqrt{3}$ Phase current,
- Voltage across each element is equal to the line voltage,

it can be concluded that if measurements are done differently and the currents
through load element are measured, then equations (8.6) to (8.8) can take the form

$$\text{Active power} = 3 \times \text{Phase voltage} \times \text{Phase current} \times \text{pf} \qquad (8.9)$$
$$\text{Reactive power} = 3 \times \text{Phase voltage} \times \text{Phase current} \times \sin\theta \qquad (8.10)$$
$$\text{Apparent power} = 3 \times \text{Phase voltage} \times \text{Phase current} \qquad (8.11)$$

EXAMPLE 8.11:

A 380-V, three-phase heating system has three 11-Ω elements. Find the apparent power, the active power, and the reactive power if the heater is connected to
a line using (a) a wye connection and (b) a delta connection.

Solution: In three-phase systems, unless mentioned otherwise, the line voltage
is given. So, in this example 380 V refers to the line voltage. Also, since the
device is a heater, one can conclude immediately that the power factor is unity
(pf = 1) and as a result the active power and apparent power are numerically
the same, and the reactive power is zero.

a. Wyw connection:
   The voltage applied to each element is the phase voltage, which is

   $$\text{Phase voltage} = \text{Line voltage} \div \sqrt{3} = 380 \div 1.73 = 220\,\text{V}$$
   $$\text{Phase current} = \text{Line current} = 220 \div 11 = 20\,\text{A}$$

   Using equation (8.6), thus leads to

   $$\text{Active power} = (\sqrt{3})(380)(20) = 13164\,\text{W} = 13.164\,\text{kW}$$

   The same answer could be found by equation (8.9), but the round-off
   errors can make it slightly different.

   $$\text{Active power} = (3)(220)(20) = 13200\text{W} = 13.2\,\text{kW}$$
   $$\text{Reactive power} = 0\,\text{VAR}$$
   $$\text{Apparent power} = 13.2\,\text{kVA}$$

line current
The current in the
supply lines of a
three-phase electrical
system.

b. Delta connection:
The line voltage (380 V) is applied to each element, thus

$$\text{Phase current} = 380 \div 11 = 34.55 \text{ A}$$

And from equation (8.9) we have

$$\text{Active power} = (3)(220)(34.55) = 22800 \text{ W} = 22.8 \text{ kW}$$
$$\text{Reactive power} = 0 \text{ VAR}$$
$$\text{Apparent power} = 22.8 \text{ kVA}$$

We can see that if a three-phase load is connected in delta, it draws more current from the circuit and consumes more power (the power consumption is more by a factor of $\sqrt{3}$). Also, note that although the active and apparent powers have the same numerical value, for each one the appropriate unit must be used.

> The voltage value mentioned for a three-phase system refers to the line voltage.
>
> If a three-phase load is delta-connected it consumes $\sqrt{3}$ times more power than if it is wye-connected. The current is accordingly $\sqrt{3}$ times higher.

8.2   Grid power requirement

Now that you know all that is concerned and the necessary terms in three-phase AC, we can consider the conditions and requirements for connecting a wind turbine to a grid. These conditions are not required for an isolated wind turbine. In connecting a wind turbine to a grid it is necessary that (see also section 6.6.2):

a. The voltage of the turbine output is the same as the grid voltage.
b. The frequency of the generated current is the same as the grid frequency.
c. The turbine output is synchronized with the grid.
d. The three-phase orders (sequence) of the turbine and the grid are the same.
e. In addition to these primary requirements, it is necessary that the power factor of the grid be maintained within certain values. Therefore, sometimes it is necessary that, if possible, the power factor be improved, or at least not be decreased by the introduction of a turbine to the grid.

substation
An electricity utility consisting of a number of transformers and other equipment to increase or decrease and regulate the voltage of incoming three-phase electricity.

Taking care of all the preceding conditions is not manual work and is done by a computer that monitors and supervises all the functions of a turbine.

The voltage of a turbine generator is normally stepped up by a transformer to a higher voltage corresponding to a **substation** before it is raised again to that of a grid. A substation is an electric unit in electricity transmission and distribution where the line voltage is either increased to a higher value or

© Cengage Learning 2012

FIGURE 8.10 *A substation mainly consists of transformers, breakers, voltage regulators, and safety devices. It may also contain other equipment.*

decreased to a lower value. For example, the voltage of a generator can be 600 V, 900 V, or 4000 V, which is increased to 11,000 V or 13,800 V. All the wind turbines in a wind farm are connected to a substation, which in this case is called the **collector.** There may be different turbines, having generators at different voltages, but their transformer must accordingly bring up their voltage to the voltage of the collector substation. The grid voltage can be 69,000 V, and a step-up transformer in the substation brings up the 11,000 V or 13,800 V to this level.[1]

A picture of a substation is shown in **Figure 8.10.** Changing voltages in substations is performed by transformers. You learn more about transformers in chapter 12. In addition to transformers, one can find breakers and voltage regulators in substations. Breakers are protective devices that disconnect circuits in abnormal conditions.

The frequency of a generated current is a function of the operating speed of a turbine generator. This is a matter that must be considered during the design process. This subject will be further discussed with more details in chapter 11 on the control of wind turbines. The operating frequency of electricity in North America is 60 Hz. The frequency of the electricity from a turbine must be within a certain range (say 1%) of this number; that is, the maximum error (deviation) that a turbine can have is limited by local authority for electric operations and standards.

collector
A point where all the turbine outputs in a wind farm are connected to be further connected to a grid by a transformer.

---

[1] Standard transmission (line-to-line) voltages in the United States are 69 kV, 115 kV, 138 kV, 161 kV, 230 kV, 345 kV, 500 kV, and 765 kV, set by the ANSI (American National Standards Institute).

After a turbine starts from rest it is not sufficient that the voltage and frequency of the generator arrive at the specified levels. It is quite important that the up and down (sinusoidal) variation of voltage correspond exactly to those of the grid. Otherwise the turbine becomes a load to the grid, consuming energy from it instead of contributing (injecting energy) into it. Again, this is the job of the turbine controller in the modern turbine to switch on the connection at the appropriate instant, or induce a delay to bring the turbine in synchronization with the grid.

The fourth item of having the correct sequence for the three phases is taken care of during installation of a turbine and making cable connections. The wrong order can damage the generator and/or the transformer. It is like wrongly connecting two generators together, which can overload one or both and damage them.

Connecting the phases in the wrong sequence is like turning a generator in the wrong direction. The directions of rotation of a turbine and its generator are determined in the design stage of a wind turbine. However, during installation it is quite important that the cables are connected correctly according to the manufacturer's schematics.

AC loads are normally inductive rather than capacitive. In other words, generally speaking, the loads in a grid consist of lighting, heating, and motor. We shall see in chapter 11 that some generators also consume reactive power as they work and, thus, contribute to lowering the power factor of a grid (which is not desirable). As discussed in this chapter, capacitors cancel the current of inductive loads and, thus, can be employed to correct the power factor to some degree. On the other hand, while in operation, some types of generators can perform in such a way that they inject capacitive reactive power into the grid, thus improving the power factor. This subject is further discussed in the modes of operation of generators in chapter 11.

8.3   Harmonics and power quality

Out of the basic requirements in section 8.2, items c and d are maintained after a turbine has been connected properly to a grid. The voltage, frequency, and power factor are subject to variations due to either load change or turbine performance. Load variation is a fact for any grid, since there is no control on the users and their continuously fluctuating demand for power. In wind turbines, the wind itself has variations of an order of less than a second. As a result, it is not possible to have the voltage and frequency at exact desired values and, consequently, there will be small fluctuation in both voltage and frequency. These fluctuations show up in flickers and represent a not clean current, which is undesirable as well as harmful for many devices. Power quality is defined by how near a voltage waveform is to a sinusoidal function and how clean it is from spikes and flickers.

There are two aspects in any random fluctuation in voltage and frequency. One is the duration of any such spike and one is the deviation in magnitude from the required value. The smaller are these two, the cleaner is the current. As a measure of power quality, the allowed ranges for all of these variations are currently defined by electric authorities in each country or region. For instance, the frequency fluctuation must be within ±0.01 and for a duration of less than 200 msec. In this sense, the power factor is not a measure of quality, but can lead to a loss of quality, as well as extra cost for the power-producing companies. It is the utility companies, therefore, that avoid having low power factors.

In addition to the voltage and frequency, **harmonics** of the electric current also can play a role in reducing the power quality. The harmonics of a 60-Hz current have frequencies of 120 Hz, 180 Hz, 240 Hz, and so on, which are the second harmonic, the third harmonic, and the fourth harmonic, and so on, respectively, for the fundamental frequency of 60 Hz.

The harmonics of a 120-V alternating current electricity are usually at a much smaller level than 120 V, for instance 1 or 2 V. Nevertheless, they can alter the shape of a waveform and deviate it from being a sinusoidal wave. This can cause problems in certain devices, and therefore, the harmonics are not desired to coexist with the fundamental frequency. In this way, the harmonics strengths (magnitudes) are to be within a certain limit in order not to degrade the power quality, in a manner similar to the voltage and frequency.

harmonics
AC signals with twice, three times, four times, and so on the frequency that can become mixed with the base AC (with fundamental frequency) electricity in a circuit.

Wind turbine as a discrete generator   8.4

Now that we have learned the relationships for power in electricity we can consider the effect of a load when connected to a generator. First, we consider the case of one generator supplying power to a number of loads, as shown in **Figure 8.11**. All the loads together can be assumed to be one load to the generator, demanding a current I at a phase angle θ. The current and phase angle are not constant and can vary from one instant to the next.

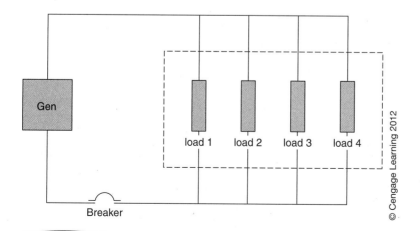

© Cengage Learning 2012

FIGURE 8.11 *A generator supplying power to a number of loads.*

The generator is expected to provide the demanded current at each instant based on the given condition (phase angle) maintaining the voltage and frequency at their nominal values. In other words, the loads altogether require an active power and a reactive power according to equations (8.2) and (8.3) based on the values of voltage, frequency, and current. The active and reactive powers are related to each other according to equation (8.3). The preceding facts reveal that the generator must provide power to maintain the current in the circuit, while keeping the voltage and frequency as required. This implies three separate control actions for voltage, frequency, and current. If the load current is suddenly lowered by turning off equipment, then the immediate result before the control action takes place is an increase in frequency and in voltage, which leads to an increase in current for other equipment.

In the case of multiple generators, which applies to all the grid-connected wind turbines, the loads are distributed among various generators. But the effect on each individual generator is the same, meaning that in each generator the frequency, the voltage, and the current must be continuously controlled.

For wind turbines, in comparison with other means of electricity generation, there is one more item to bring into consideration in the control action. That is the available wind (measured by wind speed) at each instant of time. If there is not enough wind for the demanded power, then in no way a turbine can keep the magnitudes of the three variables frequency, voltage, and current as expected. Thus, the current is controlled based on how much power is available.

8.5   Islanding

All the turbines in a wind farm are connected to the high-voltage grid through a collector substation. For example, there can be 15 turbines in a wind farm that must be connected to a grid. Unless only one turbine exists for the farm, no turbine is individually connected to a grid. The control system of each individual turbine monitors the safe performance of the turbine and its connection or disconnection from the substation. This includes synchronization of the voltage with the grid.

In addition to the breakers for each turbine, a substation has its breakers, which can isolate a whole wind farm from the grid. In situations like a lightning strike such a breaker can trip and disconnect the wind farm from the grid. Suppose that the wind farm has some load of its own, and this situation does not lead to overspeeding and eventual disconnection of each turbine (more on this in later chapters). If the disconnection from the grid is prolonged, then the group of turbines in the farm can continue working together, while due to a drift in frequency go out of synchronization with the rest of the grid. This is called **islanding.** Islanding is not a desirable condition and can damage equipment, due to voltage sag or overloading when attempting to reinstate the grid connection. **Figure 8.12** shows a simple case. While the two generators are connected, each one contributes to powering the circuit. However, if the connection at a point such as point A is lost, then after some time, the synchronization between

islanding
A state of isolation of part of an AC network from the rest, thus resulting in the possibility of frequency and voltage drift while operating.

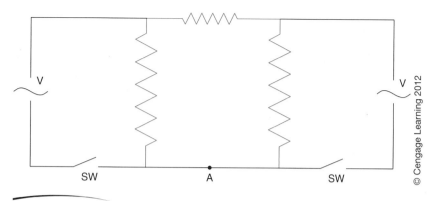

FIGURE 8.12 *A simple example of islanding.*

the two generators may get disrupted, and it is not a good idea to connect the two parts together just by closing the circuit at point A.

Islanding can also happen in any part of a grid when one or more generators provide power to the loads in that part of the grid, while because of a malfunction or tripping of a breaker they become isolated from the rest of the grid.

Dealing with islanding is not an easy task and a number of islanding schemes have been suggested and are practiced by utility companies. The solution, however, depends on each individual case, and it may take hours to correct the situation.

Chapter summary   8.6

- Electric power is always proportional to the product of voltage and current.
- Resistive load is a load due to resistors and anything independent of frequency.
- Capacitors and inductors exhibit resistance to the flow of electricity, which depends on frequency.
- The frequency-dependent resistance from capacitors and inductors is called reactance.
- Capacitors in AC circuits are capacitive loads. They have capacitive reactance.
- Inductors in AC circuits are inductive loads. They have inductive reactance.
- Capacitive and inductive loads can cancel each other's effect.
- In DC electricity, there is only one type of power, corresponding to resistive load.
- In AC electricity, there are three types of power: active power, reactive power, and apparent power.
- Active power corresponds to resistive loads in AC electricity.
- Reactive power corresponds to inductive and capacitive loads.

- Inductive and capacitive loads cause a phase difference between voltage and current.
- The phase difference is measured as an angle (say in degrees) and can be between -90 to $+90$.
- "Current leading voltage" means a current waveform reaches its maximum (or minimum) value before voltage reaches its maximum (or minimum) value.
- AC electric motors put both inductive and resistive load in a circuit.
- Capacitors in parallel with motors reduce the inductive load.
- "Current lagging voltage" means current reaches its maximum or minimum after voltage.
- Active power is the real power that converts to heat or work.
- Reactive power is the power stored in each half cycle and then sent back to the circuit.
- Apparent power is what is required to be provided for maintaining a current I in a circuit of voltage V. A generator in a circuit must provide the apparent power.
- The preferred value for the phase angle in a circuit is that it be as close to zero as possible.
- The power factor is the value of the cosine of the phase angle. Its preferred value is 1.
- In order to connect a turbine to a grid it is necessary that the turbine output and the grid (1) have equal voltages, (2) have the same frequency, (3) have synchronized waveforms (reach their minimum and maximum simultaneously), and (4) have similar phase sequences.
- A turbine must not lower the power factor of the grid.
- The output from all the turbines in a wind farm first goes to a collector substation. The substation raises the voltage from the turbines to that of the grid.
- Power quality involves that the voltage in an AC circuit has a perfect sinusoidal form and there are no fluctuations in the voltage and frequency magnitudes. Any variation in the form of the voltage signal and presence of flickers and spikes represent a lack of quality.
- Harmonics are low-voltage signals that have frequencies such as two times, three times, and so forth of the grid frequency. These can mix up with AC electricity and cause higher currents and undesirable results. They decrease the power quality of a circuit or grid.
- Islanding happens when part of an AC network, such as a wind farm, becomes isolated from the rest of the network, which can lead to a drift in voltage and frequency. It is not a desirable situation and must be avoided.
- In connecting a wind turbine to a grid, there are three fundamental control actions to be carried out: control of the voltage, control of the frequency, and control of the current based on the available power from wind.

ADVANCED LEARNING

Capacitive Reactance

Since a capacitor constantly charges and discharges every time that an alternating current completes a cycle, current continues to flow in a circuit that contains a capacitor. This fact leads to the following results:

1. For a voltage with a sinusoidal waveform, the current will also be sinusoidal, but the maximum current does not happen when the voltage is maximum, but when the change in voltage is maximum, which is at zero value of the voltage (change is depicted with the slope to the sinusoidal curve at each point). This is why the current in a circuit that contains only capacitive load leads the voltage by 90°.

2. We cannot expect that the current in such a circuit is infinity (that is, there is zero resistance to the flow of current). So, there is some opposition to the flow of current that can be defined in terms of resistance and measured in ohms. This entity is called reactance.

3. Frequency has an effect on what happens in an AC circuit, since the charge and discharge of the capacitor takes place faster as the frequency goes up.

The formula for calculating the reactance of a capacitor is as follows:

$$X_C = \frac{1}{2\pi fC}$$

where X_C is the capacitive reactance, measured in ohms, 2π is constant, f is the frequency of the circuit measured in Hertz (Hz), and C is the capacitance of the capacitor, measured in farads (F).

Example 1:

How much is the reactance of a 3-μF (microfarad) capacitor in an AC circuit whose frequency is 60 Hz?

Solution: The first step is to convert the value of the capacitor into farads, by dividing it by 1 million.

$$C = 0.000003 \text{ F}$$

Then from the formula

$$X_C = \frac{1}{(2)(3.14)(60)(0.000003)} = 884.2 \ \Omega$$

(continued)

Example 2:

If the frequency of the circuit is 400, instead, how much is the reactance?

Solution: This can be done either by plugging numbers into the formula, or since the frequency has increased by the ratio 400/60, and the frequency f is in the denominator, showing a division; we may conclude that the new value of X_c is

$$(884.2) \div (400/60) = 132.6 \ \Omega$$

Inductive Reactance

The mechanism that an inductor introduces an opposition to the flow of current in an AC circuit is based on the magnetism that is generated. This magnetic field, however, opposed the current generating it. Thus, at the time the voltage is maximum, there is the highest opposition to the flow of current, and the current is the lowest in magnitude (zero). The magnetism thus generated acts as a reserve of energy in the inductor and it will be given back to the circuit after half a cycle. It is given back to the circuit when voltage value is low around zero. Similar to the case of a capacitor, we can conclude the following for a voltage of sinusoidal waveform:

1. The current waveform is also sinusoidal, but the maximum current corresponds to when the voltage value is about zero; that is, there is a 90° phase difference between the voltage waveform and the current waveform.
2. The current in this case lags the voltage (it follows the voltage with 90° delay).
3. The frequency has an effect on the reactance of an inductor.

   The formula to calculate the reactance is

$$X_L = 2\pi f L$$

where X_L is the inductive reactance of the inductor measured in ohms, f is the circuit frequency, and L is the inductance of an inductor measured in henrys (H). We can see from this relationship that as f increases, the opposition magnitude to the current flow increases (opposite of what happens in capacitors).

Example 3:

A coil has an inductance of 20 mH (millihenrys) and its resistance is negligible. How much is its reactance when placed in a circuit with a frequency (a) 50 Hz and (b) 400 Hz?

Solution: From the formula

a. $X_L = (2)(3.14)(50)(0.02) = 6.28 \ \Omega$
b. $X_L = (6.28)(400/50) = 50.2 \ \Omega$

Review questions

1. Why must a turbine be connected to a grid?
2. What are the conditions that must be satisfied before a turbine can be connected to a grid?
3. What happens if the voltage from a turbine is not the same as the voltage of the grid?
4. Why is concern about power important when connecting a wind turbine to a grid?
5. What is the expression for power in DC electricity?
6. Why is AC power different from DC power?
7. What are the three terms for power in AC electricity?
8. What is an inductive load?
9. What is a capacitive load?
10. In AC circuits, what is reactance?
11. What is the difference between active power and reactive power?
12. Can one add the active and reactive powers together? How?
13. What is phase angle?
14. Does the power factor have anything to do with the phase angle?
15. How much is the best value for the power factor?
16. What kind of load causes the power factor to go down?
17. Can the power factor be larger than 1? Why or why not?
18. What is meant by "improving" a power factor?
19. What is meant by "correcting" a power factor?
20. What is meant by "power quality"?
21. What are harmonics? Are they good or bad?
22. Why is electricity generation and distribution accomplished in three-phase?
23. What is the difference between a Y connection and a Δ connection?
24. In a three-phase system with a Δ connection, which one has a larger value, phase current or line current?
25. In a wind farm, the outputs from all turbines are connected to a collector. Why is each wind turbine not connected to the grid separately?
26. Has islanding anything to do with wind turbines? Is islanding good, or should it be avoided? Why?

Problems

1. A DC motor is rated at 90 V and is used for a job requiring 5 horsepower. What is the current in the motor circuit?
2. A 1.5-kW, 220-V, 50-Hz, single-phase AC motor has a power factor of 0.75 at full load. How much is current for the line connecting the motor when it is operating at 100% capacity?
3. An AC circuit has a single-phase applied voltage of 240 V. The load consists of a resistive load of 48 Ω and a reactance of 56 Ω. What is the reactive power of the circuit?

4. In a single-phase AC circuit the resistive load is 82 Ω, the capacitive reactance is 124 Ω, and the inductive reactance is 50 Ω. Find the active power and the apparent power, if the applied voltage is 180 V.

5. In the circuit of problem 4, what is the power factor? How can this power factor be corrected?

6. What is the phase angle in the circuit of problem 4? Does current lead or lag the voltage?

7. If the motor in question 2 was three-phase and connected to the line with a Y connection, how much will the current in the line be?

8. A three-phase motor draws 10 A of current from a source. If the line voltage from the source is 220 V, how much is the apparent power of the motor?

9. If the power factor of the motor in question 8 is 0.82, how much horsepower do you expect from this motor?

10. A generator powers three motors, one rated 1.5 kW and a pf of 0.65, one rated 2.8 kW and a pf of 0.78, and one rated 3.5 kW at a pf of 0.9. How much is the current through the generator if the operating voltage is 600 V?

PROJECT

Make a circuit as shown in the following figure. Make sure you connect the batteries with the correct polarity, as shown. The two lights are similar; and make sure they are good for the voltage of the batteries. Close all three switches. If the batteries are similar and both are new or have the same age, then the two lights must have the same brightness. Check that as far as two of the switches are kept closed, no matter which pair, then both lights are ON.

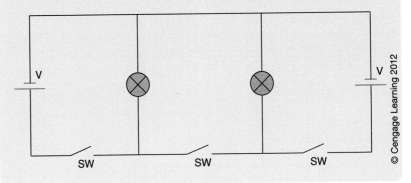

V   V

SW   SW   SW

Wind Turbine Mechanical Systems

KEY TERMS

abrasion
alignment
arm
backlash
bevel gear
carrier
cavitation
drive gear
driven gear
driving gear
efficiency
epicyclic gear
fatigue
gear train
helical gear
idler gear
internal gear
misalignment
output
pinion
pitch circle
pitch diameter
planet (planet gear)
planetary gear
ring gear
scuffing
spur gear
sun gear
worm gear

OBJECTIVES

After studying this chapter, you should be able to:

- Recognize the role of a gearbox in a wind turbine.
- Learn the structure of gears used in wind turbines.
- Understand the relationship among speed, torque, and power.
- Realize the weakest point of a gear system.
- Understand how wear and failure can happen in a gear system.
- See that part of energy converts to heat and the gearbox oil must be cooled.
- Know about efficiency and that there is always some power loss.

9.0   Introduction

One of the major components of a wind turbine is the gearbox.[1] A gearbox in machinery serves to match the speeds of the driving member and the driven member. Normally, the prime mover, a motor or an engine, which is on the driving side, is faster than the driven members and a gearbox is used to reduce that speed to a suitable value. In wind turbines, however, we have the exceptional case of increasing the low speed of a rotor to match the higher speed required by a generator. While the subject of gears and gear trains is much broader than what is discussed here, this chapter serves as a quick reference to gearing technology, with emphasis on the application in wind turbines.

9.1   Gearbox fundamentals

In machinery it is often necessary to change the rotational speed of a shaft to a desirable or more useful value. Mechanically this can be done, in general, by employing both belts and chains, or by using gear systems. If the distance between the two shafts is relatively large, then a belt or chain can be used; but if the space does not allow, the only solution is using gears. This is illustrated in **Figure 9.1.**

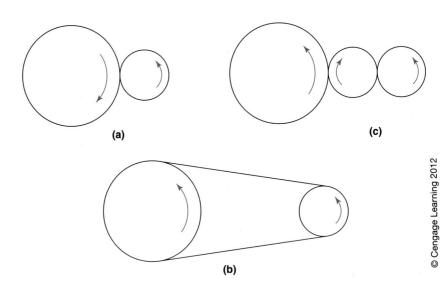

(a)

(c)

(b)

© Cengage Learning 2012

FIGURE 9.1 *Comparison of chain/belt drive with gears.*

---

[1] Future turbines may not have a gearbox, but until then a gearbox remains to be a vital part of a wind turbine.

While distance is a primary factor, the required strength or power requirement is often more important, and in this regard, chains and belts do not have the same performance as a gearbox. For belts two sets of pulleys are used (sprockets for chains) and in gears two gears can be meshed together. In either case, the ratio between the diameters (or radii) of the two parties determines the ratio of conversion of the rotational speeds. In addition to the rotational speed, the direction of rotation is also a parameter that must be taken into account. While in using a belt or chain the direction of rotation does not change, in using a pair of gears there is a change in the direction of motion. A second gear can be employed to counteract this directional change, as shown in figure 9.1(c). In industry it is customary to call the smaller and larger gears the "pinion" and the "gear," respectively. In this text, we refer to them as the **driving gear** and the **driven gear,** as appropriate. A driving gear is also called a **drive gear.**

The scenario in figure 9.1(c) is not the only way to resolve the direction change. Another solution is to use an **internal gear.** Also, note that the radii of the gear employed to reverse direction does not have to be equal to that of either of the two gears. These themes are shown in **Figure 9.2.** In an internal gear, the teeth are inside the rim. Comparing parts (a) and (b) in figure 9.2 also shows the difference in the distance between the two shaft centerlines. The **idler gears** in figures 9.2(c) and (d) are free gears and their shafts do not carry a load.

The load-carrying shafts are input and **output** of the gear system. In most cases, the smaller gear is the input and the larger gear is the output, and the gear is used as a speed reducer. Theoretically, the input and output placement can be interchanged in most of the gears.

pinion
The smaller gear in a pair of gears or the output gear in a drive train.

driving gear
Drive gear, a gear that gives energy to its mating gear.

driven gear
A gear that receives energy from its mating gear.

drive gear
Also called driving gear, a gear that transfers its energy to the mating gear.

internal gear
A gear in the form of a ring with its teeth on the inside rather than outside.

idler gear
A gear standing between two main gears whose role is to change direction of rotation as well as filling the gap.

output
The product or result of processing of a machine, system, and so on. The output of a gearbox, for example, is a rotational motion with a required speed and torque.

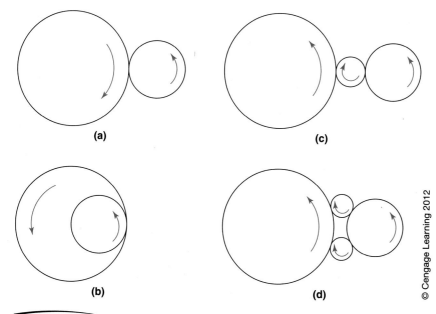

(a)   (c)   (b)   (d)

© Cengage Learning 2012

FIGURE 9.2 *Maintaining the direction of rotation.*

9.1.1 Gear types based on gear axes

worm gear

A pair of gears with a high gear ratio, and used only to reduce speed. The axes are at 90°. A worm gear has the property that it is impossible for the driven gear to force the drive gear rotate; thus, if the drive gear stops it also locks the driven gear.

bevel gear

A type where a pair of gears have their axes at 90°.

In the aforementioned gear arrangement, the axes of the two shafts onto which gears are mounted are parallel to each other. However, there are other gear arrangements in which the shafts are not parallel. **Figure 9.3** shows the types of gears where the shaft axes are perpendicular to each other. When two lines are perpendicular to each other, depending on the case, we may or may not be able to pass a plane from the two axes. In other words, the two axes either intersect or they do not intersect. Figure 9.3(a) is a schematic of a **worm gear** and figure 9.3(b) shows the schematic of a **bevel gear.** The ratio of output rpm to input rpm can be much smaller in a worm gear compared to the other gear types. Another difference of the worm gear compared with other gear types is that whereas theoretically one can select either gear to be the input, in a worm gear it is not possible to have the worm as the output.

Since the teeth in a pair of gears must mesh together, the shape of teeth must accordingly satisfy certain profile requirements. For instance, in bevel gears, the teeth are at 45°, as illustrated in **Figure 9.4(a).** The teeth in these gears cannot have the same form as those shown in figure 9.2. **Figure 9.4(b)** shows a variation of the bevel gear called the spiral bevel gear, with a still different tooth profile.

9.1.2 Gear types based on tooth form

spur gear

A gear in which the teeth are parallel to the axis of rotation.

helical gear

A type of gear in which the teeth are angled with respect to the shaft axis in order to reduce the backlash compared to spur gears.

In the simplest form, the teeth are parallel to the axis of the gear or a plane may be passed through a tooth and the gear axis. This type of gear is called **spur gear** and is shown in **Figure 9.5(a).** In spur gears, only one pair of teeth is engaged at any time. As can be seen later in this chapter, this is not the best design. In order to improve the performance of gears, the geometry of teeth is modified so that at each instant of time more than one gear is in mesh with its counterpart. This type of gear is called **helical gear,** and it is shown in **Figure 9.5(b).**

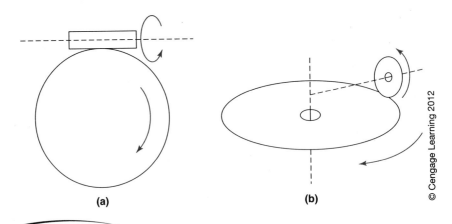

(a)   (b)

© Cengage Learning 2012

FIGURE 9.3 *Worm gear and bevel gear schematics.*

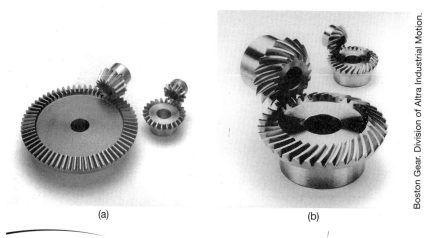

Boston Gear. Division of Altra Industrial Motion.

(a) (b)

FIGURE 9.4 *(a) Bevel gears and (b) spiral bevel gears.*

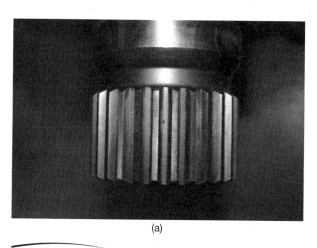

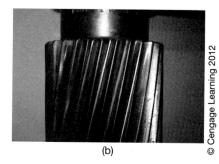

© Cengage Learning 2012

(a) (b)

FIGURE 9.5 *(a) Spur and (b) helical gears.*

Other types of gears 9.1.3

As the preceding two classifications of gears are combined, they lead to various gears for different purposes and of different profiles. This is irrespective of material used, which also plays an important role in the very many kinds of gears that exist. Information about these can be found in other books.

In wind turbines gears are used mainly in the yaw system and in the rotor shaft. The blade pitch control system may also involve gears. All the gears in a wind turbine are made of steel. The yaw system usually has internal spur gears as shown in figure 9.2(b). The main shaft uses a different type of gear as discussed in the next section.

9.1.4   Planetary gears

planetary gear
A set of gears arranged in a special form, consisting of a sun gear in the middle, an outer ring gear with internal teeth and number of (usually 3) planet gears engaging between the two. The planet gears are mounted on a bracket called an arm (or a carrier) that can turn independently of the sun and the ring. A planetary gear can accept two input speeds.

epicyclic gear
Another name for planetary gear.

The type of gear employed in wind turbines is the **planetary gear,** known also as an **epicyclic gear.** An epicyclic gear has certain advantages over other gears that make it more appropriate for wind turbines as well as other applications. Consider the combination of gears shown in figures 9.2(b) and (c); that is, consider that in the internal gear, the smaller gear is used as an idler for another gear, as shown in **Figure 9.6(a).** The result is that the idler gear just transfers motion to the gear in the middle, whose shaft is alongside the same shaft as the outer (ring) gear. The arrangement is, thus, concise and takes less space. Notice, as well, that the idler gear shown does not have to be only one. In fact, there can be any number of idlers, if space permits, as depicted in **Figures 9.6(b)** and **(c).** The most common number of idler gears is three. **Figure 9.7** shows a typical planetary gear and the associated definitions.

As it can be understood from figures 9.6 and 9.7, the idler gears cannot be left without a shaft and a mounting to hold their shafts. The arrangement for holding them in the correct position with respect to each other and to support their shafts is shown in figure 9.7.

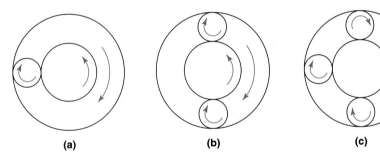

(a)　　　　　(b)　　　　　(c)

© Cengage Learning 2012

FIGURE 9.6 *Principle of construction of planetary gear.*

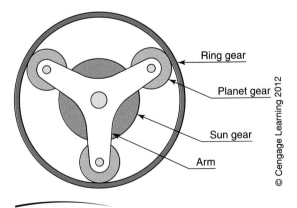

Ring gear

Planet gear

Sun gear

Arm

© Cengage Learning 2012

FIGURE 9.7 *Planetary (epicyclic) gear.*

In a planetary gear, the gear in the middle is called the **sun gear,** the idlers are called **planet** or **planet gear,** the outer gear with internal teeth is called the **ring gear,** and the bracket holding the planet gears is called the **arm** or the **carrier.** If there is only one planet gear, there is insufficient balance in the system as far as forces are concerned. If there are too many of them, then there is a lot of loss due to friction. The best design has three planet gears. The name planetary gear stems from the fact that the planets rotate around the sun gear. One advantage of planetary gears, which makes them more attractive for wind turbines, is that the input and output shafts are coaxial. In practice, however, if a planetary gear is combined with other gears, then the positional relationship between the input and output shafts depends on the way the gears are put together.

One of the aspects of a planetary gear is that except for the planet gears any of the three other components (sun, ring, and arm) can be used as input or output. More interestingly, we may have two inputs and one output. For example, we can turn the ring at a given rpm and turn the arm at another desired rpm, and the sun gear turns accordingly. More on this is discussed in the next section (see section 9.2.3). In most applications of the planetary gear (including wind turbines) the ring is kept fixed (that is, zero rpm); this makes an easier application of the planetary gears.

sun gear
The innermost gear in a planetary gear around which the planet gears rotate.

planet gear
Any of the gears in a planetary gear that stand between the sun gear and the ring gear.

ring gear
A gear in the form of a ring with inside teeth that engages with planet gears in a planetary gear.

arm
Part of a planetary gear that holds the planet gears together. Also called carrier.

carrier
Part of a planetary gear that holds the planet gears together. Also called *arm*.

Gear ratio   9.2

The main role of a gear system in a machine or device is to change the angular speed of a shaft connected to the driving element (a motor, for instance) to a desired value that is more appropriate for a particular application. A second shaft on which the driven gear is mounted takes the desired speed. The ratio between the two angular speeds is governed by the number of teeth on each gear. The purpose of a pair of gears can be primarily to change the torque on the shaft. However, the torque and speed are always related, as discussed in section 9.3. In this section, we study the speed relationships between a pair of gears and between a number of gears when put together in tandem.

Gear ratio and speed ratio in two gears   9.2.1

Figure 9.8 shows a pair of gears in mesh. For two gears to work together, the size of the teeth must be the same. The size of the teeth is defined by a number of parameters that characterize their dimensions and their other properties. For instance, gears in a clock are a different size than the gears used in a wind turbine. Similar to threads on nuts and bolts, gear size specifications are standardized, such as in the metric system or in American standards. The associated information is beyond the scope of this book and is not discussed here.

In two gears in mesh, depending on the type of gears, two or more teeth are in contact. The teeth on the driving gear push the teeth on the driven gear to transfer the motion. This pushing, nevertheless, must take place while the teeth

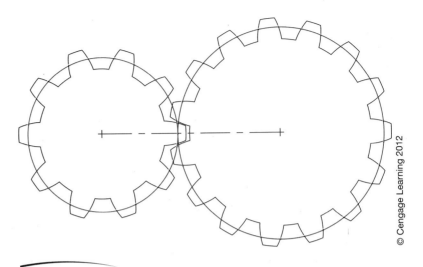

FIGURE 9.8 *A pair of gears in mesh.*

surfaces smoothly roll on each other without any sliding. Also, the two gears must not lock to each other and prevent smooth motion. In this sense, the teeth must have a well-defined geometry and be well machined to that geometry; otherwise problems will arise.

When two gears are in mesh, the top surface of the teeth of one gear (called the top land) does not touch the valley of the mating gear. Therefore, there is a gap between the two gears. This gap improves the continuity of smooth motion as well as serving as a reservoir for lubricating oil, so that the surfaces remain lubricated during motion.

A graphical representation of gears can show each gear by three circles with the same center. This is shown in **Figure 9.9.** The outer circle corresponds to the outer size of the gear (top of teeth) and the inner circle corresponds to the root of the teeth. The third (middle) circle in each gear represents the size of a cylinder such that if the two gears are replaced by two rubber cylinders, they roll on each other and transfer the motion. This circle is called the **pitch circle.** For two meshing gears, the pitch circles are tangent to each other. The diameter of pitch circle is called **pitch diameter** (PD).

If the number of teeth in the pair of meshing gears are denoted by N_1 and N_2, and their angular speeds (say in revolutions per minute, rpm) are n_1 and n_2, respectively, the following relationship holds:

$$\frac{n_2}{n_1} = \frac{N_1}{N_2} \tag{9.1}$$

Furthermore, if their corresponding pitch circle radii are r_1 and r_2, since the number of teeth is proportional to the circumference in each gear and the circumference is proportional to radius, we can write

pitch circle
An imaginary circle in gears representing a gear size. If two meshing gears are replaced by two rubber cylinders that roll on each other and transfer the motion with the same gear ratio, the circle representing each cylinder is the *pitch circle* of that gear. For two meshing gears the pitch circles are tangent to each other.

pitch diameter
The diameter of the *pitch circle.*

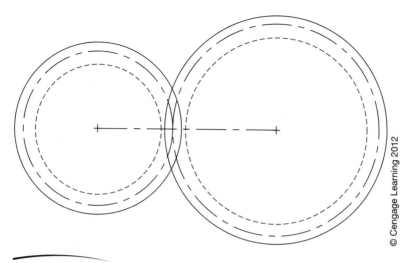

FIGURE 9.9 *Representation of a pair of gears.*

$$\frac{r_1}{r_2} = \frac{N_1}{N_2} \qquad (9.2)$$

It follows, thus, from equations (9.1) and (9.2) that

$$\frac{n_2}{n_1} = \frac{r_1}{r_2} \qquad (9.3)$$

EXAMPLE 9.1:

If the numbers of teeth in a pair of gears are 20 and 48, what is the ratio of speeds between them? If the smaller gear rotates at 900 rpm, what is the speed of the larger gear?

Solution:

From equation (9.1) we can see that

$$\frac{n_2}{n_1} = \frac{N_1}{N_2} = \frac{20}{48} = \frac{1}{2.4} \quad \text{or} \quad 1:2.4$$

That is, the angular speed of the larger gear is 2.4 times smaller than that of the smaller gear. The speed of the larger gear is thus,

$$900 \div 2.4 = 375 \text{ rpm}$$

EXAMPLE 9.2:

The motor shaft in a machine has a speed of 1800 rpm. What gear ratio is required to reduce this speed to 400 rpm? If a selected driving gear has a radius of 3 in., what should be the diameter of the larger gear? If there are 12 teeth on this 3-in. gear, how many teeth are there on the larger gear?

Solution:

The gear ratio is found from the ratio between the two speeds.

$$1800 \div 400 = 4.5$$

The gear ratio is 1:4.5. Since the speed is to be reduced, the driven gear must be 4.5 times larger than the driving gear.

$$\text{Diameter of the large gear} = (3)\,(4.5) = 13.5 \text{ in.}$$

$$\text{Number of teeth on the larger gear} = (12)(4.5) = 54$$

Note that the number of teeth must always be an integral number. If the number of teeth was 11 instead of 12, then we would have

$$\text{Number of teeth on the larger gear} = (11)(4.5) = 49.5 \rightarrow 50$$

9.2.2   Gear ratio and speed ratio in gear trains

gear train
A set of gears
arranged together to
obtain a desired gear
ratio.

When more than one pair of gears are put together, the assembly is referred to as a **gear train**. As we saw in figure 9.1, the role of the third gear can be to change the direction of rotation of the output shaft. It can be seen that the number of teeth of a third gear does not play a role in the gear ratio. **Figure 9.10(a)** illustrates three gears; the input shaft is attached to gear 1, and gear 3 is the output. Suppose that the number of teeth for each gear and the angular speeds are denoted by N_1, N_2, and N_3 and n_1, n_2, and n_3, respectively. It follows from equation (9.1) that

$$\frac{n_3}{n_1} = \frac{n_2}{n_1} \times \frac{n_3}{n_2} = \frac{N_1}{N_2} \times \frac{N_2}{N_3} = \frac{N_1}{N_3} \tag{9.4}$$

Supposing now that a fourth gear is added, by the same token one can see that the relative speed between the input and output shafts depends only on the gear ratio of the first and last gears. For the arrangement in **Figure 9.10(b)**, we have

$$\frac{n_4}{n_1} = \frac{n_2}{n_1} \times \frac{n_3}{n_2} \times \frac{n_4}{n_3} = \frac{N_1}{N_2} \times \frac{N_2}{N_3} \times \frac{N_3}{N_4} = \frac{N_1}{N_4}$$

or

$$\frac{n_4}{n_1} = \frac{N_1}{N_4} \qquad (9.5)$$

The intermediate gears could be used to transfer the motion between shafts, as well as to maintain the desired direction of rotation. Imagine if only two gears were to be used instead of the arrangement in figure 9.1(b), that the diameter of each gear would have to be much larger in order to fill the space between the shafts and maintain the gear ratio and the direction of rotation. This would require unnecessarily heavy and expensive gears.

If the desired speed ratio in a device is large, say 40, this implies a large gear ratio. From various technical viewpoints such a large gear ratio is not desirable and must be avoided. In such a case, the large number of the gear ratio must be broken into smaller values and gear pairs of practical ratios must be combined. For instance, for a speed ratio of 40, two speed ratios of 5 and 8 are much more practical. As a rule of thumb, gear ratios higher than 10 must be avoided.

Figures 9.10(c) and **9.10(d)** show examples of such an arrangement, which is often employed in gear trains. In these arrangements, some gears have the same shaft and rotate together. In these gear trains, equation (9.5) is not valid, since the two gears on the same shaft have the same speed, but a different number of teeth. For instance, for the arrangement in figure 9.10(c) there are four gears, the same as in figure 9.10(b). However, equation (9.5) can be applied to the latter, but not to the former.

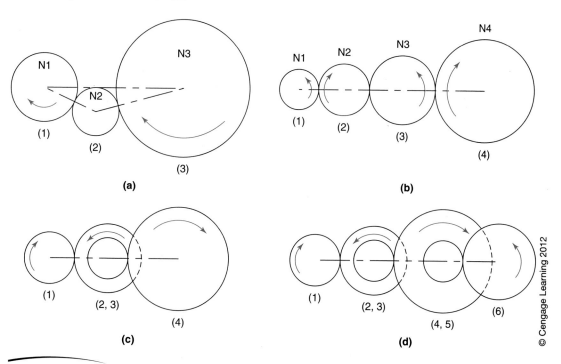

FIGURE 9.10 *Gear trains.*

© Cengage Learning 2012

> From various technical viewpoints, a large gear ratio for a single pair of gears is not desirable and must be avoided. As a rule of thumb, gear ratios must not exceed 10.

In order to find the gear ratio for the arrangement in figure 9.10(c) one can follow the same approach as used for equations (9.4) and (9.5), but keeping in mind that some of the angular speeds are equal. The speed of gear 4 (or, in other words, the speed ratio of gear 4 to gear 1) in figure 9.10(c) can be found by combining the speed ratio of gear 3 to gear 1 and the speed ratio of gear 4 to gear 3. Noting that gears 2 and 3 have the same speed, the speed ratio of gear 4 to gear 1 can be found by applying equation (9.1) for gears 1 and 2, and for gears 3 and 4, as follows

$$\frac{n_4}{n_1} = \frac{n_2}{n_1} \times \frac{n_4}{n_3} = \frac{N_1}{N_2} \times \frac{N_3}{N_4} \tag{9.6}$$

The same reasoning can be used for any combination of gears in a gear train. For example, in the arrangement in figure 9.10(d) there are six gears. The first four gears have the same construction as that in figure 9.10(c), and gears 4 and 5 have the same speed. In order to find the speed of gear 6 with respect to gear 1 we can multiply the speed of gear 4 by the gear ratio of gear 6 to gear 5. The former was already found in equation (9.6). The speed ratio of gear 6 to gear 1, therefore, is

$$\frac{n_6}{n_1} = \frac{n_4}{n_1} \times \frac{n_6}{n_5} = \frac{N_1}{N_2} \times \frac{N_3}{N_4} \times \frac{N_5}{N_6} \tag{9.7}$$

EXAMPLE 9.3:

The gear on the left in the gear train shown in figure 9.10(d) is on the input shaft and has a rotating speed of 1800 rpm. If $N_1 = 20$, $N_2 = 25$, $N_3 = 17$, $N_4 = 38$, $N_5 = 14$, and $N_6 = 28$, find the speed of the output shaft (gear 6).

Solution:

Equation (9.7) can be directly applied to this problem; thus,

$$\frac{n_6}{n_1} = \frac{N_1}{N_2} \times \frac{N_3}{N_4} \times \frac{N_5}{N_6} = \frac{20}{25} \times \frac{17}{38} \times \frac{14}{28} = \frac{17}{95}$$

and the speed of the output shaft is

Speed of gear 6 = (1800)(17/95) = 322 rpm

Note that the answer is rounded to an integral number.

Speed ratio in planetary gears

In the same way that gears can be combined together, as discussed in the previous section, planetary gears can be combined with the other gears in a gear train in order to obtain a desired gear ratio. This is the case in most wind turbines. The speed ratio in a planetary gear, by itself, will be discussed separately, since this gear is different from the other gears.

As mentioned before, with a planetary gear it is possible to run either two of the ring gear, the sun gear, or the arm at any speed, and the third one will rotate at a speed that can be determined from the associated relationships. Here we consider only the case that the ring gear is stationary, as in the case of wind turbines. If the number of teeth in the ring gear and the sun are denoted by N_R and N_S, respectively, when the ring is kept stationary, we have the following relationships:

a. The sun gear is input, and the arm is output:

$$\frac{\text{Output speed}}{\text{Input speed}} = \frac{N_S}{N_S + N_R}$$

(9.8)

b. The arm is input, and the sun gear is output:

$$\frac{\text{Output speed}}{\text{Input speed}} = \frac{N_S + N_R}{N_S}$$

(9.9)

As we can see, the number of teeth of the planet gears does not come into effect. However, that number cannot be arbitrary, since the space between the sun and the ring gears must accommodate them; and this space depends on the radii of the sun and ring gears, which in turn dictate their numbers of teeth.

It follows from equations (9.8) and (9.9) that, in order to increase the speed (which is the case for wind turbines), the arm must be the input and the sun gear must be the output of the planetary gear. When desired, it is possible to combine a planetary gear with other planetary gears or other gears in order to obtain a higher gear ratio. In a wind turbine, it is not uncommon to have gear ratios of 70 to 90. In such a case, two sets of gears are combined.

The planetary gear discussed in this chapter is the simplest form of planetary gears. It is possible to combine two or more planetary gears in one casing to create a high ratio of speed reduction (or exceptionally, speed increase). One of the advantages of planetary gears so arranged is that the input and output shafts have the same axis (they are alongside each other). A comparison with the arrangements in figure 9.10 illustrates the compactness of planetary gears compared to other gears. They are also more stable since the power is divided (evenly) between a number of planet gears. This is why planetary gears are employed in wind turbines.

EXAMPLE 9.4:

In a planetary gear, the number of teeth in the ring gear and the sun gear are 89 and 16, respectively, and there are 36 teeth on the planet gear. What is the gear ratio of this gear?

Solution: Based on equation (9.9),

$$\frac{\text{Output speed}}{\text{Input speed}} = \frac{16 + 89}{16} = \frac{105}{16} = 6.5625$$

The number of teeth on the planet gear does not come into the calculation. Nevertheless, based on the numbers for the sun gear and the planet, one can see that the planets in this gear are larger than the sun gear, which is quite acceptable. This gives a high speed ratio.

9.3   Speed and torque relations

In chapter 7, you learned that power in a mechanical system is the product of angular speed and torque [see equation (7.3)]. This relationship is very important. Try to master understanding its application in gear systems. In a gear assembly, the input shaft has a rotational speed n_{in} that, through the gears, is modified to the rotational speed n_{out} on the output shaft. If for the moment the losses due to friction are ignored (discussed later in this chapter), the maximum power that the system can handle, assuming the mechanical structure of the gearbox to be strong enough, is equal to the power of the driving device. If this power is denoted by P, and the torque of the input shaft is represented by T_{in}, we have

$$P = n_{in} \times T_{in} \tag{9.10}$$

Similarly, if the torque on the output shaft is T_{out}, since the power P is not lost, as assumed, on the output shaft we also have

$$P = n_{out} \times T_{out} \tag{9.11}$$

A comparison of equations (9.10) and (9.11) reveals that

$$n_{in} \times T_{in} = n_{out} \times T_{out} \tag{9.12}$$

Equation (9.12) implies that if the gearbox is used as a speed reducer (that is, $n_{in} > n_{out}$), then the output torque is larger than the input torque by the same ratio. More particularly, from equations (9.1) and (9.12) it follows that for a pair of gears in mesh,

$$\frac{T_2}{T_1} = \frac{N_2}{N_1} \qquad (9.13)$$

And, for a set of gears in a gear train,

$$\frac{T_{out}}{T_{in}} = \frac{n_{in}}{n_{out}} \qquad (9.14)$$

EXAMPLE 9.5:

In the speed reducer shown in figure 9.10(d), $N_1 = 20$, $N_2 = 25$, $N_3 = 17$, $N_4 = 38$, $N_5 = 14$, and $N_6 = 28$. Find the output torque if the input torque is 100 ft-lb.

Solution: The output-to-input speed ratio was already found in example 9.3 to be 17/95. According to equation (9.12) or (9.14),

Output torque = (100)(95/17) = 588.8 ft-lb

An increase in torque and a decrease in speed (or vice versa) go hand in hand, and cannot be altered. Sometimes the primary reason for using gears is to match the torques, rather than matching the speeds. For instance, if a motor does not have sufficient torque to move a load, by using a speed reducer the torque can be augmented. In this way, the motor power has not been changed, but its torque has been increased. The trade-off is a reduction in the speed of moving the load.

In a gear system, if the torque is to be increased, the speed will be reduced. We cannot increase (or decrease) both the speed and the torque.

Force loads on gears

9.4

In chapter 7 we learned about force, torque, and the torque of a force. It is recommended that you review that chapter to refresh yourself on the subject, before you begin this section.

When a gear is rotating at an angular speed n_1 and its shaft carries a torque T_1, it transfers power to another gear. If the second gear has a rotational speed n_2, the torque on its shaft is

$$T_2 = \frac{n_1}{n_2} \times T_1 \qquad (9.15)$$

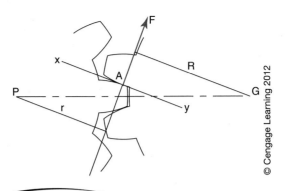

FIGURE 9.11 *Force transfer between two meshing teeth in a pair of gears.*

in light of equation (9.14). This transfer of power is through the teeth, which, as mentioned earlier, roll on each other while the driving gear teeth push the driven gear teeth. In spur gears, one tooth of the driving gear is in contact with one tooth of the driven gear. In helical gears, more than one pair of teeth are engaged at each instant of time; nevertheless, for the design of the gear it is assumed that only one pair of teeth are engaged.

Figure 9.11 shows a pair of teeth in contact. In the figure, P and G are the centers of the driving gear and the driven gear, respectively. At the instant shown, a force is exerted from the driving gear on the driven gear. This force is normal to the two surfaces of the gears that roll on each other.

The value of the force F depends on the torque transfer between the gears. Since power is defined by the product of torque and angulr speed [see equation (7.3)], the force F depends on the power transfer of the gears. When the tooth of the driving gear pushes the corresponding tooth of the driven gear, it is subject to the same force exerted on it from the driven gear tooth. This is to say that the teeth of each gear are subject to a force that depends on the power handled by the gears in a gearbox.

From the preceding discussion, we can conclude that all the power in a gear is transferred by each tooth and, therefore, the teeth are the weakest part of a gear. In other words, in the design of gears, the teeth are sized to take the entire load, and in case of overload they are subject to fail.

In wind turbines, because of continuous variation of wind speed, the load on a gearbox is not uniform and gears are subject to impulsive loads and shocks, which are not good for the gears.

> In a wind turbine gearbox, the load is not uniform and continuously varies because of wind speed fluctuation. The gears are subject to unfriendly impulsive loads and shocks, which are not good for them.

Friction and wear   9.5

Ideally the two contacting teeth of a pair of meshing gears roll on each other during motion. This, however, may be not what happens in practice. Gear teeth are designed such that the two surfaces roll on each other, but imperfections can occur during manufacturing of gears as well as during their assembly in a gearbox. As a result, the teeth may have slight sliding on each other due to not having the perfect cut, or due to **misalignment.** Misalignment means that the shafts of two gears that must be parallel to each other, or along the same axis, have an angle with each other. This angle may be very small, but as far as it is not zero, there is a misalignment.

    Errors in manufacturing arise from the machinery that does the cutting job, as those machines are subject to wear and loss of precision. These errors can be minimized, but they still exist. Also, an operator's experience can play a role in augmenting the imperfections. In the presence of all these imperfections in the geometry of teeth and misalignments in the shafts, there is friction between the contacting surfaces. This friction causes the two gears to wear. Also, it generates heat.

    Even if the two mating gears are perfectly aligned, there is always some wear between two gears, particularly when they are new. Wear implies one or both gears being rubbed away. If the wear is excessive, a lot of noise and heat are generated and the chance of failure increases tremendously. If the wear is at an acceptable level, heat and fragments of eroded metal are still generated during operation of gears.

    Vibration is another (severe) cause of wear in gears, because it results in a hammering action between the gear teeth. Misalignment causes vibration, and vibration is the worst cause of wear and tear in machinery.

    Since the wear in the teeth of a gear starts at the teeth surface, in order to reduce wear the surfaces of gears are hardened. In this way, the resistance of the metal to wear increases greatly, say, by a factor of 20. Hardening is customary for steel, and in particular, it is frequently used for gears. Hardening can be performed by a chemical or heat treatment process and it is possible to harden only the outer surface of a metal part, or the whole metal part.

misalignment
The state of not being aligned for two matching shafts or gears that must be aligned (i.e., their axes make a straight line).

Lubrication and cooling   9.6

The heat generated in a gearbox must be removed from the gears; otherwise, they can easily become overheated by the accumulation of heat. This is particularly true for high-power machinery, having high speeds and heavy gears, including wind turbines. Also, gears must be lubricated to reduce the friction and ease the motion.

    In gears, liquid lubricants are used that do both cooling and lubrication jobs. A paste-type lubricant, such as grease, is not appropriate for such a job, since it does not carry out the cooling function. This is why gears are housed in a gearbox, which allows for a liquid lubricant, as well as protecting gears from contamination and ensuring the safety of the operators and the environment.

As the lubricant contacts the gears, it takes away the heat from the hot surfaces of the teeth and also washes away the debris generated by the rubbing action of the gears. Thus, it is imperative that the lubricant has a continuous circulation to become cooled. In larger machinery, including wind turbines, a radiator is employed for this purpose, and the circulation of a lubricant is assisted by a pump. The inclusion of a filter in the circulation path is quite important. The lack of a filter can lead to severe damage from the floating debris having a sanding action on the teeth and other surfaces in the gearbox.

9.7   Efficiency

So far in this book we have not considered the effect of losses that happen during the transfer of energy and power. This is a good time to bring this matter to the surface and discuss more realistically what happens to the energy grasped from wind by the rotor of a turbine before it shows up in the form of electrical energy to be injected into the grid.

The heat generated in the gearbox, as discussed in section 9.6, is part of the mechanical energy input to the gear that has turned into heat. This heat is, obviously, lost because it cannot be recovered in the form of mechanical energy. As a result, there is a difference between the input and the output powers of the gearbox, and that difference is the corresponding power that is converted to heat. In this sense, we can always say $P_{out} < P_{in}$. Alternatively, this can be written as

$$P_{out} = P_{in} - P_{lost} \qquad (9.16)$$

efficiency
The measure of how much of the energy given to a device or machine is consumed by it in order to function, when modifying a specification of that energy; efficiency is the ratio of the output energy to the input energy.

The ratio $\dfrac{P_{out}}{P_{in}}$ is a number smaller than unity and without dimension (that is, no units should be specified for it) that represents the **efficiency** of a device. Normally efficiency is given in percentages. This is done by multiplying the number calculated by $\dfrac{P_{out}}{P_{in}}$ by 100. For example, the efficiency of a gearbox can be 94%, meaning that the output power is only 0.94 of the input power. The higher the efficiency of any piece of equipment the better it is, because this implies that a smaller portion of the power has been lost, and a higher portion of the power (or energy) can be put to work. In a gearbox, the loss of energy occurs not only because of gears. In all rotating machinery, the shaft is mounted on bearings and, although small, some power is lost in the form of generated heat due to friction.

In all machinery, a better design implies a higher efficiency and a poorer design implies a lower efficiency. By a better design we also mean more accurate manufacturing and more precise workmanship. Obviously, the difference appears in the price, as well. Measuring the efficiency of a gearbox (or other equipment) is not a simple matter. It has to be carried out by measurements and by laboratory experiments. Usually a manufacturer carries out these experiments and shows the results for efficiency in the machine data.

In the same way that a portion of input power is lost in a gearbox (which is a mechanical system) in the form of heat, in a generator, a portion of power also turns into heat. This heat is due to both the bearings and the current in the windings of the generator. As a result, an efficiency term is associated with the generator. In general, any motor, generator, gearbox, and so on functions with an efficiency that is less than 100%. Also, the efficiency of a device is not necessarily constant. It can change with the power, meaning that at a particular power the efficiency is at its maximum value, and at other power values it decreases. Usually, the efficiency is maximal at the rated value of any device.

It is interesting now to see the value of the total efficiency of an assembly of various systems that are put together, as in a wind turbine. If a system is made up of a number of components that are cascaded together, as in a wind turbine, then the efficiency of all the components must be multiplied together to find the total efficiency. If the efficiency of each component is shown by eff_1, eff_2, eff_3, and so on, then

$$\text{Efficiency}_{\text{total}} = \text{eff}_1 \times \text{eff}_2 \times \text{eff}_3 \times \ldots \qquad (9.17)$$

> Usually, the efficiency is maximal at the rated value of any device.

EXAMPLE 9.6:

In a wind turbine, the rotor and bearings, together, have an efficiency of 97%; the gearbox and generator efficiencies are 95% and 98%, respectively; the generator cables before the transformer have a loss of 1%; and the efficiency of the transformer is 96%. If at a given time, the power capture from wind is 1000 kW, how much is the power after the transformer?

Solution: Here we have five components, both electrical and mechanical, before the point at which we want to measure the available power. Although the cables are different from other devices, we can consider them as a component with a given efficiency. The loss in the cable is given as 1%. We first convert it to the form of efficiency by subtracting this percentage loss from 100%. The efficiency of this component, thus, is 99% (i.e., 100% − 1%). Now we have the efficiencies of all the components and we can apply equation (9.17):

$$\text{Efficiency}_{\text{total}} = (0.97)(0.95)(0.98)(0.99)(0.96) = 0.858 \rightarrow 85.8\%$$

As we can see, only 85.8% of the power reaches a point just after the transformer; that is, 14.2% (100% − 85.6%) of the energy is lost.

$$\text{Lost power} = (1000)(0.142) = 142 \text{ kW}$$

$$\text{Useful power} = (1000)(0.858) = 858 \text{ kW}$$

EXAMPLE 9.7:

In a gearbox whose efficiency is 95%, the output shaft has 1.2 MW of power. How much is the input power?

Solution: Note that in all problems like this one, the output power is known, and the input power must be more than the given value. Here, the output is 95% of the input. We are looking for a value 95% of which is given. We can directly use the definition of the efficiency. Thus,

$$\frac{P_{out}}{P_{in}} = 0.95 \quad \text{or} \quad \frac{P_{out}}{0.95} = P_{in}$$

which gives

$$P_{in} = \frac{P_{out}}{0.95} = \frac{1.2}{0.95} = 1.263 \text{ MW}$$

9.8   Backlash in gears

Backlash
The free motion between two meshing gears due to noncontact free space between their teeth.

It is very important in studying gear systems to know what backlash is and to know that *it is an unwanted feature*. **Backlash** is the free play between two meshing gears. If two meshing gears are made with large tolerances when manufactured, the gear tooth surfaces do not roll on each other as they should. Instead, there is always a small gap between the teeth, and the gears can move with respect to each other freely without transferring motion. This free motion can be very small, say, smaller than 1° of rotation, but it is larger than acceptable for a good pair of gears. The effect of backlash is more undesirable in precision machinery, as in machine tools where precise machining is expected. In addition to loss of precision, backlash causes vibration and noise in gears.

In wind turbines, since the wind speed is not constant and has fluctuations, backlash leads to a hammering action between the teeth of two meshing gears. This hammering action, although small, in the long run can result in gear damage and failure.

In order to reduce backlash in gears, manufacture of gears (by machining) must be performed with more precision and tighter tolerances. However, manufacturing inaccuracies are unavoidable and backlash can also develop after the gears run-in (after they have been put to service and have run for a number of hours). One common way to avoid backlash in gears is to employ helical gears, where at each instant more than one pair of gears are engaged (see figure 9.5). Helical gears are more expensive to manufacture, but in many situations, such as in wind turbines, helical gears should be used to eliminate any backlash.

Helical gears are more expensive to manufacture, but in many situations, such as in wind turbines, helical gears should be used to eliminate any backlash.

General problems of gears and gear failure

9.9

Gearboxes, in general, are subject to a number of problems during their operation. These problems can be enhanced or accelerated if faulty situations happen. Faulty situations are undesirable conditions that may arise for various reasons. For example, the lubricating oil leaks and the oil level drops, or improper oil is used. As a result of any of these situations insufficient lubrication and cooling takes place and parts become heated or hot spots develop, or other damage can occur. Any damage can lead to tooth breakage and the permanent failure of a gearbox.

There are two particular issues with wind turbine gearboxes. The first one, as mentioned earlier, is that in wind turbines the gearbox is not used as a speed reducer, as is the case in the majority of applications. The second issue is that, because of the nature of wind, the gearbox is subjected to frequent and sudden changes of the power it handles. Such power fluctuations, which in turn translate to load variation on the teeth, is not a desirable situation for the gearbox.

In this sense, the chances of damage and failure in wind turbine gearboxes are more than in many other applications. Moreover, changing a gearbox in a wind turbine is extremely expensive, since the operation is not on the ground.

In summary, the failure in a wind turbine gearbox can be due to (a) design and manufacturing, (b) mounting and usage, or (c) operation and maintenance.

Poor or wrong design definitely can lead to problems during operation. Similarly, defects in manufacturing create unnecessary stresses, wear, heat, and **fatigue.** In mounting a gearbox, **alignment** with the rotating shaft, in both the input side and the output side, is absolutely necessary. Alignment implies that the axes of two mating gears are completely parallel to each other. Misalignment causes the teeth to go under fluctuating stress (see the text in this chapter about stress), meaning that the stress in parts of a tooth continuously changes from compression to tension and vice versa. This is called fatigue and, compared to a part under the same magnitude but constant stress, can tremendously decrease the useful life of a machine part. Proper mounting of a gearbox implies taking care that the alignment remains within the recommended tolerance. Bad usage implies if a wrong gearbox is selected for an application; for instance, when a gearbox cannot handle the power or cannot transfer the torque it is applied on. This leads to overload on the gear teeth.

Examples of faults in operation were just mentioned. These are essentially based on the lubrication oil, overheating, and malfunctioning of an associated

fatigue
Undergoing repeated alternative tension and compression stress.

alignment
Having the axes (of two shafts, for instance) along a straight line.

device such as a pump that must circulate the oil. Other examples are contamination of oil, say if humidity gets to the oil; and if the oil filter clogs and oil pressure drops. Faults in maintenance refers to any mistake concerning prolonged delays in oil change, the wrong oil type, and so on.

Typical gear damage or failures are:

1. Fracture
2. Bending
3. Wear
4. Fatigue
5. Cracking
6. Scuffing

Fracture implies that a piece of tooth breaks apart. In such an occurrence, not only does the tooth with the broken part become weaker, the broken part has a grinding action on all other parts, because it can move around with oil into all the other teeth, until it is stopped by the oil filter.

Gear tooth bending is a deformation of teeth in the direction of load (direction of force on the teeth; see figure 9.11). Note that when two objects (here the gear teeth of the two meshing gears) push each other, the force on one is the opposite of the force on the other (see the discussion on force and stress). This can happen due to overloading and/or high temperatures. High temperatures can weaken a metal part in terms of lessening its strength to loads. Thus, the part can bend or deform more easily.

Wear is normally in the form of **abrasion,** where two matching parts can grind against each other. In the case of two gear teeth, this happens if there is not sufficient room for smooth rolling motion between the teeth. This can also happen due to contamination and debris in the oil. A physical abrasion action is normally called erosion, and if it happens due to chemical reactions (acid in the oil, for instance) it is called **cavitation.**

Fatigue, as mentioned earlier, occurs when a machine element is subject to frequent change in the applied force. As a result, the part is forced to compress and then pulled to extend. Consider a gear tooth that is pushed up, then pushed down, on a permanent basis. This causes fatigue. Vibration in gears causes such a phenomenon. Fatigue can occur in the tooth roots due to the bending of teeth (which can be called bending fatigue), or it can happen at the points of contact where two teeth press each other and then are relieved. Note that we also have thermal fatigue, which happens in parts of devices that are subject to temperature fluctuations (becoming very hot and then very cold, repeatedly). This does not normally happen in gears.

Cracking is self-explanatory. It is the starting point for eventual fracture, and it is due to localized stress. Cracks can start under the surface, and eventually expand to the surface at different parts of a tooth body.

Scuffing is the term used when tiny pieces of metal tear from one gear tooth and due to localized heat adhere (weld) to the body of the matching teeth. This causes the surfaces to become rough, instead of being smooth, which in turn causes the development of more stress in the teeth and vibration in the gears.

abrasion
Wear in machinery components where two matching parts can grind each other.

cavitation
An interaction between a metallic rotating piece and a liquid, such as in a gearbox or a pump, which causes the metal to be eroded and cavities to form.

scuffing
Transfer of metal particles by tearing from a tooth and adhering to another tooth by welding (due to heat and pressure).

FORCE, PRESSURE, AND STRESS

By now you have a good understanding of force in a mechanical system. Also, in chapter 3 we learned about the relationship between force and pressure and between force and torque. Here, we see the relationship among force, pressure, and stress.

Both pressure and stress are the force per unit area and for both the unit of measurement in the English system is "pounds per square inch" (psi). Nevertheless, pressure and stress should not be mixed up, and they cannot be used in place of each other. Whereas pressure is used for the effect of a force on an area, particularly for liquids, where pressure is transferred in all directions, stress is used to describe the strength of a machine part and to determine its load-taking capacity. It is pressure at the molecular level, where the integrity of a piece is concerned.

The first figure here shows a valve under a load. Suppose that the pressure in the vessel is 240 psi. The valve serves to keep the opening closed. The cross sections of the valve on the top and in the middle are 1.5 in.^{2} and 0.16 in.^{2}, respectively. This 240-psi pressure pushes the valve with a force of

$$F = 240 \text{ psi} \times 1.5 \text{ in.}^2 = 360 \text{ lb}$$

This force is transferred along the body of the valve to the other end. In the narrow part of the valve, the force is still 360 lb, but the cross section area is only 0.16 in.^{2}. We can now say the stress (and *never the pressure*) in the body of the valve is

$$360 \text{ lb} \div 0.16 \text{ in.}^2 = 2250 \text{ psi}$$

This stress must not exceed the allowable stress of the metal for the operating temperature, taking into account the necessary safety factor.

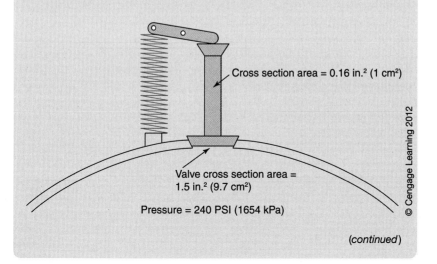

Cross section area = 0.16 in.^{2} (1 cm^2)

Valve cross section area = 1.5 in.^{2} (9.7 cm^2)

Pressure = 240 PSI (1654 kPa)

© Cengage Learning 2012

(continued)

The kind of stress just described is called *normal stress*. It can be in the form of compression (compressive stress), as in the preceding example, or in the form of tension. If a rod or cable, for example, is pulled from both sides with some force, the resulting stress is not compressive; it is called *tensile stress*. Compressive and tensile stress is shown in parts (a) and (b) of the second figure here.

In addition to normal stress, which tends to break (or deform) a piece by either pulling it or compressing it, as shown in part (c) of the second figure here, *shear stress* tends to break (or deform) a piece by cutting it. This is what happens when a material is cut by a pair of scissors. These three types of stresses occur in a part that is under load. For instance, a shaft that is subject to torsion is under a load that tends to twist the shaft. Under such a load the body of the shaft undergoes shear stress. Also, if an object such as a metal bar is forced to bend [see part (d) of the second figure here], this leads to formation of compressive and tensile stress in the body of the object. In this case, parts of the body are under tension and parts are under compression. In the case shown, the upper half is under compression and the lower half is under tension. When a body undergoes bending due to force, it is subject to shear stress also. The force wants to cut the piece across.

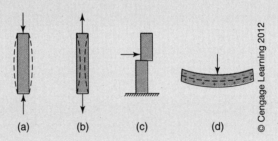

(a)　　(b)　　(c)　　(d)

© Cengage Learning 2012

9.10　Chapter summary

- Gears are used to change the speed and torque between two shafts.
- At least two gears are necessary. More can be employed in a gear system.
- The driving shaft is the input and the driven shaft is the output.
- Gears are usually used as speed reducers. In such a case, the torque on the output shaft is larger than the input torque.
- In wind turbines, gears increase the speed of the rotor. This is necessary to match the higher speed of electric generators.
- Depending on the requirement, various configurations can exist for a pair of gears. These can be categorized in terms of the distance and angle between the two (input and output) shafts. In spur gears, the shafts are

parallel, whereas in bevel gears and worm gears, the shafts are perpendicular to each other.

- Gears can also be categorized based on the angle of the teeth with the body of the gears. Helical gears have teeth that are cut at an angle. Their advantage is reduction or elimination of backlash.
- Backlash is the free play between matching teeth in a pair of gears. Backlash is not a wanted feature and can lead to vibration and noise.
- In wind turbines planetary gears are used. Planetary gears are made of a set of gears with a specific arrangement. These are the sun gear, the ring gear, the planet gear, and the arm. In many applications, including wind turbines, the ring gear is stationary and does not rotate.
- In wind turbines, the arm is the input and the sun gear is the output.
- Gears can be combined together in a gearbox. In a wind turbine gearbox, the planetary gear is combined with ordinary gears.
- Lubrication is very vital in the operation of gears. It serves for reducing friction, cooling, and moving dirt and debris away to be filtered.
- Alignment is also crucial for proper operation of gears. Misalignment leads to vibration and quick deterioration of gears.
- Any damage in gears eventually leads to permanent failure. Damage can be a result of overload, overheat, misalignment, and vibration.
- Gear failure can start in the form of fracture, bending, wear, fatigue, cracks, and scuffing.

ADVANCED LEARNING

In addition to the basic types of gears mentioned in this chapter, other types of mechanisms work for speed reduction in machinery. Some of these can be specialized, but there are two types of mechanisms worth mentioning here. These are not necessarily used in wind machines and probably will never be.

Harmonic Drive

A harmonic drive is a type of speed reducer with no backlash, a high gear ratio (could be up to 100:1), and a high torque capability. Also, similar to planetary gears, the input and output shafts are coaxial. Harmonic drive gears work based on a wavy motion of a flexible metallic membrane. They can be used for increasing speed, too. The figure below shows a harmonic drive. The piece in the middle can be connected to the input shaft. This piece rotates with the input speed and creates a wavy motion in the membrane. The membrane rotates while it changes form to conform to the shape of the inside piece. The membrane has smaller number of teeth than the outside ring. In its rotation the membrane causes the outer ring to rotate slowly in the opposite direction, since the teeth have to match up.

(continued)

For speed reduction, the speed ratio in the harmonic drive is

$$\text{Reduction ratio} = 1 : \left(1 - \frac{\text{Membrane teeth numbers}}{\text{Outer ring teeth numbers}} \right)$$

For example, if the number of teeth in the membrane and the outer ring are 49 and 50, respectively, then the gear ratio is 1: 0.02 or 50.

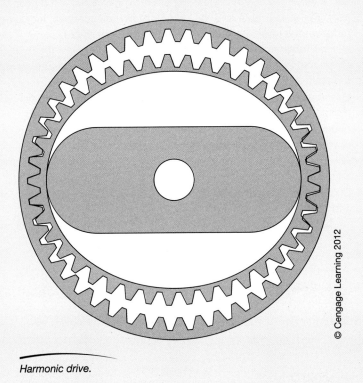

Harmonic drive.

© Cengage Learning 2012

Variable speed drives

The principle of variable speed drives is two rotating disks in contact with each other, as shown in the figure. Transfer of motion takes place by means of friction between the two disks. The two axes are perpendicular to each other, and the smaller disk can move along its shaft toward or away from the center of the other disk. This changes the point of contact along the periphery of different circles on the larger disk. As the smaller disk moves outward, its point of contact follows a larger circle. This increases the speed ratio. Similarly, moving toward the center reduces the speed ratio.

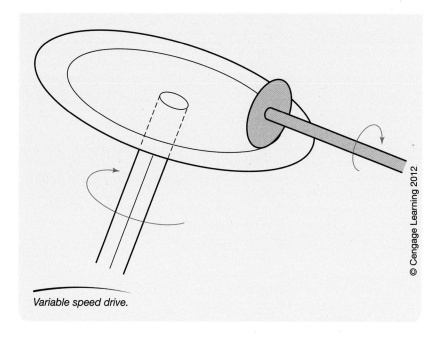

Variable speed drive.

© Cengage Learning 2012

Review questions

1. What is the main role of a gearbox in a wind turbine?
2. What is the main difference between the gearbox in a wind turbine and one in any other machine?
3. What is the difference between a spur gear and a helical gear?
4. Do all the gears have the same tooth geometry?
5. What is the gear ratio?
6. In a gearbox of a wind turbine, which is larger: the torque on the rotor shaft or the torque on the generator shaft?
7. What is the advantage of a planetary gear?
8. How many planets are usually in a planetary gear?
9. What is the other name for a planetary gear?
10. Can a worm gear be used in a wind turbine?
11. What is an internal gear?
12. What is the difference between the ring gear and the sun gear in a planetary gear?
13. In a planetary gear, can the sun gear have more teeth than the ring gear? Why or why not?
14. What is the expression for power in a gearbox?
15. What is the expression for efficiency in a gear system?
16. Can efficiency be higher than 1? Why or why not?

17. What is the specific feature of pitch circles in a pair of gears?
18. Does the gear ratio depend on the number of teeth?
19. In a pair of gears, what geometric parameters determine the gear ratio?
20. In a three-gear system, what are the effects of the middle gear?
21. In a pair of gears, if the input rpm doubles, what happens to the output rpm?
22. What is backlash?
23. Why is backlash undesirable?
24. How can we reduce backlash?
25. Which type of gear does not have backlash?
26. Why is the gearbox in a wind turbine more prone to damage than the gearbox in other applications?
27. What are the main roles of lubrication oil in a gearbox?
28. Why is it necessary to change the oil in a gearbox?
29. Can lubricating oil correct the misalignment?
30. What is overload and what is its main cause?
31. What can happen if overload happens in a gear?
32. Name three types of failure in a gearbox.
33. What is meant by scuffing?
34. What is fatigue?

Problems

1. The numbers of teeth in a pair of gears as a speed reducer are 120 and 48. If the input speed is 1200 rpm, what is the output speed?
2. In problem 1, if the output torque is 500 ft-lb, how much is the torque provided by the input shaft, assuming the efficiency is 100%?
3. In problem 2, how much is the input torque if the efficiency is 90%?
4. In problem 1, if the radius of the 120-tooth gear is 8 in., how much is the diameter of the 48-tooth gear?
5. In problem 4, if the given radius is the outer radius, is the pitch circle diameter of the 48-tooth gear larger or smaller than the value you obtained?
6. In a four-gear straight-line gear train [similar to that shown figure 9.10(b)], the smallest gear is 2.5 in. in diameter and the largest gear has a diameter of 12.5 in. If the mechanism is used for speed reduction, how much is the gear ratio between the input and output?
7. In a gear train similar to that shown in figure 9.10(d) used as a speed reducer, the first gear has 15 teeth and, the second gear has two sets of teeth. One has 30 teeth and the other has 14. The third gear has two sets of teeth, one with 32 teeth and the other with 20. The last gear has 44 teeth. Determine at what rpm the input shaft should rotate so that the output shaft has a speed of 179 rpm.
8. In a two-gear system the efficiency is 95%. What input power will provide an output power of 20 hp?

9. In a planetary gear suppose $N_S = 44$ and $N_R = 100$. If the sun is connected to the shaft of an engine whose rpm is 450 rpm, what is the speed of the arm?

10. In problem 9, suppose that there is no arm and the planet gears are restrained from rotating. This assembly of the internal ring gear, the planet gear, and the sun gear is similar to three gears with the middle gear having no role in speed ratio [see figure 9.2(c)]. How much is the ring gear speed?

11. In problem 9, suppose that the arm is the input and the sun is the output. Find the sun gear rpm.

12. If the output (the sun gear) in problem 11 is subject to 200 ft-lb of torque, how much torque must the engine provide? (Assume 100% efficiency.)

Projects

PROJECT 1

In a planetary gear, $N_S = 244$ and $N_R = 400$. The planet gear's number of teeth is always proportional to the gap between the ring and sun gears. Find out for each turn of the arm around the sun gear how many turns the ring gear rotates. (Assume that the sun gear does not rotate, to make it easier.)

PROJECT 2

Suppose that two shafts are 1.5 ft apart and you must select gears to reduce the speed by a ratio of 3:1 for these shafts. You have two choices: either use one pair of gears, or alternatively, use four smaller gears instead. Assuming that all the gears that you select have the same thickness and are made of the same material, compare the two scenarios in terms of the weight of the total number of gears required. (Hint: Use the surface area of gears to compare their weights.)

Control of Wind Turbines
Part 1: Mechanical Side

KEY TERMS

blade pitch control
crowbar
downwind turbine
look-up table
overspeed
stall
stall control
upwind turbine
wind turbine power curve

OBJECTIVES

After studying this chapter, you will be able to:

- Define the control parameters in a turbine.
- Explain how these parameters are related.
- Understand why a turbine must be controlled.
- Describe how a turbine reacts while wind speed varies.
- Explain how a turbine can fail if its controls do not function properly.
- Use technical terms such as pitch control and yaw control.
- Comprehend what stall control is.
- Differentiate between simpler and less expensive control of a turbine and the more sophisticated and more expensive control.
- Explain why a turbine must be stopped at wind speeds beyond a certain value.

10.0   Introduction

A wind turbine is a complicated system and its operation has to be continuously monitored and controlled in order that it functions as it should. It is expected from a wind turbine that

a. It adjusts itself to the wind.
b. It generates the maximum possible power under the operating condition.
c. It safely delivers the generated electricity to the grid.
d. It monitors its own safety and safe operation.

These four functions are for commercial wind turbines that are connected to a grid; that is, the large wind turbines. If a wind turbine is not connected to a grid, it is still required to perform the other three functions.

Control of wind turbines covers monitoring and executing the required action for all the components in the turbine as a whole system. **Figure 10.1** shows the interrelationship between the various parameters that are involved in the operation of a wind turbine. On one hand, the turbine must mechanically adapt itself to the wind conditions, respecting the requirement for operational speed and power. On the other hand, the generator must conform to the requirements of grid (voltage, frequency, delivered power, and power factor), taking into account the available power in addition to its speed and voltage.

The subject of connecting to a grid and the modes of operation of a generator are discussed in the next chapter. In this chapter, the matters pertinent to the controls that stem more from mechanical viewpoints are discussed.

10.1   Power accord with the wind speed

In a steam turbine or a hydraulic turbine (water turbine), normally the main control is based on the power requirement. A throttle valve is adjusted to deliver sufficient fuel to the turbine to maintain its speed. In industrial-level turbines, a synchronous generator is employed, which has to rotate at a constant (synchronous) speed. Normally sufficient steam or water is available for continuous operation of the system.

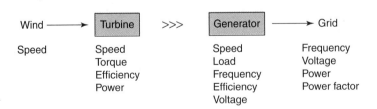

FIGURE 10.1 *Parameters involved in operation and control of a wind turbine.*
© Cengage Learning 2012

Quite the opposite, in wind turbines there is no continuous and reserved supply of wind and, as we have learned so far, wind speed can vary from minute to minute, at a larger scale, and it can even have variations from one second to the next. For this reason, control of a wind turbine is not (and cannot be) performed in the same manner as the other power plants, and it is more complicated.

A wind turbine must detect the wind direction and speed, and act accordingly. If a turbine is in operation it must continuously orient itself to the direction of the wind. In older turbines, some have been designed to have a downwind configuration (see **Figure 10.2**). A turbine having such a configuration is called a **downwind turbine.** This could be employed in small turbines and no active yaw control was required to orient a turbine to the wind direction. In today's large turbine, however, it is the yaw system that actively turns a turbine in the wind. Also, if the turbine is to be stopped it is turned out of wind, implying that if it was already in the stop position and the wind direction changed, it must be yawed out of the wind.

The other accord to be made with wind is the wind speed. Wind speed determines how much power exists in the wind (see chapter 2). Measurement of wind speed is an absolute necessity; otherwise a turbine cannot function. In a small isolated wind turbine (a stand-alone turbine that is not connected to a grid), it is possible to have a simple control system, or even not measure the wind speed. However, in today's large turbines the wind speed must be measured to be entered into the onboard computer, based on which the available power can be computed.

downwind turbine
A turbine in which the rotor is such that the wind hits the tower before reaching the blades.

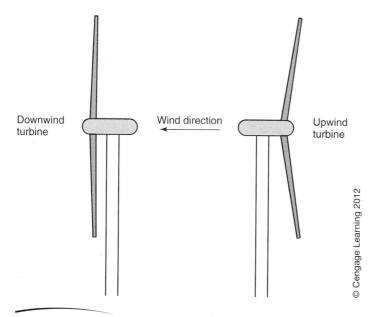

FIGURE 10.2 *Upwind and downwind configurations.*

10.2   Wind speed measurement

Each turbine is equipped with an anemometer that is normally installed on the top of the nacelle. Ideally, the measurement made by an anemometer must represent a true value for wind speed. However, since the wind stream is disturbed by the motion of blades, particularly in an **upwind turbine** (almost all today's modern turbines), the speed as measured can be deviated from the real value of wind speed. Figure 10.2 shows the meaning of upwind and downwind configurations of a turbine.

A correct measurement of the wind speed can be performed by an anemometer installed on a met tower far enough from and in front of the turbine. This is a difficult task, since it not only requires the extra cost of a separate tower, a fixed met tower cannot always be in front of a turbine, due to the change in the direction of wind. Nevertheless, a compensation for the aforementioned error in wind speed measurement can be considered within the software of the onboard computer, if a correct analysis of the aerodynamics process in the blades is made. This is a matter that depends on the manufacturer's expertise and quality work.[1]

As you may recall from chapter 6, any turbine has a cut-in speed, the wind speed below which it does not generate any power, and a cut-out speed, above which the turbine should not operate. The first thing a turbine controller must determine is whether the turbine must continue its present status (operating or braked), or if it must be switched to the opposite status. Obviously there is a gray area around the cut-in and cut-out speeds, since there may not be a sharp transition in each zone. For example, if the cut-in speed for a turbine is 4 m/sec (about 9 mph), wind could be blowing at this speed for say 40 min. Since this is the average speed, the real speed of wind is fluctuating around 4 m/sec. For situations such as this, the controller should not repeatedly turn the system on and off based on the instantaneous wind speed. Wind speed could be measured hundreds of times per second. Provision for how to deal with, or what action to take for, this situation must be thought of and embedded in the algorithm for the turbine control.

10.3   Power calculations

We can recall from chapter 2 that the rated value (installed power) of a turbine is based on the rotor size (the area that the rotor sweeps) and a given wind speed, as well as the average air density in the region and the turbine power coefficient (which depends on the wind speed and rotational speed). The rotor size is constant, and the air density depends on the temperature, but its changes are small. The two more important parameters are the wind speed and the power coefficient.

upwind turbine
A wind turbine installed on its tower in such a way that the blades are in front of the tower and wind reaches the blades before reaching the tower, as opposed to the downwind turbine.

[1] Future wind turbines can be equipped with measuring devices that can predict changes in the wind speed before it reaches a wind turbine.

Since it is always preferred that the turbine catch the maximum possible power from wind based on the available wind, the pitch angle must be adjusted as required. We refer to figure 6.10, which is repeated here for convenience in **Figure 10.3.** During operation, for any wind speed, the power that a particular turbine can provide, taking into account the cut-in and cut-out speeds as well as the maximum generator capacity, is determined from a **look-up table,** which is based on this curve. The necessary calculations are performed at the design stage to create a curve similar to that in figure 10.3 and its associated look-up table. A look-up table is the corresponding numbers in the form of a table, which resides in the onboard computer memory. Such a table may look like **Table 10.1.**

A similar table can be used for the set point of the blade pitch angle, based on which the pitch angle control mechanism adjusts the pitch angle at different stages of operation and different wind speeds.

look-up table
A table of calculated values of a variable that is used to determine a numerical value without doing the calculations. Using such a table can be much faster.

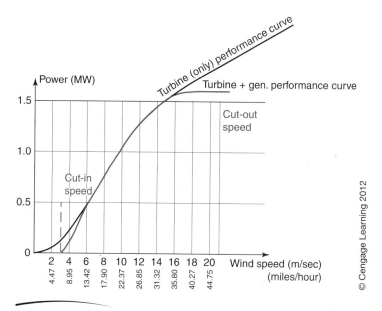

FIGURE 10.3 *A typical wind turbine power curve.*

TABLE 10.1

Look-up table of power curve in figure 10.3									
Speed, m/s	3	4	5	6	7	8	9	10	11	12
Power, MW	0	0.1	0.35	0.49	.62	.78	.9	1.05	1.18	1.27

Speed, m/s	13	14	15	16	17	18	19	20	20.5
Power, MW	1.38	1.44	1.51	1.56	1.57	1.59	1.6	1.6	1.6

© Cengage Learning 2012

> The blade pitch angle set point used for adjustment of blades during turbine operation can be put in a look-up table in the onboard computer.

10.4 Output power quality and requirement

The requirements for a generator to be connected to a grid and share the supply of electric power reduces to three parameters that must match those of the grid, as discussed in chapter 6. These are the voltage, the frequency, and the power factor. Obviously, the voltage and frequency of the generated power must be within the tolerance as defined by grid standards. Moreover, the power factor of the entire grid must be kept within a range to satisfy the requirements set forth by utility and distribution companies. In this sense, the reactive power by a generator must not exceed the allowed limit.

The three aforementioned parameters are electric variables, but they influence the "power" that must be provided by the turbine. Thus, the power captured from the wind affected by the efficiency of the gearbox and the efficiency of the generator must be delivered to the grid while satisfying the conditions for voltage, frequency, and power factor. At each instant, the available power at the shaft of the generator can be determined based on the wind speed and other parameters. This is a mechanical power. Based on this available power and the voltage, the current can be calculated. Depending on the type of generator in use and the modes of operation (discussed in chapter 11), if this current is not monitored and controlled, either the frequency or the voltage or both can go beyond the expected range of values.

Thus, the controller in a wind turbine must work based on continuous measurement of wind speed, voltage, current, frequency, and power factor. In addition, all the other operating variables, such as temperature and pressure, associated with safe and healthy operation of a turbine must be monitored. While it is not possible to show the control block diagram for all turbines, we can show a typical block diagram for the control of wind turbines based on only the principal variables. In practice, the commands associated with operating conditions are added to this diagram to shut down, put in stand-by, restart, report faults, and issue alarms. **Figure 10.4** shows a typical block diagram for control of a wind turbine.

> The controller in a wind turbine must work based on continuous measurement of wind speed, voltage, current, frequency, and power factor.

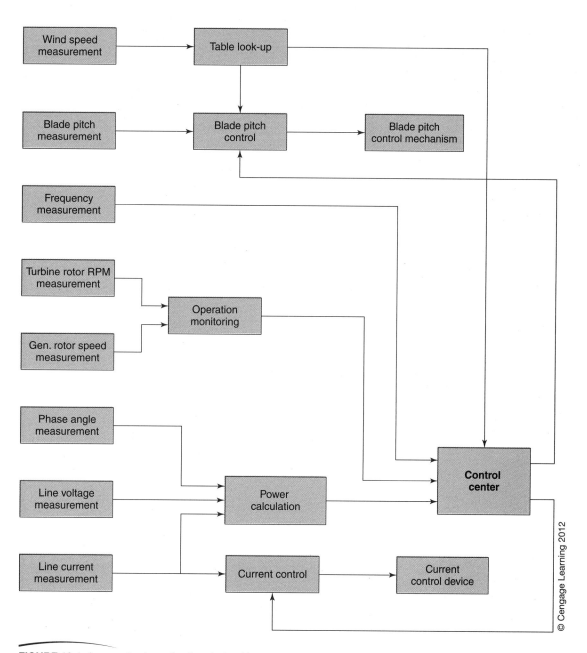

FIGURE 10.4 *A general schematic of a wind turbine controller.*

© Cengage Learning 2012

10.5   Blade pitch control

blade pitch control
The capability of turning a turbine blade about its axis with respect to the hub, thus changing the blade pitch angle.

The **blade pitch control** mechanism rotates the blades about their longitudinal axis with respect to the hub. This mechanism necessarily must operate from inside the hub. This can be done by an electric or a hydraulic actuator. Normally, the three blades must be adjusted to the required value simultaneously.[2] Pitching the blades alters their angle of attack; or better said, it changes the angle of attack of each segment in a blade by the same amount. As a result, the wind power capture capacity of the turbine changes (see chapter 3). As can be realized, adding this capability involves more components and more cost. Nevertheless, all of today's megawatt size turbines are equipped with blade pitch control capability.

By pitching blades the range of power capture capacity of blades can be altered dramatically from minimum to maximum. Thus, for each wind speed, one may catch between zero and the maximum possible power from wind. The variation of pitch angle, in fact, alters the power coefficient (C_p) of the turbine. When a turbine is not in production, the blades are feathered by pitch control.

All megawatt-size turbines are equipped with blade pitch control.

stall control
The effect of decrease in the wind power capture at higher wind speeds (due to stall) is a regulating mechanism to prevent them from going into overspeed or overload. A turbine using this feature does not use blade pitch control, and is called a stall-controlled turbine.

stall
The effect of a decrease in the lift force with an increase in the angle of attack in airplane wings and wind turbine blades. This happens at some higher values of the angle of attack. Before stall takes place, the lift force increases with an increase in the angle of attack.

Older turbines were **stall controlled,** instead. When **stall** occurs, the lift force decreases and the drag force increases. This can also happen in airplane wings, which is not desirable. At a certain angle of attack, the local circulation of air behind the airfoil creates a vacuum, as a result of which the drag force increases and the lift decreases. **Figure 10.5** illustrates a stall condition. This condition does not simultaneously happen for the whole blade, since the angle of attack is not the same along the length of a blade in a wind turbine. Since the angle of attack depends on the wind speed and the rotational speed of the blades (that is, the relative speed of air with respect to blade), stall condition can occur in part of a blade, depending on the twist angle. Then the entire lift and drag forces on a blade can change, which leads to the change of harnessed power.

A blade can, thus, be designed (by varying the twist angle) such that after a certain wind speed, the power grasp capacity does not increase (and even decreases). The behavior of an airfoil designed in this way is such that at higher wind speeds stall happens. Therefore, unless the angle of attack is reduced, the lift force on the airfoil decreases and the drag force increases. This by itself is a regulatory action that decreases the power capture of a turbine and prevents it from overspeed at higher wind speeds.

When the drag forces on the blades increase, however, this extra drag force must be resisted by the whole structure of a blade and then by the tower. This is one of the disadvantages of stall-controlled wind turbines that the turbine

---

[2] Some newer turbines may have three separate pitch control systems for each of the three blades. The technology is developing and new ideas come into action based on research and development.

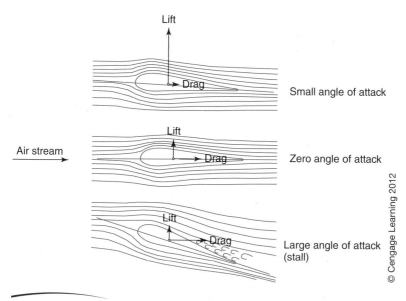

FIGURE 10.5 *Stall in an airfoil.*

FIGURE 10.6 *A turbine with fixed blades.*

must withstand increasing thrust loads as the wind speed increases beyond the turbine power capacity. **Figure 10.6** shows a turbine with fixed blades.

In some older (and smaller) turbines, an auxiliary braking mechanism came into action at higher rotational speeds of a blade. This mechanism would reduce the aerodynamic lift forces and prevent a turbine from speeding.

It consisted of a normally hidden and locked plate that comes off the blade and makes more air resistance for blade rotation. **Figure 10.7** depicts such a mechanism. The idea was a passive self-regulatory action, but this often did not work properly and the plate retraction into hidden position failed. So, putting the blade back to normal had to be done manually, which is time consuming.

Instead of this braking system, a tip brake was implemented in which the tip of each blade was a separate piece. In the normal working condition, the tip was aligned with the rest of the blade, forming one uniform blade. In the case of overspeed, the centrifugal forces on the tip section of the blade would trigger a mechanism to open. The blade tip then would rotate by 90°. This action modified the aerodynamic behavior of a blade and, as a result, the blade slowed down. This type of brake is shown in **Figure 10.8.** This mechanism could also suffer from failure to go back to its normal position.

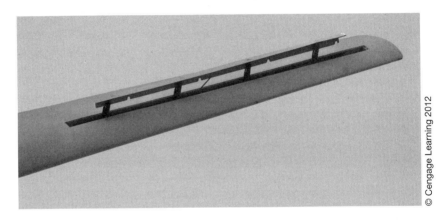

FIGURE 10.7 *Braking mechanism to prevent a turbine from overspeed.*

FIGURE 10.8 *Tip braking system.*

Turbine operating power curve   10.6

Figure 10.3 illustrates a typical characteristic curve of wind turbines based on which a turbine is to adjust its parameters and set points for the operation at various wind speed conditions. It is interesting to understand how this curve has been generated. We recall from chapter 7 that the performance (power coefficient and power capture) of a turbine changes with the tip speed ratio and that for a specific rpm of a turbine the tip speed ratio depends on the wind speed (see figure 7.6). On the other hand, we would like a turbine to always work, as much as possible, with maximum power coefficient, that is, around the peak of its characteristic curve at each wind speed. If the maximum points of the characteristic curves for each wind speed are connected together, the resulting curve shows the desired points of operation of a turbine at various speeds. This curve, when blended with a cap for the maximum capacity of the turbine generator, defines a curve based on which a turbine is scheduled and controlled. That is the curve introduced in figure 10.3; it is referred to as a **wind turbine power curve. Figure 10.9** illustrates the process and the resulting curve is highlighted.

> The power curve of a turbine is built by connecting the maximum points of individual characteristic curves at various blade pitch angles, taking into account the maximum capacity of the turbine generator.

In figure 10.9, each curve in the family of similar looking curves represents a different pitch angle for the blades of a turbine. The corresponding value of the pitch angle for each wind speed can be added to the look-up table, similar to table 10.1, for blade pitch angle adjustment.

wind turbine power curve
A curve specific to a wind turbine that determines the output power versus the wind speed. This curve is used by the wind turbine controller during operation to adjust the blade pitch.

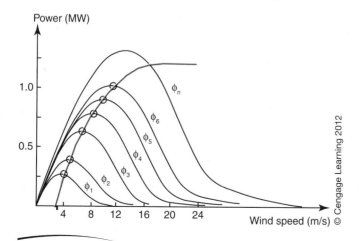

FIGURE 10.9 *Development of the power curve of a turbine.*

10.7   Yaw control

When working, a horizontal-axis wind turbine must orient itself with the wind direction. This is one of the automated functions of a turbine based on determining the direction of wind at each instant. The control loop for yaw motion can be independent of the other functions. A number of motors, for example, eight, working in parallel are employed for this purpose. The arrangement can be different from one turbine to another. Normally, however, this mechanism consists of a large ring gear (approximately the size of the tower top), and the yaw motors through a set of gears turn on the yaw ring gear [see figure 9.2(b)]. This turns the nacelle with respect to the tower with a slow motion. **Figure 10.10** shows the yaw gear-motors for a Vestas 82 wind turbine. The cables in the picture are the generator output, which extend to the bottom of the tower.

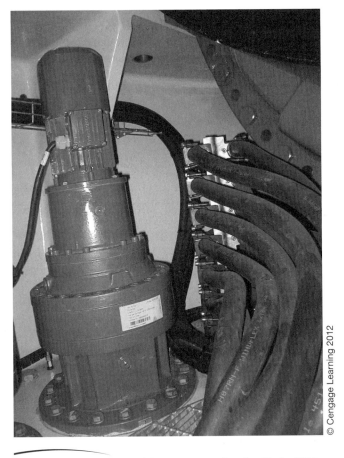

© Cengage Learning 2012

FIGURE 10.10 *One of the eight yaw gear-motors in a Vestas V-82, seen from inside the nacelle.*

Parking brake   10.8

When a turbine is not in operation, that is, when a turbine is turned off, it must be kept in a fixed position and stopped from moving. In other words, the turbine must be parked. This is very important for any maintenance job on the turbine. It is very dangerous and can be fatal to work on a turbine if it is not braked. Also, during operation, whenever a turbine must be stopped, for example, in wind speeds below the cut-in and over the cut-out speeds, a turbine must be prevented from rotating. This is achieved by the turbine brake system, which can be similar to the brakes in a car, usually the disk brakes, installed on the high-speed shaft. **Figure 10.11** shows an example of a typical brake on a small (350-kW) turbine.

In addition to a disk and pads for parking the turbine in a fixed position, some turbines are equipped with a pin and a hole in a convenient position. This is an extra safety measure that is used during maintenance work. During the work, the pin is inserted manually into the hole by a technician who is going to work on the turbine. This makes sure that the rotor cannot rotate. It is also helpful to keep the three blades in one or more predefined position. For certain jobs on the hub or the blades, the turbine must be kept in a particular position.

> It is very dangerous and can be fatal to work on a turbine if it is not braked.

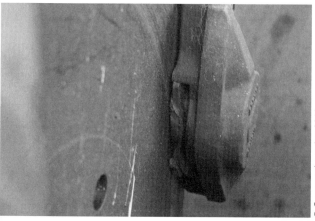

© Cengage Learning 2012

FIGURE 10.11 *Example of the brake system in a small turbine (in larger turbines this arrangement may be under a cover for better safety).*

10.9   Fault control

One of the roles of a monitoring and control system in a turbine is that if some fault is detected, the proper action is taken. A fault can be any malfunctioning of a component, including damage and breakage; high or low operating temperature or pressure; or a value outside of the allowed tolerances for a mechanical or electrical parameter such as speed, voltage, current, and the like.

Depending on how serious the fault is, a turbine must be shut down if the fault cannot be corrected. If a turbine has to be shut down, a process of shutting down the turbine must be followed. Some faults may lead to a temporary action that must be checked a number of times to see if it persists.

The number of faults that can occur is large and it is not practical to list them, since faults can be different from one machine to another. Some of the more important mechanical faults that can happen are mentioned here:

a. Excessive vibration in various parts of the turbine; this can be in any of the main bearings, in the nacelle, in the tower, and in the gearbox.
b. High temperatures of various fluids in the gearbox, bearings, oil cooling system, or any other hydraulic system.
c. Low level or low pressure of various liquids in various parts of the system.

Also, some faults are not detectable by a turbine controller, unless redundant systems are used. Redundant means that one parameter or variable is measured by two devices, so that if one fails the other shows a (different) value. For instance, suppose that the anemometer malfunctions or freezes in a freezing rain. The reading from the anemometer then shows a zero speed for wind, whereas there is wind. In such a case, the controller interprets this as lack of wind and, thus, the turbine is put into halt and with the brakes on. This is a typical scenario that calls for a technician to act.

overspeed
Speeding up and reaching speeds that are beyond the designed values.

One of the major faults that can happen in a wind turbine and can lead to disastrous results is when a turbine speed increases beyond the allowed limit. This is called **overspeed.** When a turbine rotates, there are dynamic forces that act on blades and all the other related mechanical components. The magnitude of these forces depends on the speed of rotation. If the speed goes up, the magnitudes of these forces increase as well, and can go beyond the values for which the components are designed. If this happens, these components can break. And if one component in a mechanical system breaks and the cause is still present, this can lead to other component failure and serious damage.

In particular, if the speed of a wind turbine increases beyond its designed value, the blades can fly away and the whole turbine can fall apart in a short time. If during the normal operation of a wind turbine, the load is suddenly taken off the turbine, this can lead to overspeed. This can happen if the generator is disconnected from the grid, for instance, in the case of lightning tripping the overhead breakers and disconnecting the circuit. Even if the turbine controller starts shutting down the turbine in such a condition, the momentum in the rotor, as a result of its existing speed, can speed up the turbine to overspeed. In order to prevent this from happening to the turbine, modern turbines are equipped with a set of resistors that can connect to the generator and become a temporary load, until the situation is corrected, or the turbine has slowed down. These resistors, called **crowbars,** are not normally connected, but in the case when such a fault occurs they kick in and connect to generator stator. As a result of taking a large load from the generator, these resistors become hot and their heat must be taken from them as fast as possible. For this reason, they are placed outside of the nacelle where they can be cooled by fresh air.

crowbar
A set of electrical resistors that come into operation only when a generator is disconnected from a grid and its load has vanished. These resistors form a temporary load for the generator so that the turbine does not overspeed.

Chapter summary   10.10

- A wind turbine is a complex electromechanical system that cannot work properly without control.
- The role of a turbine controller is to monitor the safe operation of a turbine in addition to taking the correct action based on the parameter measurement for correcting faults, changing the electrical load on the turbine, and shutdown and start-up of a turbine.
- At each instant, the mechanical and electrical parameters must match for balance of power in a turbine.
- Mechanical parameters are rotor speed, torque on the rotor, and power. Also in wind turbines equipped with blade pitch control, pitch angle can be regarded as one of the mechanical parameters.
- Electrical parameters are the generator speed, the voltage, the frequency, and the current of the generated electricity, and the electric power (including the active power, reactive power, and the power factor).
- A turbine may be designed to work in an upwind or a downwind configuration. See figure 10.2 for the difference. A downwind turbine has the property of self-alignment with wind (passive yawing).
- In modern turbines, active yawing is employed. The turbine is yawed into wind by rotating the nacelle with respect to the tower.
- For each modern turbine, an operating characteristic curve determines the power output at each wind speed. This is provided as a look-up table in the turbine control system.
- Wind speed is measured by an anemometer.
- Smaller wind turbines can work based on stall control. In stall control, the power capture capacity of a turbine lowers as the wind speed increases beyond certain value. This limits the maximum power.
- A turbine must be kept in a park position when not operating. This is done by the turbine brake. Turbine brakes are similar to car disk brakes.

ADVANCED LEARNING

Control Systems

As we have learned so far, many variables must be controlled in a wind turbine, such as blade pitch angle and nacelle yaw angle. Also, on the electrical side, there are voltage, current, frequency, and other variables that must be controlled to a *desired value*. Here, we elaborate on how an entity, in general, is controlled. The discussion here applies to any variable that needs to be controlled. But for sake of clarity, we may use the terms for blade pitch control, when necessary.

Suppose that the value of pitch angle for blades at a certain time during the operation of a turbine must be 35°. This value is from a reference point for measurement of blade angle that one may physically count 35°. Moreover, this value is not fixed, since a minute later, it could be altered to, say, 38°. But, for any given instant the figure defines the desired value.

A control action is normally taken in a *control loop.* In order for better control, *feedback control* is used and the value to be controlled is continuously checked and compared to the desired value, and corrections are made. In the control loop the desired value is the *set point.* An action is taken on the difference between the *actual value* (that is, the measured value) and the set point. This difference is called *error.* The figure below shows a feedback control loop.

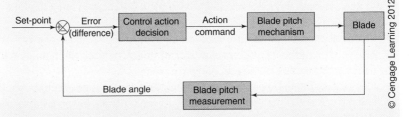

The action to be taken based upon the error is performed by the *actuator.* For a pitch control system, the actuator is an electric motor or a hydraulic piston that turns the blade. The *command* to the actuator for how much must it rotate the blade can be found by a control law. There are various control laws; but the most common and traditional ones are proportional control and PID control (PID stands for *proportional, integral,* and *derivative*). Each element in a PID control has a different role. Proportional control introduces an action proportional to the value of the error. Its role is for stabilization of a system. The integral action is intended to take care of a drift in the controlled value. This drift is called *offset* or *steady-state error.* If not taken care of, by integral control, the value reached can be slightly different from the desired value. Derivative control has the effect of looking forward and predicting what can be expected to happen, and adding the necessary action.

Review questions

1. Why must a turbine be controlled?
2. What is the main parameter based on which a turbine should be controlled?
3. What are the parameters that must be respected associated with a turbine operation and control?
4. What is blade pitch control?
5. Do all wind turbines have blade pitch control?
6. What is the purpose of blade pitch control?
7. What is yaw control?
8. Why should a turbine be yawed?
9. What are the meanings for upwind and downwind turbines?
10. In terms of yawing, what is the difference between an upwind turbine and a downwind turbine?
11. What happens if a turbine cannot yaw?
12. Why does a turbine have a characteristic curve for its operation?
13. How is the turbine characteristic curve generated?
14. What is a look-up table?
15. What is meant by stall in an airfoil, or a turbine blade?
16. What is stall control?
17. What can be done to prevent a turbine from overspeed in strong winds and storms?
18. Does a turbine generate more power in storms? Why or why not?
19. Why do not all turbines have stall control?
20. What is the purpose of the parking brake?
21. Why must a turbine be parked when not working?
22. Is it safe to work on a turbine that is not properly parked? Why or why not?
23. If a turbine runs out of control, what can happen?
24. Give some examples of faults that can happen in a wind turbine.
25. What happens if a turbine is suddenly disconnected (for some reason) from the electric grid?
26. What precautions are made to prevent a turbine from overspeed?

PROJECT

The following figure shows the operating curve of a turbine. The range of variation of the pitch angle is between 74° and 105°, as shown. Your task is to make a table for the values of the blade pitch angle for each wind speed between 6 and 18 m/sec with the intervals of 2 m/sec. For the values of the pitch angles you should use your judgment for the best matching value, rounded to an integral number (for example, 88, 89, 90, etc.)

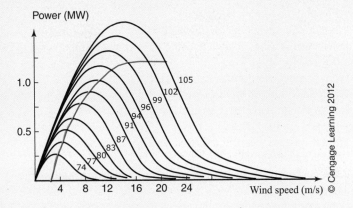

Control of Wind Turbines
Part 2: Electrical Side

KEY TERMS

asynchronous generator
direct drive mode
doubly fed induction
  generator (DFIG)
dynamo
fixed speed mode
free wheel
frequency converter
induction machine
inverter
rectifier
rotating magnetic field
slip
slip speed
slip-ring
synchronous machine
synchronous speed
variable slip mode
variable speed mode
wound-rotor induction
  machine

OBJECTIVES
After studying this chapter, you should be able to:

- Explain the ways various three-phase machines work.
- Differentiate between three-phase machines.
- Define the relationship between a generator rpm and the alternating current frequency.
- Describe the characteristics of operation of various generators used in industry.
- Define how a direct current generator is used in small turbines.
- Describe the ways an isolated turbine can operate.
- Describe the various modes of operation of wind turbine generators in connection to a grid.
- Define the advantages and disadvantages of each operating mode.
- Explain the ways that a particular type of three-phase generator can be employed in wind turbines.
- Understand and use the new terms employed in industry.

11.0   Introduction

As you have learned so far, a wind turbine is a small power plant that generates electricity. However, unlike conventional power plants, run by a diesel engine for example, the control strategy has to take into account the nature of wind. Large power plants employ steam turbines or hydro turbines as the prime mover of a generator. The capacity of a steam turbine can be 500 MW. This is why a diesel-driven turbine is considered small. Similarly, even the largest wind turbine available today, with a capacity of 2–4 MW, is considered small. In a conventional power plant, a synchronous generator is used, whereas in most of the modern wind turbines the type of generator used is the induction generator. An induction generator is also called an **asynchronous generator.**

asynchronous generator
Another name for induction generator; a generator that is not and does not work based on the same principle as a synchronous generator.

Since, in general, wind turbines can vary in size, from really small (generating tens of kilowatts of electricity) to small (generating hundreds of kilowatts of power) as well as the newest ones (at a few megawatts capacity), not all can use the same type of generator. In very small turbines a DC generator is more likely the type to be used.

In chapter 6 you learned about the principles of electric generators and how they work. Also, you learned about the necessary conditions that generators can be connected together. In chapter 10, you learned the mechanical side of the controls in a wind turbine. In this chapter, we further the study of wind turbine controls from the electrical side viewpoint. In doing so, we need first to study more about generators and the differences between them, and the way they function. Then we describe how they are employed in wind turbines, and how the generator in a wind turbine can be controlled.

11.1   Direct current generators

Although at the level of industrial power generation there is no place for a direct current generator, a small isolated turbine can be connected to a DC generator. By small, we mean a turbine with a maximum capacity of around 25 kW. Practically, the output of a DC generator is good for charging batteries. Such a scenario can happen only in small-scale applications. For instance, for powering a house in a remote area, a set of batteries can be charged by a small wind turbine. If such a house is similar to a normal house, using equipment such as a refrigerator and a TV, which are run by AC electricity, then the DC electricity stored in the battery must be converted to AC with the required voltage and frequency through an **inverter.** An inverter is an electronic device that converts DC to AC. The quality of the converted signal depends on the inverter. A cheap inverter delivers a square-wave AC signal, whereas it is possible to get a nearly sinusoidal signal from a properly designed inverter. You learn more about inverters in chapter 12.

inverter
An electronic device to generate AC electricity from a DC source.

A direct current generator is similar to a DC motor. If it is turned, it generates electricity, and if it is connected to a DC source with the proper voltage, it turns. In other words, it converts mechanical energy to electrical energy, or vice versa. Alternatively, a DC generator can be a small three-phase alternator the voltage of which is converted to DC through a **rectifier.** Such an arrangement is similar to what you find in almost all cars today. The reason for using an alternator and a rectifier in cars is because of smaller size and lesser weight compared to a **dynamo** (the name for a DC generator).

The voltage of the electricity generated in a dynamo depends on the structure of the machine and the speed at which it is rotated. In this sense, a small turbine, which is not equipped with blade pitch control, and whose rotational speed depends entirely on the wind speed, does not generate electricity at a fixed voltage. This may not affect charging of batteries, but if it is directly connected to a load, such as lights, the fluctuating voltage has a noticeable effect on the load. A voltage regulator between the load and the generator can remedy the problem.

At commercial-level generation, a DC generator is never used, for the same reason mentioned for the car dynamos and the cost. **Figure 11.1** illustrates the simple cases when a small turbine can be used for generating power in an isolated application. The third part of the figure resembles what happens in a car when an inverter is used to obtain AC electricity from a battery; for example, if you power a TV set with your car battery, in a camping or in a recreational vehicle.

rectifier
An electronic device that converts AC electricity to DC.

dynamo
A direct current generator.

> Commercial-level electricity generation is never done in DC.

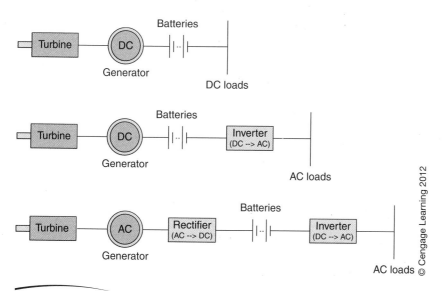

FIGURE 11.1 *Ways of isolated application of wind turbines.*

© Cengage Learning 2012

11.2 Synchronous generators

The standard type of generator for all power generation at the commercial level, except in wind turbines, is a three-phase synchronous generator. As we can see in this chapter, depending on the mode of operation, they can also be used in wind turbines. The specific difference between a synchronous generator and an asynchronous generator is that a synchronous generator must be rotated at a constant rpm in order to be connected to a grid. Otherwise the frequency of the generated electricity, which depends on the rpm, is not going to be constant and fixed at a required value. It is impossible to have different frequencies in a grid. Indeed, it is impossible to connect an AC generator to a grid with a frequency other than that of the grid. At a lower level, it is impossible to interconnect two AC generators having two different frequencies.

> It is impossible to have more than one frequency in a grid.

This does not mean that a synchronous generator cannot provide electricity when it is rotated at different rpm. In an isolated application where frequency is not important, a synchronous generator can be used. Suppose that in a remote location a single wind turbine is to provide electricity for heating and lighting. Heating and lighting are less affected by frequency. In such a case, a three-phase or single-phase synchronous AC generator can be employed.

In a synchronous generator the frequency depends on the number of magnetic poles per phase. The number of poles per phase can be as small as 2 (minimum), which form a north-south pair, to higher numbers. In order to have a certain frequency, the speed at which the generator must be rotated (in rpm) can be found from

$$N_S = \frac{60 f}{p} \tag{11.1}$$

where f is the required frequency, and p is the number of pairs of poles per phase.

This equation can be equally represented by

$$N_S = \frac{120 f}{P/3} \tag{11.2}$$

where P is the total number of poles in the stator ($P = 6p$).

As we can see, the lower the number of poles, the faster the generator must be turned. In this respect, generators that are energized by a steam turbine or a gas turbine have a smaller number of poles compared to those that are run by a water turbine. This is because steam and gas turbines have a higher speed than hydraulic turbines. For instance, in a water turbine-driven generator the total number of poles can be 96 ($P = 96$); that is, 16 pairs per phase ($p = 16$). In order for the electricity to have a frequency of 60 Hz, the turbine must run at 225 rpm.

The number of poles in a generator determines its structure and diameter size, due to the fact that all the poles must be accommodated in a circular area that forms the stator of the generator. The rotor rotates inside the stator. Accommodating more poles necessarily implies a larger diameter. For the same power capacity, generators with a fewer number of poles are slimmer and longer, whereas those with a higher number of poles are larger in diameter, but shorter in length. **Table 11.1** shows the corresponding rpm based on the number of poles for 50- and 60-Hz frequencies.

We will see later in this chapter that in one of the modes of operation, synchronous generators are employed in wind turbines. As of necessity then, the problem of frequency must be resolved.

As similar to all generators, a **synchronous machine** can serve as a motor or as a generator (this is why we call it a machine). If it is rotated, it generates electricity, and if it is connected to electricity it behaves as a motor. All the three-phase AC machines work based on the **rotating magnetic field** (see the text on *rotating magnetic field*, in this chapter). Single-phase AC machines have the same structure as the three-phase machines except that they have only one phase winding, and can be connected only to single-phase electricity.

Whereas in DC generators the magnetic field is developed by the stator winding, in three-phase AC generators the magnetic field rotates in the middle surrounded by stator windings in which the electric voltage is generated. Except in small generators this magnetic field is created by a DC machine embedded inside the rotor of the generator, and rotating with it.[1] The generated electricity is taken directly from the stator winding and there is no need for slip-rings. For more details on rotating magnetic field and the operation of synchronous generators see the text in this chapter on three-phase machines.

synchronous machine
A synchronous motor or generator.

rotating magnetic field
A magnetic field that rotates about an axis. This exists inside the stator of AC motors and constitutes the basis of their operation.

In a synchronous generator the magnetic field rotates inside the stator windings. The generated electricity is taken from the stator windings, which are stationary.

TABLE 11.1

Synchronous speed in rpm for various number of poles										
Number of pole pairs per phase	1	2	3	4	5	6	8	9	10	16
Synchronous rpm for 60 Hz.	3600	1800	1200	900	720	600	450	400	360	225
Synchronous rpm for 50 Hz.	3000	1500	1000	750	600	500	375	333	300	187

© Cengage Learning 2012

[1] Recently some wind turbine generators with permanent magnets have been developed. Such a generator has the characteristics of a synchronous machine. For a machine with a permanent magnet, the magnets can be on the outside of the stator winding, instead of inside.

11.3   Asynchronous or induction generators

The principle of operation of induction generators is quite different from that of the synchronous generator, although both work on the basis of a rotating magnetic field. The difference lies in how the magnetic field in the rotor is made. In this sense, the structure of stator is the same, and only the rotors differ from each other. That is to say, in principle, one can interchange the rotors of two similar (in size and power) synchronous and asynchronous generators.

There are two types of induction machines. Again here the term machine is used to emphasize that motors and generators have the same structure (for each category of machine), and "machine" can address either one. The two types are named based on the structure of the rotor; one is called a squirrel-cage machine and the other is called a **wound-rotor induction machine.** As the name implies for the latter, the rotor has windings, and these windings must be connected to electricity. The arrangement for windings on the rotor that rotates to be connected to outside sources that are stationary is done through slip-rings. **Slip-rings** are metallic circular rings mounted and rotating with the rotor and connected to the rotor windings. Then *brushes*, made out of carbon or similar manufactured electrically conductive material, are in contact with the surface of the rings. Brushes are pressed to the surface by springs to have a good contact, but they do not rotate. They are connected to the outside circuitry.

A squirrel-cage machine gets the name from the current-carrying bars that form a closed circuit. These bars arranged around a cylinder (rotor) and connected at the two ends form a cage shape. The reason for having bars instead of wires is the high current that must flow through them. More on the structure of these rotors can be found in this chapter under "Squirrel-cage induction machine."

wound-rotor induction machine (WRIM)
A type of induction motor or generator whose rotor contains windings, as opposed to squirrel-cage machine that has no windings in the rotor.

slip-ring
A metallic ring on the rotor of certain types of AC electric machines where brushes slip on the slip-rings to transfer electric current.

11.3.1   Induction generator characteristics

induction machine
A term that embraces both an induction generator and an induction motor.

synchronous speed
The constant speed of a synchronous machine. This speed depends on the frequency of the supply electricity.

The characteristic behavior of an **induction machine** is that if its rotor is turned slower than its **synchronous speed** it is a motor, and if it is turned faster than the synchronous speed it is a generator. The synchronous speed is governed by the grid to which the stator is connected (see the rotating magnetic field text in this chapter). The characteristic curve of an induction machine, indicating the relationship between speed and torque, is shown in **Figure 11.2.** The horizontal axis is the speed and the vertical speed is the torque. This is for a three-phase machine. The operating region is *only* in the linear region, the area around the synchronous speed where there is an abrupt slope. Outside the linear region, no matter as a motor or as a generator, both torque and speed values drop, meaning that there is a great loss of power. In that sense, an induction generator must operate around its synchronous frequency; otherwise the energy is wasted in the machine, itself, in the form of heat.

Focusing on the generator side of the curve, in the linear region if speed increases, so does the torque (and thus, the power), and vice versa. But, the variation of speed is in a very narrow range, whereas the variation of torques

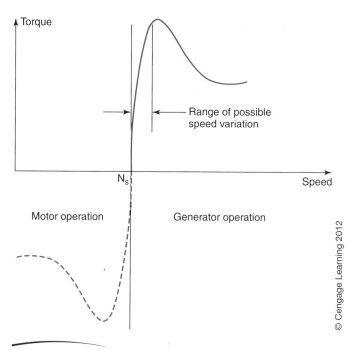

FIGURE 11.2 *Torque-speed characteristic curve of an induction machine.*

is significant. This implies that the generator has the intrinsic property of self-regulating for variation of speed, meaning that as the speed tends to increase, the torque increases as well, thus limiting the speed increase, while more power is delivered. This property makes induction generators quite attractive for wind turbines where the speed cannot be changed by opening and closing a valve, but is dictated by wind.

When wind speed increases, more power is available from wind. The increase in power can speed up the turbine, but since the operating point must follow the characteristic curve, any increase in speed must accompany some increase in the torque. This means that more power is available for harnessing. The turbine controller increases the power delivered to the grid. This keeps the speed from any further increase. The reverse happens when the wind speed decreases. It is to be noted that even if the induction generator runs at a speed above the synchronous speed, the generated electricity has a frequency associated with the synchronous speed (and not higher). For example, if there are six pairs of poles in the stator winding (two per phase), the synchronous speed is 1800 rpm for 60 Hz (see table 11.1). If the induction generator runs at 1850 rpm, it does not mean that the frequency of the generated electricity is more than 60 Hz.

If the generator is run exactly at the synchronous speed, it does not deliver any power, and if the speed drops below the synchronous speed, the machine automatically becomes a motor. With the increase in wind speed, more power is generated until the maximum capacity of the generator is reached. Beyond that, the power capture of the turbine has to be limited. This is where the role of the mechanical control comes in (see blade pitch control in chapter 10). Otherwise, the generator falls into the wrong region of its curve, which leads to a malfunction and failure. In the nonoperational region of the curve, the speed tends to increase, but small torque can be produced. However, if already a large torque (due to its load) is on the generator shaft, then that torque acts like a brake on the generator, stopping it altogether. All the parts of both the turbine and the generator undergo a lot of stress if this happens.

If the wind speed gradually drops, at one point the generator speed becomes lower than the synchronous speed. The generator turns into a motor, and instead of giving energy from the grid, it becomes a load. At this time, the turbine controller must prevent this from happening by disconnecting the generator from the grid. In this condition, until the wind speed catches up, the turbine rotates freely and has a **free wheel** status.

The linear region of a typical induction generator characteristic curve extends only within only a few percent, for example 3% of the synchronous speed. The graph in figure 11.2 is exaggerated for clarity. In this regard, the variation of speed is only within 3% of the nominal value. This deviation from the synchronous speed is called **slip.** Later in this chapter we see that one of the modes of operation of wind turbines is slip control. More about slip can be found in the discussion under "induction machines."

free wheel
A standby state of a wind turbine when insufficient wind is available to turn the turbine.

slip
The act of not rotating at synchronous speed in induction motors and generators.

11.3.2   Doubly fed induction generator

doubly fed induction generator (DFIG)
An induction generator in which the rotor is made of windings and requires slip-rings to connect to electricity.

A **doubly fed induction generator (DFIG)** is a wound-rotor induction machine used as a generator. You must read the section in this chapter about how a wound-rotor induction machine works in order to understand this subject better. A large number of the most recent wind turbines employ a DFIG because of its advantages in terms of harnessing more power. This extra power grasp is mostly at the higher end of the operating wind speed range (between cut-in and cut-out wind speeds), where the operating curve is flattened by imposing a limitation on the turbine power catch (see figure 10.3 in chapter 10).

Compared to a squirrel-cage induction generator, which must run only a few percent above the synchronous speed, a DFIG can run at a higher speed; that is, it can extend the slip (the percentage of speed over the synchronous speed) and still generate power without falling in the nonlinear part of its characteristic curve. For example, if the synchronous speed is 1200 rpm for a machine and the maximum slip is 3%, a squirrel-cage generator can have a speed of up to 1236 rpm, whereas a similar size DFIG can work up to a speed of say 1320 rpm (1200 + 10%).

Similarly, on the lower side of the wind speed range, where a squirrel-cage generator turns into a motor if the speed drops below the synchronous speed, a DFIG can still work as a generator. In other words, the range of speeds at which the generator can generate power is extended at both ends. Whereas the amount of extra power at the lower side is small (because it corresponds to wind speeds of around 3–4 m/sec), on the higher side it becomes very significant (since it corresponds to wind speeds of over 20 m/sec.)

In addition to the preceding advantage, by employing a DFIG there is a control on the power factor and reactive power, which is very attractive from the viewpoint of grid connection. This extra capability over the squirrel-cage machine is at a cost. The cost is due to the necessary electronic devices that must be incorporated in the system. Nevertheless, compared to a synchronous generator (which must operate in the variable speed mode), the additional cost is again attractive, as we can see in section 11.4 under "modes of operation."

What happens in a DFIG is very interesting. Since the rotor winding terminals are accessible, not only can one connect the rotor winding to an external resistor, it is also possible to connect it to a voltage source. If the rotor winding is connected to a resistor, then when the machine works as a generator, a low-frequency AC current passes through the resistor. The frequency of this current corresponds to the difference between the rotor speed and the synchronous speed. For instance, if the rotor has a speed of 1860 rpm for a synchronous speed of 1800 rpm, the frequency of the rotor current is 60 rpm or 2 Hz [since there are two pairs of poles per phase; see equation (11.1), 60 rpm corresponds to 2 Hz. If there was only a pair of poles per phase, then 60 rpm would imply 1 Hz].

This low-frequency current is a source of electricity that is connected to an external (to the generator) resistance (see the figure on the subject of Wound-Rotor Induction Machine, in this chapter). The resistance is in fact the load to this source of electricity. In wind turbines, this extra power from rotor winding is first converted to DC and then converted back to an alternating current with the desired frequency and voltage.

Suppose now that an AC current is injected into the rotor by connecting the rotor terminals to an AC source, instead of connecting to a load. The current from this source interacts with the current already in the rotor.

If the injected current is such that it has the same frequency and the same intensity as the current from the rotor, but with a phase difference of 180°, the two signals cancel each other and the net rotor current will be zero. If this happens, then although the rotor is rotating above the synchronous speed, no voltage is generated in the stator, because the magnetic field of the rotor has diminished. Now assume that the injected current has a larger intensity and, as result, it not only cancels the rotor current but forces the current in the rotor to be in the reverse direction with the same low frequency of, for example, 2 Hz for the aforementioned numbers. This action forces the synchronous speed of the generator to be lowered (by 2 Hz).

In practice (in wind turbines), this injected rotor current is performed stepwise, based on the wind speed. When the synchronous frequency is shifted to a lower value, the speed torque characteristic curve is shifted from its original location on the speed axis to a new position. **Figure 11.3** shows the shift of the synchronous speed for a number of cases of providing the rotor winding with an AC current with different frequencies.

The net result of this phenomenon is that even if the machine rotates at a lower speed than the synchronous speed, it can still generate electricity; that is,

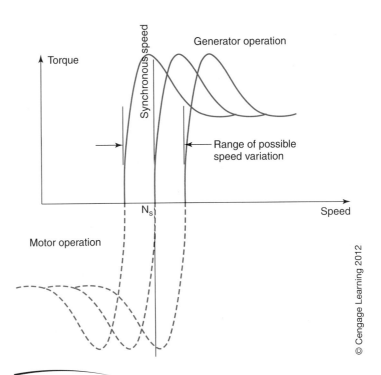

FIGURE 11.3 *Shift of the characteristic curve in a doubly fed induction generator.*

it still behaves as a generator and does not become a motor. That is to say, the synchronous speed of the machine has shifted by the same amount as the frequency of the injected current. There is a catch here, however, for the balance of the powers. Injecting a current into the rotor implies giving some power to the rotor in the form of electrical energy. This energy received by the rotor is converted to mechanical energy and is added to the total energy of the rotor. Thus, the machine has more energy than if no current was injected to its rotor. In a way, one can say that the machine borrows electrical energy in the rotor, but gives it back in the stator. The net production is the difference between the energy generated in the stator and the energy given to the rotor.

In conclusion, with a doubly fed induction generator one can get some extra power from the rotor winding when the rotor speed is higher than synchronous speed, and can inject some electrical power to the rotor in order to continue power generation when the rotor speed falls below the synchronous speed.

> In a doubly fed induction generator one can get some extra power from the rotor winding in addition to the power obtained through the stator winding.

ROTATING MAGNETIC FIELD

Consider three pairs of similar windings that are physically placed 120° from each other, as shown in **Figure 1**. If these windings are connected to three-phase electricity, in each winding a magnetic field develops the strength of which varies according to the variation of the AC current in the windings. That is, each magnetic field traces a sinusoidal pattern, continuously changing from zero to a maximum and back to zero, and then changing direction, repeating the same pattern in the opposite direction. Since the currents in a three-phase system are 120° out of phase, the variations of the three fields are not coincident, meaning that their maximum and minimum values do not occur at the same time. If all the magnetic fields developed this way were parallel to each other, then the resulting magnetic field would have the strength of all the individual fields added together in the same direction. However, since the windings are at 120° from each other, their fields will be added together as vectors (see chapter 3), each one having components in two perpendicular directions. Therefore at each instant of time the resultant magnetic field has a different direction within 360°. This is partially shown in **Figure 2** for 30° intervals. As we can see, the resulting magnetic field rotates with time.

For three windings, as was the case here, for each cycle of the current, the resulting magnetic field rotates by one revolution. The same arrangement can be practiced with two sets of windings (six pairs) connected to three phases in a correct order and each one being physically at 60° degrees from the neighboring windings. In such a case, we can see that for each full cycle of current variation, the magnetic field revolves half a revolution.

By increasing the number of windings and connecting them in a proper manner the number of revolutions per second of the magnetic field can be modified. This is what is done in AC machines in order to change the operating rpm. As we have seen elsewhere in the book, the relationship between the number of poles (each pair of windings counts for one set of north and south poles) and the rpm is

$$N_s = \frac{60f}{p}$$

In this equation f is the line frequency, and p is the number of *pairs of poles per phase*. That is, for six windings as shown in figure 1, for instance, p is equal to 1. The subscript s in N_s represents the *synchronous speed*. **Figure 3** shows the stator winding of a three phase generator. **Figure 4** illustrates a cut-away of a three-phase machine;

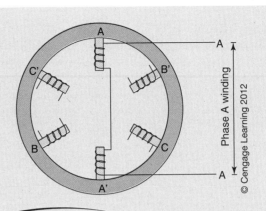

FIGURE 1 *Three pairs of windings 120° from each other.*

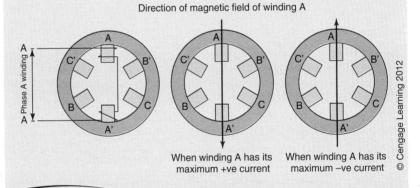

Direction of magnetic field of winding A

When winding A has its   When winding A has its
maximum +ve current   maximum −ve current

FIGURE 2 *Direction of magnetic field of one phase.*

notably it shows the rotor that sits inside the stator winding (magnetic field) by a number of bearings, also shown. This machine is much smaller than that in figure 3, and all the inside components are on the same frame. Most machines have fins at one end of the rotor, which function as a blower when rotating, forcing air across the hot components in order to carry the heat away. This is also shown in figure 4.

Note that in a three-phase system there are three connections to be made. If (any) two connections are interchanged, the direction of rotation of the magnetic field reverses. This can be used to reverse the direction of rotation of three-phase motors.

FIGURE 3 *Stator windings of a three-phase generator creating a rotating magnetic field. The magnetic field is inside the winding, where the rotor is inserted. The three windings are distributed in 360 degrees, so that the magnetic field is uniform all around.*

Stator winding

Connection to electricity

Bearing

Fan

FIGURE 4 *Cut-away of a three-phase machine.*

SYNCHRONOUS MACHINES

A synchronous machine has a stator, consisting of a number of windings connected together in an order, to create a rotating magnetic field, plus a rotor consisting of pairs of magnets, which is placed in the middle of the stator. The terminals of the stator winding are to be connected to electricity. For better efficiency the rotor can have the same number of poles as the stator. The stator of a synchronous machine is shown in figure 3 for a rotating magnetic field, and a rotor with two pairs of poles is shown below.

Rotor of a synchronous machine.

In order to understand how a synchronous machine works, we first consider a motor. When the stator of a three-phase machine is connected to a three-phase power supply with a particular frequency, a rotating magnetic field is generated (see the section on the rotating magnetic field in this chapter). Suppose that the frequency is 60 Hz, and the stator has two sets of poles per phase. According to table 11.1 this implies that the generated magnetic field rotates at an angular speed of 1800 rpm. Suppose that the windings are such that the axis of rotation of the magnetic field is vertical. Now, think of placing a compass in the middle of the stator windings, that is, in the middle of the magnetic field. Maybe you expect that the compass rotates with the same speed of 1800 rpm. Well, it does not. The compass will stay stationary, since the speed is too much for the compass to follow and orient itself with the magnetic field.

Now, however, assume that you manually turn the compass, so that its speed reaches near the speed of the magnetic field. Now, the compass is able and will follow the rotation of the magnetic field and will rotate with it; that is, with the same speed that the magnetic field rotates.

This is the basis for the operation of a synchronous machine, when used as a motor. It needs a magnet in the rotor, inside the stator winding. Compared to a DC motor, you see that in the DC machine the field is provided by the stator, whereas in the synchronous AC machine the magnets are in the rotor. They can be permanent magnets (PM) in the smaller machines, or electromagnets in the larger machines.

Consider now a stand-alone synchronous machine. When the rotor is manually revolved, due to the interaction of the magnetic field of the rotor with the windings of the stator, a voltage is induced in the stator. This voltage has a sinusoidal form and can be connected to a three-phase load. The frequency of this generated voltage depends on the speed of rotation of the rotor. It follows the relationship mentioned for the rotating magnetic field, as seen in table 11.1.

THREE-PHASE WOUND-ROTOR INDUCTION MACHINE

We recall from chapter 6 that if a wire moves inside a magnetic field, a voltage is induced in it, and if a wire inside a magnetic field carries a current, then a force is exerted on it. This is the basis for any electrical machine. The difference between various machines is how this natural phenomenon is put to work. The motion of a wire and a magnetic field is relative; that is to say, the wire can be stationary while the magnetic field moves.

Consider the following figure in which a one loop wire (for simplicity) is placed inside a rotating magnetic field.

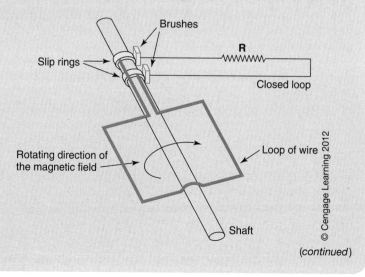

© Cengage Learning 2012

(continued)

Notice that the single wire loop is connected to a resistor and together they form a closed loop. Also notice that the resistor is external to the wire loop and its connection to the wire loop is through slip rings that are mounted on the shaft holding the loop. The loop ends are fixed to the slip rings and two brushes make the connection between the slip rings and the external circuit.

Rotation of the stator magnetic field is equivalent to moving the wire in a stationary field. The following lines describe what happens as a result:

1. Because the magnetic field is moving, it induces a voltage in the wire.
2. Because the wire ends are connected to the resistor and form a closed circuit, a current is developed in the loop (including the resistor), proportional to the induced voltage.
3. Because of the current in the loop, a force is generated that pushes each side of the loop wire in opposite directions, thus creating a torque. This torque makes the wire loop and its shaft rotate.

This is what happens in a wound-rotor AC induction motor. There is no connection to the electricity for the rotor winding (the loop wire). But the rotor windings make a closed circuit through the external resistor.

We see that the torque develops if there is a current in the rotor windings. If there is no current the torque diminishes. In other words, as far as there is a current in the rotor winding the motion exists. This current exists when there is a relative motion between the magnetic field and the rotor winding. If the rotor runs at the same speed as the magnetic field, then there is no relative motion. As a result, the rotor always runs slower than the magnetic field.

For simplicity and clarity of the figure, in the preceding description there was only one loop (two connections), which corresponds to a single phase machine. But the discussion is equally valid for three-phase systems.

SQUIRREL-CAGE INDUCTION MACHINES

Alternating current induction machines are either of the "wound-rotor" or the "squirrel-cage" type. Both work on the same principle as described in this chapter under three-phase wound-rotor induction machines. The difference is in the structure of the rotor. As the name implies, the rotor of a wound-rotor machine consists of wire windings that end up in three slip rings for 3-phase and two for single-phase, mounted on the shaft of the rotor. The slip rings are then connected through brushes to three resistors that can be outside the machine.

Suppose now that the outside resistors are just a piece of wire with no resistance. This immediately causes the current in the rotor windings to go up considerably. In order to compensate for this current increase the thickness of the wires in the rotor windings must be increased accordingly. In addition to the wire thickness increase, we observe that in such a case, then, there is no need for a set of slip rings, since the external resistors do not exist anymore and, thus, the wire loops can be closed inside the rotor rather than outside the rotor. In fact, this latter reality has been the basis of the squirrel-cage machines in which the extra cost of the slip rings and brushes have been eliminated.

The windings of a squirrel-cage machine are a number of thick bars of copper or aluminum (in order to carry very high currents) that are shorted together (forming parallel components) at both ends. Since the metal bars must go around the rotor in a circle, with the end rings connecting them together they form a cage, as shown below; thus, the name "squirrel cage."

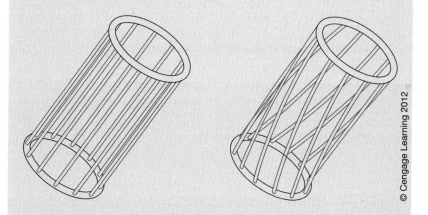

© Cengage Learning 2012

This is the simplest form of the cage. The bars can be slanted (like a twisted cage), or they can have a different cross section rather than being circular. These are the variations of models of squirrel-cage machines for various reasons or purposes. The aluminum or copper cage is embedded in a ferrous material that is laminated (like the metals in a transformer) in order to prevent or reduce eddy currents, thus reducing heat loss. The cage, therefore, is not empty and there is no place for a squirrel. The following figure shows a picture of a squirrel-cage motor, illustrating a typical squirrel-cage rotor.

(continued)

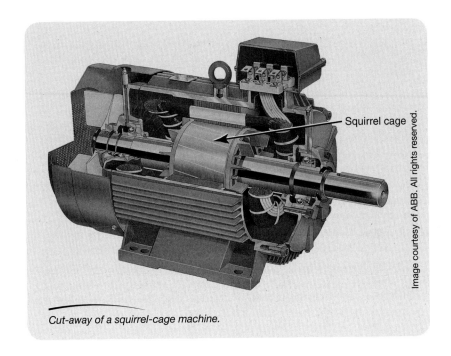

Squirrel cage

Cut-away of a squirrel-cage machine.

CHARACTERISTICS OF THREE-PHASE INDUCTION MACHINE

slip speed
The difference between the actual rotating speed of an induction motor or generator and the corresponding synchronous speed. Slip speed is not constant and can change based on the load on the machine shaft.

The difference between the speed of the magnetic field and the rotor speed is called **slip speed** and the ratio of this difference to the synchronous speed (the magnetic field speed) is called *slip*. The magnitude of slip is usually small, say below 3%. It can be determined from

$$s = \frac{N_s - N}{N_s} = 1 - \frac{N}{N_s}$$

where N_s is the synchronous speed and N is the rotor speed.

The magnitude of N (thus the slip) depends on two parameters, the resistance in the rotor circuit and the load on the rotor shaft. The figure below shows this variation for two different values of the rotor external resistance. The effect of the load on the shaft is represented in each curve by the resistive torque of the shaft; that is, the torque of the load that must be overcome by the motor in order to continue rotation.

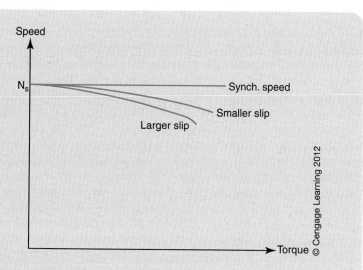

The above figure shows only the useful operating region of the entire curve for an induction motor. The following figure shows the whole curve for the machine functioning as a motor.

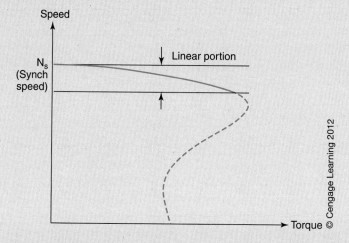

An examination of the preceding figure and comparison with figure 11.2 shows that the two figures represent the same thing; only the axes are swapped. This is a part of the whole graph without showing the rest corresponding to generation. Moreover, in figure 11.2 our concern is more on the generator portion of the graph, which is the continuation of the same curve shown above on the other side of the speed axis.

If instead of letting the machine rotate as a motor, the rotor is given energy so that its speed increases and eventually exceeds the synchronous speed, then it behaves as an induction generator. The energy given

(continued)

to the rotor will be converted to electric energy, which can be taken from the stator winding. If the machine rotor is turned exactly at the synchronous speed, no electricity is generated. The rotor must be revolved with a speed higher than the synchronous speed, so that it functions as a generator.

As a generator, the speed-torque characteristic of the machine is based on the same graph as the motor. In fact the two parts of the graph (as a motor or a generator) are mirror images of each other. The performance of the machine is affected by the magnitude of the resistance of the rotor winding circuit (including the external resistance). This is illustrated in the following figure.

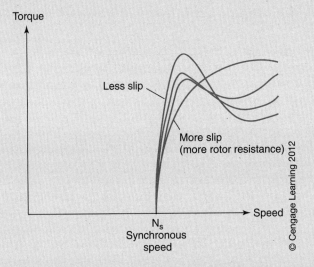

The effect of changing the value of the external resistance is to change the slope of the linear part of the speed-torque curve. This implies a larger range for the speed variation as the torque changes. It may sound like a desirable feature to be able to change the speed of the machine when required. Nevertheless, increasing this resistance implies throwing away useful energy in the form of heat.

11.4   Modes of operation of wind turbines

The way a wind turbine is connected to a grid depends on the type of the generator employed and the design specifications of the turbine. This determines what other equipment is necessary and what control strategy may be utilized. We call the various ways practiced by wind turbine manufacturers the modes of operation.

Modes of operation fall into two categories in the first place: those that use synchronous generators, and those that use asynchronous generators. In the first category the modes are: **direct drive mode,** fixed speed mode, and variable speed mode. In the second category, there are also three different modes that wind turbine generators are put to work. In one of them, a squirrel-cage generator is used, and in the other two, the generator is a wound-rotor induction machine. Thus, we can refer to these modes as (a) squirrel-cage generator mode, (b) **variable slip mode,** and (c) variable slip with doubly fed induction generator mode.

Since the technology is still developing, new designs can be expected.

direct drive mode
Connecting a turbine directly to a generator, without a gearbox in between.

Direct drive mode

As the name implies, in this mode of operation the turbine and generator are directly connected to each other, and there is no gearbox between them. This is shown in **Figure 11.4.** A three-phase transformer stands between the generator and the grid, to increase the voltage from the generator to the grid voltage. A difficult condition to be satisfied here in such a mode is the necessity for the turbine and the generator speed to be the same. Such a requirement implies that the design speed of a turbine be high and a generator with a large number of poles be used. As we can recall, a generator with a large number of poles (similar to those used in water turbines) has a relatively large diameter. Also, operating a turbine at a high speed creates a lot of noise, which is not desirable.

New designs try to overcome the problem of speed match by other possible means such as hydraulic converters, or using permanent magnet generators. Employing permanent magnets eliminates the need for higher speeds of a generator in order to provide the rotor magnetic field. However, the frequency of the generated AC electricity is low and, therefore, a frequency converter (see section 11.4.3) must be employed to raise the frequency to that of the grid. Most recently Siemens introduced a permanent magnet (synchronous) generator for wind turbines used in variable speed mode (see section 11.4.3), but connected directly to the turbine rotor. This generator has other design features such as placing the permanent magnets on the outside of the stator winding. **Figure 11.5** shows this new design in a 3-MW wind turbine.

11.4.1

variable slip mode
A way of operation control of wind turbines with an induction generator for connecting to a grid. In this mode, a wound-rotor induction generator is to be used and some power is extracted from the rotor winding. The turbine must have an almost constant rpm.

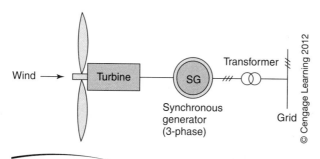

© Cengage Learning 2012

FIGURE 11.4 *Direct drive wind turbine.*

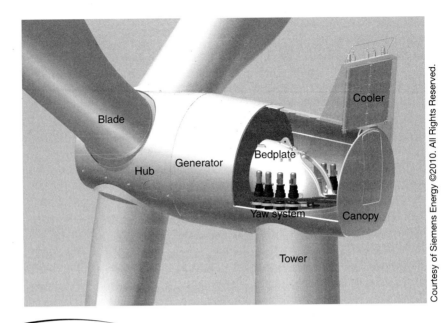

FIGURE 11.5 *Turbine with permanent magnet generator.*

11.4.2 Fixed speed mode

fixed speed mode
A way of operating
a wind turbine
where the turbine
speed must be kept
constant.

In **fixed speed mode** a synchronous generator is connected to the turbine by means of a gearbox (**Figure 11.6**). The gearbox increases the slow speed of a wind turbine to a high speed, matching that of the generator. In this sense, a smaller size generator can be used compared to the direct drive, but still the necessity of running the turbine at a fixed speed is there. This mode was practiced before the induction generators found their place in wind power generation.

Since there can be a considerable difference in the amount of power at high wind speeds compared to low wind speeds, in some turbines two generators were involved, a small generator for low wind speed and a large generator for

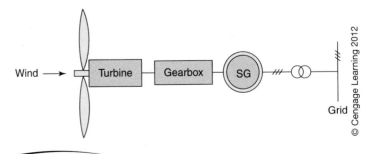

FIGURE 11.6 *Fixed speed mode of operation.*

high wind speeds. A switching mechanism was required to disengage one and bring in the other at a certain wind speed. This arrangement is no longer utilized presently in today's more sophisticated and more powerful wind turbines.

Variable speed mode

In order to capture wind power at a wider range of wind speeds the **variable speed mode** is utilized by some turbine manufacturers. This method relaxes the tight condition of turbine speed control. Variable speed turbines use electronic converters to change the generated AC electricity (which can be at any frequency, depending on the wind speed) to DC and then convert the DC back to three-phase AC. With such a design there is no restriction on the speed at which a turbine runs, except that it should not overload the generator and the electronic components.

Converting AC to DC is performed by a rectifier, and converting DC to AC is carried out by an inverter. The two of them together, as utilized in the variable speed mode, is sometimes called a **frequency converter,** since the initial AC is converted to a second alternating-current electricity with a different frequency. This arrangement is depicted in **Figure 11.7.**

One of the disadvantages of the variable speed turbine is the initial cost of the electronic devices involved, since these devices must be able to handle the full power of the turbine. This cost can constitute a considerable percent of the total cost, and can be prohibitive.

Squirrel-cage generator mode

Squirrel-cage machines are the most economical type of alternating current machines. Also, they are less expensive to operate. Before doubly fed induction generators found their place in wind turbines, squirrel-cage generators were employed in most of the modern wind turbines. Even now, because of the lower cost, a great number of wind turbines utilize squirrel-cage generators.

As mentioned earlier, a squirrel-cage machine should operate in the linear portion of its speed-torque characteristic curve. This governs the range of speeds that a turbine can operate. By measuring the wind speed, a turbine controller takes care of the necessary action for engaging a turbine to production

11.4.3

variable speed mode
A way of operation control of a wind turbine for connection to a grid. In this mode, a synchronous generator is used, but it is not necessary to run the turbine with a constant rpm, as opposed to fixed speed mode.

frequency converter
A device that takes AC electricity with a given frequency as input and delivers AC electricity with a (different) desired frequency.

11.4.4

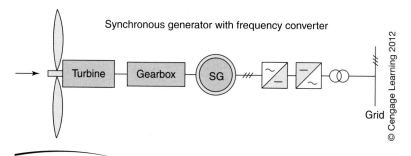

FIGURE 11.7 *Variable speed mode.*

or putting it on free wheel, and regulating the power output in order not to overload the generator.

The schematic of a wind turbine utilizing a squirrel-cage generator is the same as shown in Figure 11.6; only the type of turbine is different.

11.4.5   Variable slip mode

From the discussion of the induction machines in this chapter it can be seen that if the slip of a machine is altered by introducing external resistance in series with the rotor winding it is possible to better adapt the operating point of an induction generator to the maximum point of the turbine characteristic curve. This is illustrated in **Figure 11.8.** This method has been implemented in some turbines in order to have the turbine operate more efficiently and near the maximum point of its characteristic curve. In figure 11.8, the straight vertical line represents the case if a synchronous generator was employed. The slanted lines represent the characteristic curves of an induction generator with five different resistors in the rotor winding. The outermost curve to the right (with the smallest slope) corresponds to the highest resistance in the rotor circuit, or the least rotor current. As we can see, for higher wind speeds, this curve intersects the turbine characteristic curves nearer to the maximum values as compared with the curves with higher slope. Thus, the turbine has a better efficiency.

Instead of a variable resistance in the rotor circuit it is possible to utilize the rotor voltage to power a load, or even to increase its voltage to the grid voltage level and inject it into the grid. Obviously the cost of the hardware to carry out this job is going to be an additional expense to the turbine cost.

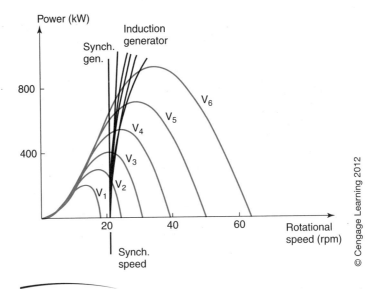

© Cengage Learning 2012

FIGURE 11.8 *Variable slip mode of turbine operation (with wound rotor induction generator).*

Variable slip with doubly fed induction generator   11.4.6

We have already had a discussion of how the doubly fed induction generators work (see section 11.3.2). Suppose that a squirrel-cage induction generator can work at wind speeds between 4 m/sec and 20 m/sec. This implies that the cut-in speed is 4 m/sec and the turbine does not have any production below this speed, and the cut-out speed is 20 m/sec and the turbine must stop if the wind speed exceeds this. Now, if a same size induction generator with a wound rotor is used instead, because the rotor winding is accessible, we can connect the rotor winding to external circuits. The squirrel-cage generator had to be stopped at a 20 m/sec wind speed because the higher speed wind would cause the generator to overload or the speed would go beyond the limit, increasing the frequency. With the wound-rotor generator, part of the available energy is extracted from the rotor winding, and at a 20 m/sec wind speed the generator has not yet reached its maximum capacity; it can continue to generate electricity, say, up to 23 m/sec wind speed.

In addition to that, at lower-end wind speeds instead of stopping production at a 4 m/sec wind speed, this generator is still capable of continuing to work until wind speed drops to, say, 3 m/sec; thus, grasping more wind power within a wind regime.

In a doubly fed induction generator, a portion of the generated electricity is from the stator. This portion can be up to about 70% of the generator power rating. The rest, reaching a maximum of 30% of the generator power rating, comes from the rotor winding. The portion generated by the stator is formed by the same principle as the squirrel-cage generator and the performance of the generator and its circuitry are similar to those of a squirrel-cage generator. On the other hand, the frequency and the voltage of the portion of electricity from the rotor are not compatible with those of the grid. In order to feed this portion into the grid both the frequency and the voltage must be changed. This is done in a similar manner to what was discussed for the variable speed mode of operation. The alternating current electricity from the rotor is first converted to DC with a rectifier and, in turn, the DC voltage is converted back to AC with the required frequency and voltage.

Figure 11.9 shows the schematic for the setup in a doubly fed induction generator. As we can see, the electricity from the stator is connected directly to the grid, whereas the electricity from the rotor must pass through the back-to-back converters for AC-to-DC-to-AC conversion. The back-to-back converter works both ways. It consists of two programmable converters that can convert AC to DC or DC to AC, based on the requirement, according to the wind speed. The converters are connected together by a capacitor, which performs as a DC source. The voltage of this capacitor is kept at a constant value by one of the converters performing as a source while the other converter acts as a load.

At higher wind speeds where the generator rotor rpm is higher than the synchronous speed (based on the grid frequency), the converter connected to the rotor winding acts as an AC-to-DC converter and supplies the DC source.

Wind turbine with doubly fed induction generator

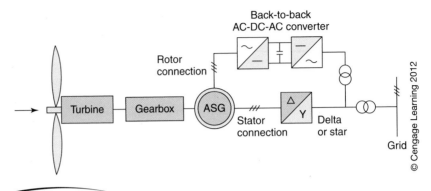

FIGURE 11.9 *Doubly fed induction generator schematic.*

At lower wind speeds, when the generator rotor rpm drops, the actions of the two converters are switched. An alternating current with appropriate frequency is generated by the converter attached to the rotor winding and is injected to the rotor windings. In this case, the converter attached to the rotor winding becomes a consumer (a load) and the other converter becomes a source.

The advantage of this mode over the variable speed mode, described earlier, lies in the fact that the cost of the electronic components in this mode is much lower than in the variable speed mode, since they must be rated for only 30% of the power. This is a remarkable advantage of doubly fed induction generators used in wind turbines compared to variable speed mode using synchronous generators.

We should know, however, that a doubly fed induction generator has slip-rings for connection to the rotor winding. The brushes that make the connection (slide on the slip rings) have to be changed after some time of being in service. In this respect, a doubly fed induction generator needs more maintenance than a squirrel-cage generator.

11.5   Control schemes of wind turbines

In chapter 10 we learned about control of wind turbines from the mechanical side and the role of mechanical parameters. In this chapter, we learned about the electrical side of a wind turbine and the electrical parameters that come into play. Since a wind turbine is an electromechanical system, a control must involve all the parameters, especially because they are very much interrelated through the amount of power that is grasped from wind in mechanical form and is delivered in the form of electrical energy. Figure 10.4 in the previous chapter shows a general control block diagram. In this figure there are eight principal parameters for control purpose that are to be measured.

The variables in a wind turbine and the common symbols to denote them are turbine rotor speed (ω_t); turbine rotor torque (T_t; a control system might use the rotor torque); mechanical power (P_t, a product of ω_t and T_t); blade pitch angle (θ); grid frequency in hertz (f); stator frequency in rpm (ω_s), corresponding to f; generator rotor rpm (ω_r, corresponding to turbine rotor speed ω_t); stator voltage (V_s, must be the same as grid voltage); and stator current (I_s).

In addition to the preceding, if a doubly fed induction generator is used, other variables that need to be measured are rotor winding voltage (V_r) and current (I_r); rotor current frequency (f_r); phase angle between V_s and I_s (φ); generator active power (P); and reactive power (Q).

Other parameters such as the gearbox ratio, the blade twist angle, and so on are part of the design and are fixed. Of the above variables, the stator voltage and frequency must be equal to those of the grid. That is, f (and thus, ω_s) and V_s must be kept constant.

All the generator output electrical variables must be controlled based on the available power from wind, as well as any requirement for phase angle φ. These are P, Q, and I_s.

Figure 11.10 illustrates a general schematic for control of a wind turbine in which the reactive power can be controlled. Control of the reactive power

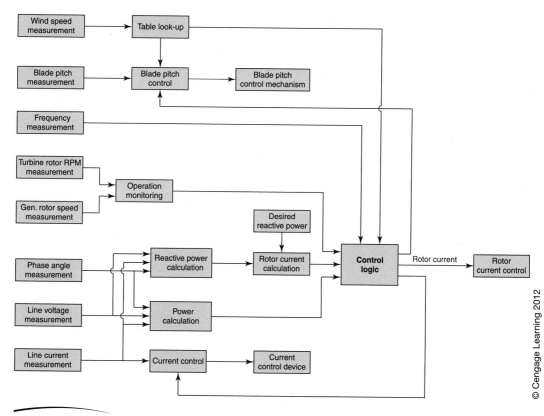

FIGURE 11.10 *Schematic of a general control system for a wind turbine with reactive power control.*

corresponds to altering the magnetic field of the rotor by having an appropriate current in the rotor winding (If it is not a permanent magnet in the case of a synchronous generator, in which no control action on the reactive power can be made).

11.6   Chapter summary

- Direct current generators are used only in small isolated turbines.
- The electricity from a turbine with a direct current generator can be stored in batteries.
- Industrial turbines always use three-phase alternating current generators.
- There are mainly two types of three-phase AC generators: synchronous generators and asynchronous generators.
- An asynchronous generator is also called an induction generator.
- A motor and a generator have the same structure.
- Synchronous machines must operate at the synchronous speed.
- Induction machines do not run at synchronous speed; motors run at a slightly lower speed than the synchronous speed, and generators must run at a slightly higher speed. This is dictated by the characteristic curve of the machine.
- A generator can be connected directly to a turbine (without a gearbox). This is called direct drive.
- In most cases there is a gearbox between a turbine and its generator to increase the low speed of turbine shaft to the generator required speed.
- There are many modes in which turbines to be connected to a grid can be operated.
- For a synchronous generator, we have fixed speed mode and variable speed mode.
- For an induction generator, we have squirrel-cage generators, which can be run in only one mode of operation. We also have wound-rotor induction generators. These have a rotor with windings that can be connected to an outside circuit with the help of slip rings and brushes.
- Wound-rotor induction generators can be run in variable slip mode. There is a simpler variable slip mode in which the slip of such a generator can be varied. There is also a more sophisticated mode used by many of today's modern turbines. Turbines employing this mode are simply called turbines with doubly fed induction generators.

ADVANCED LEARNING

More on Doubly Fed Induction Generators

One of the major advantages of a wound-rotor induction generator over a squirrel-cage machine is the fact that in addition to the main concerns about the voltage, frequency, and power, it is possible to independently control the two components (active and reactive) of power in such a generator. This independent control of active and reactive power is a property of synchronous generators.

When a squirrel-cage induction generator, which is the simplest generator and requires the simplest control, is connected to a grid it introduces its reactive power as well as the true (active) power that it generates. This is due to the windings of its stator that have inductive reactance. The amount of reactive power increases (but is not proportional) to the generated power. In order to reduce the reactive power, banks of three-phase capacitors are added in parallel with the generator. But, since the power generation is not constant, the values of the required capacitors must be determined based on the average power production. Power factor correction (reduction of the reactive power) at all power values requires an array of capacitors. The number of capacitors that must be inserted into or taken out of the circuit is determined based on the amount of power produced at each instant. This must be done automatically and, therefore, the necessary equipment becomes rather expensive.

If a DFIG generator is employed, it is possible to control the produced power in accordance with the required quantity of active (true) power and the amount of reactive power. These two power values are independently controlled. In other words, in a more sophisticated control loop the set values for active power and reactive power can be independently assigned. This is not possible with a squirrel-cage generator.

Since control of the active power is possible with a synchronous generator (by changing the rotor magnetic field through excitation current), we can say that a doubly fed induction generator has the advantage of working as a synchronous generator, while it does not have to be run at a constant speed. The latter is what makes a synchronous generator not so suitable, or expensive (as we have learned) for wind turbines.

Control schemes for wind turbines equipped with DFIGs can be based on power control, torque control, current control, or a combination of these. An approach commonly referenced as *vector control* is based on power control where the active and reactive powers are considered as two components of a vector to be controlled. Theoretical discussion of these methods is outside the scope of this book, but the interested reader can find numerous publications on the theory of the subject in the literature.

Review questions

1. Why are direct current generators not good for grid connection?
2. What are the advantages and disadvantages of direct current generators?
3. Which are the main types of alternating current generators?
4. For commercial production, why are only three-phase generators used?
5. What happens if a synchronous generator connected to a grid is not rotating at synchronous speed?
6. What are the three main parameters that have to be considered when a turbine must be connected to a grid?
7. What are the main differences between a synchronous generator and an induction generator?
8. Name the two types of induction generators.
9. What is the similarity between a motor and a generator?
10. Why can a motor and a generator commonly be referred to as "a machine"?
11. In what ways can a synchronous generator be used in a wind turbine?
12. In what ways can an induction generator be used in a wind turbine?
13. What is slip and what is slip speed?
14. Name the modes of operation of a wind turbine, as far as its generator is concerned.
15. Explain variable speed mode of operation in wind turbines.
16. What is variable slip mode?
17. Can a synchronous generator be used in variable slip mode?
18. What is specific about a wound-rotor induction generator?
19. What does DFIG stand for?
20. What are the advantages of a turbine using DFIG?

Problems

1. A three-phase generator has a total of 48 poles (24 pairs). What is the rpm at which it must be run to generate 60-Hz electricity?
2. A three-phase generator runs at 1200 rpm and its electricity has 60 Hz. How many pairs of poles does this generator have?
3. What is the number of poles per phase in the generator in problem 2?
4. If a 60-Hz generator runs at 912 rpm, is it a synchronous or an asynchronous generator? Why?
5. If a generator runs at 750 rpm, can you say what the frequency of its-electricity is?
6. Can a three-phase generator running at 900 rpm be put in parallel with a generator running at 1800 rpm?
7. If the answer to question 6 is yes, what conditions are required?
8. If your answer to question 6 is no, explain why.
9. A three-phase machine is generating at 1450 rpm. Describe what type of machine it is and in what mode (motor or generator) it is working.
10. Determine if you want to directly (without a gearbox) run a three-phase generator from a wind turbine rotating at 18 rpm, how many pairs of poles it needs to have.

Ancillary Electric Devices

KEY TERMS

anemometer
autotransformer
bridge rectifier
full-wave rectifier
half-wave rectifier
ideal transformer
isolation transformer
pad mount transformer
pole mount transformer
ripple
sinusoidal waveform
square-wave
step-down transformer
step-up transformer
thyristor
turns ratio
wind vane

OBJECTIVES

After studying this chapter, you should be able to:

- Describe the principles of operation of all transformers.
- Define the terms step-up and step-down.
- Select a transformer power rating based on the loads to be powered.
- Realize the practical concerns when using transformers.
- Explain the differences between single-phase and three-phase transformers.
- Understand and exercise using the terms employed for transformers, such as pad mount, pole mount, step-up, etc.
- Calculate the efficiency of a transformer based on its daily usage data.
- Learn more about how a back-to-back converter in a variable speed turbine works.
- Define the principle of operation of a rectifier.
- Realize various rectifiers and sketch the schematics of each one.
- Understand the basics of operation of an inverter.
- Make a bridge rectifier.

12.0   Introduction

Unless a very small turbine and small-scale application is concerned, various ancillary devices are needed for a wind turbine to function or for transmission of the generated power. As we have learned so far in previous chapters, the generator of a wind turbine converts mechanical energy of the turbine to electrical energy. In chapter 11, we also learned about various modes for wind turbine operation. These modes, in fact, are the ways that the generator of a turbine can be used.

In this chapter, we consider the structure and the functioning of some of the more fundamental electrical devices that are used in the operation of a wind turbine. These are the transformer, the rectifier, and the inverter.

One of the most important devices in generation and transmission of electricity is a transformer. It is used not only in wind turbines, but in all electric power production and usage. One can clearly say that without transformers it is impossible to benefit from electric power, or at most the possible applications will become limited to only a small fraction of what we are benefiting from as electricity.

A rectifier is a device that converts AC electricity to DC electricity. In other words, it is an AC-to-DC converter. On a large scale, it is used in some wind turbines; but on a small scale, it is used in the majority of devices that contain electronic circuits. That is, all the devices that need to work with a battery can also be connected to AC electricity (through what is commonly called an "adaptor").

The function of an inverter is the opposite of that of the rectifier. In this sense, an inverter is a DC-to-AC converter. The common applications of an inverter are relatively few, compared to a rectifier; however, in some of the wind turbine modes of operation, a DC-to-AC converter is a necessary part of the system. Also today, more and more industrial applications, such as motor drives, employ an inverter as part of the drive for the application.

12.1   Transformers

A transformer is a vital part of alternating power generation. Its main function is to step up (increase) or step down (decrease) the voltage of the whole or part of an AC circuit. However, in some particular applications a transformer's role can be something else.

In this section, we are going to learn the principle of operation of a transformer. For a complete and thorough knowledge of transformers, a deeper study to the various parameters and their detailed relationships is necessary, which is beyond the scope of this book.

As with AC electricity, we have single-phase or three-phase transformers. A three-phase transformer, in fact, consists of three single-phase transformers that sometimes are packed or put together for better efficiency or to save space, depending on the case. In this sense, everything that can be said for a single-phase transformer also applies to a three-phase transformer. The next discussions refer to a single transformer, but the term "single" or "single phase" is omitted without loss of generality.

In the simplest form, a transformer is made up of two windings that share a common core. The core can be of various shapes, but the shape does not play a significant role. The most common shapes are rectangular and circular. From a manufacturing viewpoint, a rectangular shape is easier to work with. **Figure 12.1** shows a transformer with a rectangular core. In this configuration, the two windings are on the opposite sides of the core. This is not necessary, and both can be on the same wing. In fact, most of the small transformers have a core as shown in **Figure 12.2,** and both windings are placed side by side on the middle wing.

The two windings are called "primary" and "secondary," depending on their role. The primary winding is connected to the supply voltage, the voltage we need to change. The secondary winding is the output, which provides the required voltage. If the primary winding has more turns than the secondary winding, the transformer decreases the voltage and it is a **step-down transformer.** If the primary winding has fewer turns than the secondary winding, the output voltage is higher than the input voltage (a **step-up transformer**).

step-down transformer
A transformer used to decrease voltage.

step-up transformer
A transformer used to increase voltage.

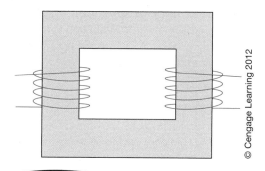

FIGURE 12.1 *Basic structure of a transformer.*

© Cengage Learning 2012

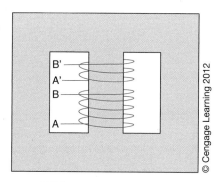

FIGURE 12.2 *Most transformers have this structure.*

© Cengage Learning 2012

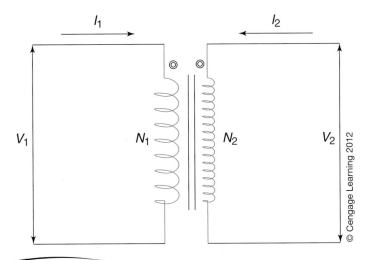

FIGURE 12.3 *Drawing symbol for a transformer; also showing the number of turns, voltage, and current in the primary and secondary circuits.*

One thing that is important in the transformer winding (in addition to the number of turns that we see shortly) is the direction of the turns. In figure 12.2, for instance, both windings have the same direction. This defines the polarity of the secondary voltage with respect to the primary voltage. That is, for this transformer if at one instant the voltage at A is positive with respect to B, the voltage at A′ is also more positive than the voltage at B′. The two windings have the same polarity. To show the polarity in the drawings of transformer circuits, a dot is used at the correct side of the transformer schematics. **Figure 12.3** illustrates the symbol for a transformer, also showing the dots representing the polarity. This figure indicates the number of turns N, the voltage V, and the current I in the primary and secondary circuits, as well, denoted by subscripts 1 and 2, respectively.

Figure 12.4 shows a very small transformer like those used in a radio or TV set. A very large industrial transformer is shown in **Figure 12.5**. While the two transformers may have the same fundamental structure (except that the latter is three-phase), the large transformer in figure 12.5 is inside an oil-filled container and has a radiator for the oil to get cooled by air.

12.1.1   Voltage relationship

Figure 12.3 represents a transformer at work. The primary winding is connected to a supply voltage V_1, and the secondary voltage V_2 is applied to a load. Based on the nature of the load (resistive, inductive, and capacitive combination), a current I_2 will flow in the secondary winding. Based on this current, the primary winding carries a current. The latter will be a (major) part of the total current in the primary winding. Consider that if the secondary winding is not

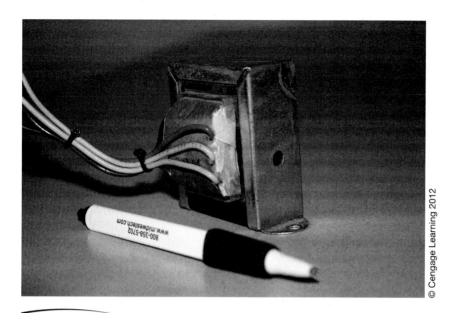

FIGURE 12.4 *A very small transformer.*

FIGURE 12.5 *A large (750-kVA) transformer.*

connected to a load, the primary winding still has a closed loop, and as a result of the resistance of the primary winding, a (small) current flows in the primary winding. Also, because of the iron core, there is always some small power consumption in a transformer that reflects in the primary winding current. Thus, in a transformer, the primary current I_1 consists of three parts, one of them due to the load connected to the secondary and the other two due to the structure (windings and core) of the transformer.

For basic relationships and simple calculations the two latter parts of the primary winding current are ignored, as if they do not exist. Such a transformer without these losses is called an **ideal transformer.** In fact, the contribution of the two ignored components of the primary current is marginal and except for large transformers and for precise calculations they can be ignored in most cases. An ideal transformer, thus, is based on two assumptions:

ideal transformer
A transformer for which all the losses are assumed to be zero and, thus, the efficiency is 100%.

a. The relationship between primary and secondary voltages is based on the ratio of the number of turns; and

b. The powers in the primary and secondary windings are equal (no losses).

The following equation is the fundamental relationship for (ideal) transformers:

$$\frac{V_1}{V_2} = \frac{N_1}{N_2} = a \qquad (12.1)$$

Here, V_1 and V_2 are the primary and secondary voltages, respectively, and N_1 and N_2 are the number of turns in the primary and secondary windings, respectively. The parameter a is called the **turns ratio.** For most transformers, the turns ratio is fixed and cannot change. However, multioutput transformers have more than one secondary winding and **autotransformers** have a variable turns ratio (see section 12.1.5).

turns ratio
The ratio between the number of turns in the primary and secondary windings in a transformer.

autotransformer
A transformer where the primary and secondary windings are parts of the same winding. The primary and secondary windings are not electrically isolated.

Equation (12.1) shows that a transformer with a given turns ratio, say, 10, can theoretically multiply the primary voltage by 10 or divide it by 10 at the secondary, depending on which winding is connected to the source. This implies an important point that in working with a transformer, special care must be taken for correct connection to the outside sources and loads. A wrong connection can easily lead to damage and injuries.

> In working with a transformer, special care must be taken for correct connection to the outside circuits.

For instance, consider a transformer with the turns ratio of 10 to convert 120 V to 12 V. If the secondary winding (with a lower number of turns) is used as the primary and is connected to 120 V, then a voltage of 1200 V will exist between the two terminals of the other winding with 10 times more turns. This is definitely dangerous for any person who may touch the terminals expecting only 12 V. It also immediately damages any device connected to the transformer.

Take note that although, as said, theoretically one can use a transformer for stepping up or stepping down a voltage, in practice design considerations come into effect and a transformer not necessarily can be connected to any arbitrary voltage or in an arbitrary fashion. We'll see more on this in section 12.1.2.

In a step-up transformer, the secondary voltage is higher than the primary voltage. In a step-down transformer it is the reverse.

Current relationship

The relationship between the primary and secondary currents for an ideal transformer is based on the power relationship; that is, the power in the primary side is equal to the power in the secondary side. The power here refers to the apparent power, that is, the product of voltage and current. Setting $P = V_1 I_1 = V_2 I_2$ leads to

$$\frac{V_1}{V_2} = \frac{I_2}{I_1} \quad or \quad \frac{I_2}{I_1} = \frac{N_1}{N_2} \tag{12.2}$$

This equation also implies that the higher current corresponds to the lower voltage and vice versa. The current I_2 can always be found if the secondary voltage and the load impedance are known. The current I_1 found in this way represents the current in the primary due to the load only. That is to say, if there is no load connected to the secondary side, this current will also be zero. So, it does not include the no-load current and the core-loss currents (due to eddy currents in the core), as was mentioned earlier.

In construction of any transformer, the size, weight, and price versus power are very important. In this respect, the wire sizes of the secondary and primary windings do not need to be the same. Always, the current in the winding with the lower number of turns is higher than the current in the other winding [this follows from equation (12.2)]. Thus, the lower voltage (higher current) side has always a thicker wire than the higher voltage (lower current) side.

For a step-down transformer, the secondary current is higher and the winding wire is thicker than that of the primary.

Transformer efficiency

The efficiency of any device or a system is the ratio of the output power to the input power; that is, the ratio of the power given to the system (or consumed by the system) and the power taken from the system. For a gearbox and a transformer, the input and output powers are of the same form, but for a motor and a generator the two powers are of different forms. Based on the definition, for an ideal transformer, the secondary power is the same as the primary power; that is, the output and input powers are equal. This implies an efficiency of 100%.

For a real transformer, however, although the efficiency is quite good and can reach up to over 98%, it is not (and cannot be) 100%. For a large transformer in service for 24 hr per day, even that 2% shortage from 100% can translate to a sizable amount of money in the long run, and cannot be ignored. For this reason, whenever necessary, the efficiency of a transformer must be determined. According to what we have discussed, the efficiency of a transformer can be defined as

$$\text{Power efficiency} = \frac{\text{Output power}}{\text{Input power}}$$

$$= \frac{\text{Output power}}{\text{Output power} + I^2 R \text{ loss} + \text{core loss}}$$

(12.3)

In equation (12.3), output power is the amount of power taken from the secondary circuit. The term I^2R denotes the power consumed in the windings of a transformer and is called "copper loss," since the wires are normally made out of copper. It is the amount of power that changes to heat in both the primary and secondary windings and is lost. The reason why the copper loss is written in this way is to indicate that it is load dependent; as the current goes up, this loss also goes up. This is contrary to the core loss, which is independent of the current. Core loss is due to local flow of electrons in the core, called eddy currents. In order to decrease the eddy current (and the core loss), the core of all transformers are laminated metal sheets, instead of being solid metal.

A more practical version of equation (12.3) is

$$\text{Energy efficiency} = \frac{\text{Output energy for a given period}}{\text{Input energy for the same period}}$$

(12.4)

since it directly measures the consumed energy and the output energy, instead of dealing with power components that cannot be easily measured. Usually a period of 24 hr is used, and the term "all-day efficiency" is employed:

$$\text{All-day efficiency} = \frac{\text{Output energy for 24 hr}}{\text{Input energy for 24 hr}}$$

(12.5)

12.1.4   Transformer power rating

A transformer is usually designed for a specific purpose with a specific power capacity and a frequency range. This determines many of the mechanical and electrical parameters involved, such as core dimensions, number of turns, current, wire size, and so on, as well as the type of cooling. A transformer can be air cooled or oil cooled. If oil cooled, it might need a radiator and other components.

Industrial transformers are rated in volt-amperes (VA). That represents the maximum capacity of a transformer. Like most machinery, a transformer has the highest efficiency if it is used at the rating for which it is designed. Although it is possible to use a higher rated transformer at lower powers, the core losses make it less efficient at lower powers.

Other useful information about transformers   12.1.5

It is possible to have multiple secondary windings on the same transformer. This provides various voltages from one source voltage, thus reducing the cost and space. This is very common in electronic devices, such as a television, that need various voltages for their operation, thus, in very small transformers. Unless it is necessary to have separate (isolated) windings, this purpose can be achieved by having various taps on the same winding; that is, there is only one winding, but multiple outputs provide different turns ratios. In a transmission line transformer, it is not normally necessary to have multiple voltages. **Figure 12.6** shows the schematics of a multioutput transformer.

When a voltage must be only slightly changed, an autotransformer can be used. For example, normally toward the end of a distribution line there is a noticeable voltage drop. In order to compensate for this voltage drop, the line voltage must be raised to its nominal value. As an example, consider the houses at the end of a street where the line voltage must be 120 V. If the voltage has dropped to 105 V, then it is necessary to increase the voltage by 15 V to bring it to the nominal level. In such a case, an autotransformer is used.

In an autotransformer, the primary and secondary windings are the same. The winding is tapped at the proper point for the desired voltage ratio. An autotransformer can be used for both stepping up and stepping down the voltage. **Figure 12.7** shows the concept.

Small autotransformers can often be used when a variable voltage is required. This is obtained by a slider that connects to different points on the winding for the secondary voltage. In larger transformers, this "tapping" can be done by the tap changer connecting to different positions of a winding.

Note that in an autotransformer the primary and secondary currents flow through the same winding. But the direction of the currents at each instant is such that the primary and secondary currents do not add together for the common part of the winding.

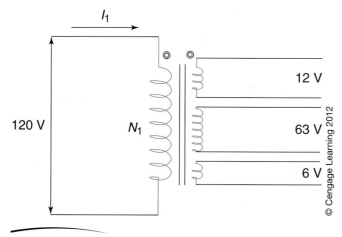

© Cengage Learning 2012

FIGURE 12.6 *Multioutput transformer.*

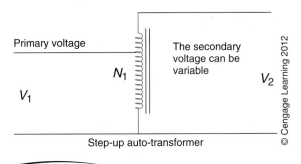

© Cengage Learning 2012

FIGURE 12.7 *Principle of operation of an autotransformer.*

Sometimes in electrical applications, it is necessary that a device be isolated from the rest of the circuit feeding it; that is, the device receives a voltage while not directly in parallel with other devices. In such a case, an **isolation transformer** is employed. In an isolation transformer, the number of turns of the primary and secondary windings is the same. As a result, the secondary voltage is the same as the primary voltage. Nevertheless, if because of a fault in the primary circuit the current suddenly jumps up, the device is not affected. The isolation transformer, thus, is used when a device must be protected from damage.

While small transformers to be used with a device (like a television) are mounted inside the device, large-size transformers are stand-alone devices and must be somehow supported in place. In this sense, transformers are divided into **pad mount** and **pole mount.** As the name implies, a pad mount transformer is mounted on a platform on the ground, whereas a pole mount transformer is more suitable for mounting on the top of distribution lines poles.

Each power-production wind turbine has a (three-phase) pad mount transformer that increases the generator voltage to the collector voltage (a collector is a switch station that connects all the turbines in a wind farm to a grid by a stepping-up transformer). Depending on the turbine, this transformer can be on the ground outside of the tower, or in many of the newer multi-megawatt turbines, it may be located in the turbine nacelle or at some point at the bottom, inside the tower.

The voltages in a generator can vary from 600 V to 4000 V. The collector voltage can be 11,000 V, or another standard low voltage (for transmission lines, 11,000 V is not considered a high voltage). A picture of a pad-mount transformer was shown in figure 12.5, and **Figure 12.8** shows three pole-mount transformers.

isolation transformer
A transformer the main purpose of which is not to change voltage, but to electrically isolate (separate) a device from the supply circuit, or isolate part of a circuit from the rest.

pad mount transformer
A transformer that can be installed on a flat base.

pole mount transformer
A transformer, normally with a round structure, more suitable for mounting on a distribution pole.

12.1.6   Three-phase transformers

For three-phase electricity transformation one can use three similar single-phase transformers. This is normally done for larger distribution transformers (such as those shown in figure 12.8) and for transmission transformers (transformers that are used for distribution and transmission lines). In smaller

FIGURE 12.8 *Pole-mount transformers.*

FIGURE 12.9 *Construction of a three-phase transformer.*

transformers, the three sets of windings are put together on a single core, as shown in **Figure 12.9.** The three primary windings are connected to the supply voltage as a three-phase load; so, they can have star or delta (Y or Δ) connection (star connection is also called wye connection). Independent of how the primary side is connected, the secondary windings can also be put together in either star or delta form. This implies that there are four ways that a three-phase transformer can be used in a circuit. The voltage ratio, thus, depends not only on the turn ratio of the windings, but also on what way the connections are (see example 2 in section 12.1.7).

12.1.7   Examples

EXAMPLE 1:

The efficiency of a 750-kVA transformer is 98%. If the average duty cycle of this transformer when in service for 24 hr, 7 days per week is 85%, find the amount of energy turned to heat each year. Do the calculation in both the metric and the English systems of measurement [1 joule (J) = 0.2388 cal; 1 J = 1 W-sec; 1 W = 3.412 Btu/hr]. If each kilowatt-hour of energy costs $0.04, determine the cost of the energy lost this way.

Solution: We first calculate the amount of electrical energy lost per year in kilowatt-hours.

Energy lost per year = (750)(0.02)(0.85)(365)(24) = 11,690 kW-hr
Cost of lost energy = (11,690)(0.04) = $4467.6
1 W = 3.412 Btu/hr → 1 W-hr = 3.412 Btu
11,690 kW-hr = (11,690,000)(3.412) = 39,886,280 Btu
11,690 kW-hr = (11,690,000)(3600) = 42,084,000,000 J = 42,084,000 kJ
42,084,000 kJ → (42,084,000)(0.2388) = 10,049,659 k.cal

EXAMPLE 2:

The turns ratio in a step up transformer is 2.65. The primary line voltage is 4160 V and the primary windings are connected in Δ. Determine the secondary line voltage (a) if the secondary windings are also connected in Δ and (b) if the secondary windings are connected in Y.

Solution:

a. Since the primary and secondary windings have the same type of connection, the secondary line voltage is multiplied by the turns ratio:

Secondary line voltage = (4160)(2.65) = 11,024 V

b. The voltage between the two ends of each secondary winding is 2.65 times that of the primary, that is, 11,024 V. Now, the line voltage is the voltage between each phase line. Thus, the line voltage for this configuration is

(11,024)(1.73) = 19,071 V

Figure 12.10 better clarifies this case.

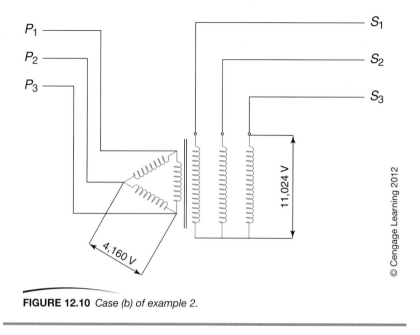

FIGURE 12.10 *Case (b) of example 2.*

ripple
The (undesirable) rapid fluctuations of voltage about the nominal value in a DC voltage

Rectifiers   12.2

Rectifiers convert AC electricity to DC. The major application of that with wind turbines, as can be recalled from chapter 11, is in the variable speed mode of operation and in wind turbines equipped with DFIG, where the AC electricity is converted to DC and back to AC with the appropriate frequency. At a smaller scale also, in very small stand-alone wind turbines, if an AC generator is employed, a rectifier can be used to convert it to DC if battery storage is a part of the setup. Nevertheless, rectifiers are used for any device that needs to be run with DC when it is powered by AC. This can be at the domestic level, such as computers and digital cameras, or at the industrial level, such as electroplating and electrolytic metal refining.

Both single-phase and three-phase AC can be converted to DC. In the simplest form, a rectifier consists of a number of diodes. The property of a diode is to let the electrical current flow in one direction by introducing very little resistance to the flow, and to block the current in the opposite direction by exhibiting a very large resistance. In this way, the alternating current changes to a DC but with a lot of fluctuations called **ripples**. Ripples are smoothed out by filters. The most common filters are capacitors in parallel with the loads in a rectified circuit.

The most common and widely used single-phase rectifier is the **bridge rectifier**. But **full-wave** and **half-wave rectifiers** can also be used, though the latter is made of only one diode and it has poor property. **Figure 12.11** illustrates the schematics of a bridge rectifier, which is made out of four diodes. The

bridge rectifier
A type of full rectifier where two sets of diodes alternatively conduct. The full voltage of the AC electricity is converted to DC.

full-wave rectifier
A type of AC-to-DC rectifier with two rectifier diodes that alternatively conduct. Each diode converts one cycle of AC to DC. Thus, all of the waveform is converted to DC.

half-wave rectifier
The simplest type of AC-to-DC rectifier, which consists of inserting one diode in the AC circuit.

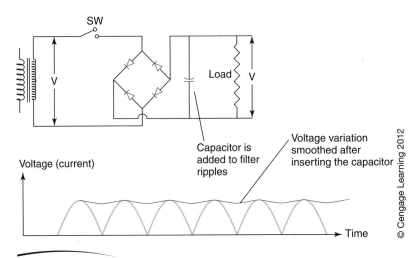

FIGURE 12.11 *Schematics of a bridge rectifier.*

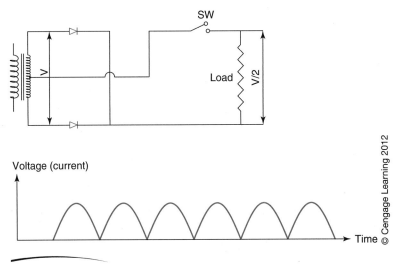

FIGURE 12.12 *Schematics of a full-wave rectifier.*

rectifier is connected to the secondary of a transformer, since this is the common case in the majority of applications. A capacitor is used to filter the ripples. The higher the capacitance of this capacitor, the better the filtering effect.

For comparison, the construction of a full-wave rectifier is shown in **Figure 12.12,** where the unfiltered output voltage is illustrated. In a full-wave rectifier the voltage obtained is only half of the supplied voltage. Also, it is required that the two half voltages be available in AC form (the center tap in the transformer). This is one of the drawbacks of a full-wave rectifier.

In practice, the power rating of the rectifier and the maximum voltage are to be considered for selection of a proper rectifier. As in DC, power is the product of voltage and current. Thus, for a particular application the rectifier diodes must withstand the applied voltage and the circuit current.

When more power is required, that is, in many industrial applications including wind turbines, a three-phase rectifier can be employed. In this way, the power is divided between the three phases, but a single DC source is obtained. **Figure 12.13** depicts the arrangement of a three-phase rectifier. The unfiltered ripples have a frequency 1/6 of the line frequency. In all the rectifiers the output voltage can be regulated by additional electronics in order to have a steadier DC source.

More powerful rectifiers are made of thyristors, instead of diodes. A **thyristor** behaves like a controlled diode (a diode with a switching action that turns it on or keeps it off) whose operation can be turned on as required. Thus, it needs additional electronic circuits for its control. The effect is that the output level (DC voltage) can be controlled. All the preceding circuits shown for rectifiers can utilize a thyristor instead of a diode; however, the cost must be justified for any application. Just for comparison, the principle of a three-phase rectifier employing thyristors is depicted in **Figure 12.14.** Note the symbol for a thyristor. The small circle denotes the thyristor gate, which acts as the control switch. As we can see, the basic construction is as the one shown in figure 12.13. Here a driver to fire (trigger the gate of) each thyristor at the appropriate instants of time is necessary (not shown). This firing is a time-based signal that causes each thyristor to conduct for a limited period of time in each cycle. The result is a truncated sinusoidal signal (compare the graphs in figures 12.13 and 12.14). The amount of truncation is based on the firing time (normally called firing angle, defined in degrees), which determines the desired voltage of the DC electricity. That is, based on the input AC voltage and frequency, the firing time (during the first quarter of one cycle of the sine-wave signal, for each phase)

thyristor
A type of transistor, used in controlled rectifiers and inverters.

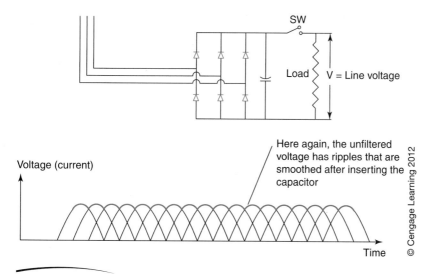

Here again, the unfiltered voltage has ripples that are smoothed after inserting the capacitor

© Cengage Learning 2012

FIGURE 12.13 *Schematics of a three-phase rectifier.*

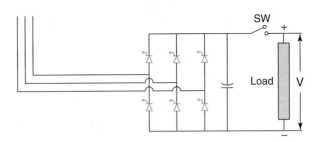

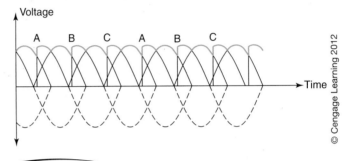

FIGURE 12.14 *Schematics of a three-phase rectifier based on thyristors.*

translates into a voltage level for DC output. The DC signal has ripples that must be filtered out.

The advantage of a thyristor and its application over a diode is higher levels of voltage and power, since it does not have a continuous current flow (it switches on and off), which can make a device hot and limited in operation. However, more discussion about a thyrsitor and the details of its operation requires a background in electronics, and is outside the scope of this book.

12.3   Inverters

An inverter is the opposite of a rectifier. It derives an AC electric supply with a desired frequency out of a DC source. Recall that in the back-to-back AC-DC-AC converter used in variable speed mode and in variable slip mode with a DFIG, we need to convert the DC electricity to AC electricity with the voltage and the frequency of the grid.

Obtaining AC from DC is possible only with power electronics devices that work based on a switching action, such as a thyristor. The required power for running the electronic circuit(s) is provided by the DC supply; then the direction of current in the device output is switched to provide an alternating current. In the simplest form, a **square-wave** signal is generated. A square-wave signal is shown in **Figure 12.15;** but this signal is good only for heating and lighting, and small low-power apparatuses. In large-scale applications and at

square-wave
A waveform each cycle of which consists of two equal rectangles, one on the positive side followed by the one on the negative side.

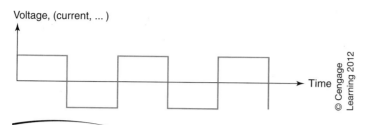

FIGURE 12.15 *Square-wave signal.*

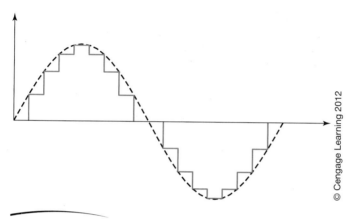

FIGURE 12.16 *Stepped waveform with a sinusoidal contour.*

high powers, for example, in wind turbines especially if to be connected to a grid, a **sinusoidal waveform** is required. A sinusoidal waveform implies that the variation of the AC voltage with time is according to a sine function.

Today, with the advancement of power electronics and with good filtering, it is possible to generate high-power and high-voltage signals that sufficiently resemble a sinusoidal form. This is accomplished with more steps of switching and other controls that initially generate an electric waveform, as shown in **Figure 12.16,** which will be smoothed out by filters in the circuit.

Obviously here again all the power as provided to the output circuitry must be supplied by the DC source. If the voltage of the DC source does not allow generation of the desired AC voltage, a transformer can be employed to step up the output voltage. For electronic components, the more the power and the higher the voltage, the higher is the price.

In order to control the frequency of the AC electricity to be as desired, normally a small (low-voltage) sinusoidal waveform with the desired frequency is generated. This is quite simple and not expensive, and is widely used in radios, TVs, oscilloscopes, and so on. Then the power electronic circuit is driven such that the switching actions follow the stepped waveform (figure 12.16) of this sinusoidal model. The cycle time determines the frequency. Firing (turning on) of the thyristors (or other transistors that can be used) and voltage variation takes place according to the time divisions within each cycle of the waveform, as in the stepped waveform.

sinusoidal waveform
Same as *sine wave*.

12.4   Anemometer and wind vane

A wind turbine cannot work without measuring the wind speed and direction. The devices to measure these two entities are an **anemometer** and a **wind vane,** respectively. These are small but very important devices for wind turbine operation.

Two types of anemometers are used. The most common anemometer is the cap anemometer, which consists of a small vertical shaft on which three cups in the form of hemispheres or a similar shape that can catch the wind are installed. With the blow of wind this device rotates and its speed of revolution depends on the wind speed. At the end of the shaft is a small DC generator that can generate a signal, the voltage of which varies. Its voltage can be calibrated for the wind speeds. This device is shown in **Figure 12.17.**

In cold regions where there is a potential for freezing rain, this type of anemometer needs to be heated in the winter. If it is not heated and it freezes, there is no reading for wind speed and, therefore, the measured wind speed reads "zero." The turbine stops generating electricity, although there is wind. In this way, a turbine can lose revenue. Any malfunction of the anemometer leads

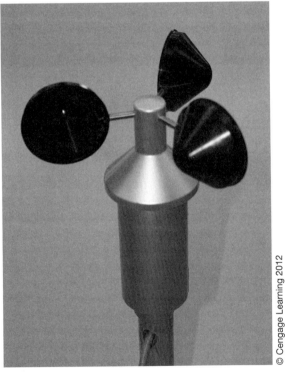

© Cengage Learning 2012

FIGURE 12.17 *Picture of a cup anemometer.*

to loss of revenue from a wind turbine. Thus, this device must always be in good order and precise.

A different type of anemometer works based on the speed of sound. The ultrasound anemometer does not suffer from freezing, like the cup anemometer, since it does not have any moving parts. The speed of sound (or ultrasound, in this case) in the air is affected by the speed of wind. Since the speed of sound in the air at a given temperature is known, any change from that value represents the speed of wind. An ultrasound anemometer has four prongs, as shown in **Figure 12.18.** Each opposite pair of prongs has a transducer and a receptor, sending and receiving an ultrasonic sound. The difference in the time measurement between the instant the sound signal is sent and the instant it is received can be employed to calculate the wind speed and its direction. The device must be calibrated for this purpose, and the output data from the device can be used in a wind turbine onboard computer.

If a wind turbine uses a cup anemometer, it is necessary to determine the wind direction using a wind vane. A wind vane is a simple device that can freely rotate about a vertical axis. It has a tail that orients it with the direction of wind if there is wind. A wind vane is shown in **Figure 12.19.**

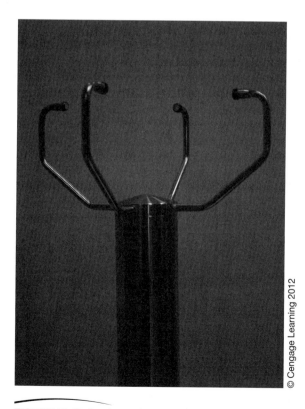

© Cengage Learning 2012

FIGURE 12.18 *An ultrasonic anemometer.*

© Cengage Learning 2012

FIGURE 12.19 *A wind vane.*

12.5 Chapter summary

- A transformer is a necessary electrical device for any wind turbine.
- A transformer steps up the voltage from the generator to the voltage level of the collector where all turbines are connected to grid.
- If a transformer increases voltage it is called a step-up transformer; if it decreases the voltage it is a step-down transformer.
- If a transformer does not change the input voltage, it is used to protect a device and is called an isolation transformer.
- Ideal transformer is a theoretical term used for transformer calculation. It assumes no losses in the transformer; that is, output power equals input power.
- The turns ratio defines the ratio between the number of turns in the primary winding to that of the secondary winding.
- An autotransformer has only one winding that serves for both input and output. The winding is tapped somewhere in order to give the desired turns ratio.
- A pad-mount transformer is suitable for mounting on a flat surface; a pole-mount transformer is suitable for mounting on the top of an electric pole.
- A rectifier is used to convert AC to DC.
- The most common rectifier is a bridge rectifier; it can be single-phase or three-phase.

- Low-power rectifiers use diodes.
- Ripple in DC is the fluctuation of voltage about a nominal value, instead of being constant.
- An inverter provides an AC voltage from a DC source; the output of a low-quality inverter has the form of a square wave.
- A thyristor is an electronic component used in rectifiers and invertors. The thyristor is a switching device.

ADVANCED LEARNING

Transformer equivalent load

While much more is necessary for a thorough understanding of the subjects considered in this chapter, the knowledge at the introductory level, as seen in this chapter, can be sufficient for working with them. Here we add another piece of useful information for those who need more insight into the work of transformers.

As usual, all the calculations are based on dealing with an ideal transformer. Suppose that a resistive load of R_2 Ω is connected to the secondary of a transformer with turns ratio a. If we want to see what the equivalent resistance in the primary is (such that it has the same effect), we can use the power relationships again.

We can assume that instead of R_2 connected to the secondary, we have a load R'_2 connected to the primary. Since the power in the primary is the same as in the secondary for this load (that is, $R'_2 I_1^2 = R_2 I_2^2$), then

$$R'_2 = R_2\left(\frac{I_2}{I_1}\right)^2 = R_2\left(\frac{N_1}{N_2}\right)^2 = a^2 R_2$$

R'_2 is said to be the secondary load, as seen from the primary.

With certain acceptable assumptions, the preceding relationship can be used for any type of load. Thus, if the secondary circuit (secondary winding inductance and the load) has a total impedance Z_2, the circuit impedance as seen from the primary is

$$Z'_2 = a^2 Z_2$$

And since the primary also has its own impedance (because of the winding), the load to the transformer is

$$Z_1 = Z + Z'_2 = Z + a^2 Z_2$$

where Z is the primary winding impedance.

Review questions

1. What is the function of a transformer?
2. What is the difference between a step-up transformer and a step-down transformer?
3. What is the relationship between voltage and current in each of the transformer categories in question 2?
4. What is meant by an ideal transformer? Is it a type of transformer?
5. Describe turns ratio and give an example.
6. Give some numbers as examples of the power ratings of small and large transformers.
7. What is an isolation transformer used for?
8. What is the ratio between the output (secondary) and the input (primary) currents in an isolation transformer?
9. What is an autotransformer and why do we need it? Why can we not use a regular transformer instead?
10. Describe pad-mount and pole-mount transformers.
11. What is the purpose of a rectifier?
12. What types of rectifier do you know?
13. What is meant by "ripple"?
14. Which one is better, a full-wave rectifier or a half-wave rectifier?
15. What is thyristor and where is it used?
16. Why do we need to use a thyristor in a rectifier? Can we use anything else instead?
17. To what does square-wave refer? With what should one compare square-wave for advantages and disadvantages?

Problems

1. A step down transformer has a 12:1 turns ratio. If the primary is connected to 120 V, how much is the voltage between the two secondary terminals?
2. A load of 40 Ω is connected to the secondary of a step down transformer. If the transformer is marked 220/110, what is the nominal secondary current?
3. The secondary of a transformer has 380 turns. If the transformer is used for 220/110 conversion, what is the number of turns in the primary?
4. The wire in the secondary winding of a transformer is thicker than the primary. Is it a step-up or a step-down transformer?
5. In a transformer the current in the secondary is 100 times the primary current. If the secondary voltage is 120 V, what is the primary voltage?
6. The secondary of a transformer has 400 turns and it is tapped at the 80th and 160th turns. If the primary has 1200 turns and is connected to 220 V, find what voltages are available from the secondary.
7. A transformer has a 400-kVA output. If the input voltage is 6.6 kV and the transformer efficiency is 0.97, find the current in the primary winding.

8. A wind turbine works 320 days per year on average. Its energy production ranges between 125,000 kWh in August to 220,000 kWh in March. If the transformer has an efficiency of 98% at full load and 97% when at reduced load, find the average magnitude of power lost in the transformer per year.

9. The total power consumption of a back-to-back converter in a turbine is 2.5%. The nominal power of the turbine is 1.65 MW, but the average generation is equivalent to 1.1 MW throughout the year. If the turbine downtime adds up to 40 days per year, how much is the yearly generation of electric energy?

10. The output voltage of a turbine is 224 V at no load and 218 V at full load. How much is the copper loss in the transformer if the current is 1000 A at full load, assuming that the entire copper loss materializes inside the transformer?

PROJECT

In this project, you are going to build a bridge rectifier. You can use your project later on as a DC source for one or more of the devices you need to buy batteries for.

For this project, you need to buy some components in addition to some equipment that you can find in your school labs. Figure 12.10 shows the components that you need to buy. These are:

One transformer, four diodes, one capacitor, a small plastic or wooden board to mount the components on, 3–4 ft of two-strand wire, one plug, and a small box to put the transformer and the circuit in.

You need a small transformer to convert 110 V to a voltage around 8 V (in order that you get a 6-V rectifier). The transformer is at your choice; you may want to have 3 V from your device, or some other value.

You need a soldering gun and some solder wire. These you can find from one of the laboratories at school.

You should make the circuit as shown in the figure. If you are not 100% sure about anything, ask your teacher or a lab technician to help you. You may want to test the device before using it. The DC signal from your rectifier can be seen on an oscilloscope. Ask your teacher for help.

Wind Turbine Mechanical Design Matters

© Cengage Learning 2012

KEY TERMS

aerodynamic force
amplitude
bedplate
centrifugal force
cone angle
fatigue
fatigue failure
flutter
frequency
inertial force
mass center
natural frequency
period
shear stress
vibration

OBJECTIVES

After studying this chapter, you should be able to:

* Describe various loads on major mechanical components in a turbine.
* Realize the reason that a turbine can break, leading not only to "not functioning" but to "disaster."
* Define the types of stress in various turbine components.
* Realize that the loads exist in both when a turbine is operating and when it is parked.
* Realize the importance of paying attention to indications of problems in turbines.
* Describe why vibration is not desirable in a turbine.
* Explain fatigue and whether fatigue stress can be prevented.
* Explain where the forces on a turbine end up.
* Determine the force from wind on a turbine tower.

13.0   Introduction

We have learned in previous chapters how the energy from wind is grasped by the rotor of a turbine and how this mechanical energy is transformed into electrical energy. However, a wind turbine cannot function in the first place if it cannot stand straight in the wind. At the same time as the energy from wind is harnessed, the wind exerts forces on the body of a turbine that push the tower and can topple the turbine.

We know that the power capacity of a turbine has a direct relationship with its size. Nevertheless, as the size grows, so does the magnitude of the force from wind. A wind turbine, thus, must be capable of withstanding the forces it is subjected to. In the same way, all the mechanical parts in a turbine must stand the forces and loads[1] that can break those parts if they are not sufficiently strong to handle the forces and loads to which they are subjected.

The forces exerted by wind on a turbine and its components can be different in magnitude and nature when a turbine is running and when it is stopped. These forces can also induce vibrations in the blades and in the turbine tower. Vibrations are not desirable in any machine and can damage or break a vibrating component much earlier than expected.

In this chapter, we study the loads on the major components of a turbine, other mechanical concerns such as vibration and fatigue, as well as some design characteristics of the components. This is only an explanation of the loads, without any stress analysis or attempt to go beyond the basics and get into the design of components. The discussion here is only intended to bring to the reader's attention the reasons why a turbine can mechanically fail and how important are mechanical issues such as vibration and fatigue.

13.1   Various operational states of a turbine

The loads that a turbine undergoes depend on whether it is working or it is stopped, and when the turbine is working, the loads depend on the status of operation. At any given time, a turbine can have one of the following statuses:

a. Normal operation.
b. Start-up and shutdown maneuvers.
c. Extreme conditions (gusts of wind, formation of ice, and so on).
d. Abnormal and fault situations.

In addition, for transportation, erection, and installation work a turbine component is subject to a different loading condition that, although relatively short in duration, should not cause an overload to the component. The corresponding loading must not be ignored.

---

[1] A load can be due to a force or a torque. But since in most of the cases under study the main cause of a load is a force, sometimes we use only the word "force."

During the design stage of a turbine each component must be designed for the worst-case scenario, that is, for the case with the highest load.

Normal operation is when a turbine is generating power. This is the state of a turbine most of the time. During operation for most turbines, the operating speed is constant and the blade pitch angle may be altered slightly by the controller, but the wind speed can change. During start-up and shutdown maneuvers, the rotor speed changes and the pitch angle is also changing. For extreme conditions, each case must be considered separately. A gust of wind applies a short-term but high-magnitude force to the parts. This may translate into a shock for certain components. An abnormal or fault situation implies cases that may arise in practice, such as rotor speeding. Rotor overspeed can happen if the electric load is suddenly taken off the turbine. Prolonged overspeed can lead to rotor breakdown and disaster.

Loads on the rotor   13.2

The rotor consists of the blades and the hub. The blades are subject to a number of loads as we'll see in this section. The hub transfers all of the loads from the blades to the nacelle. Thus, the variation of the loads on the hub is directly dependent on the loads from the blades. The hub has a shell-like structure (is not solid) and accommodates the mechanisms for blade pitch control. From a mechanical point of view, all the rotor forces must ultimately be transferred to the ground through the nacelle and the tower. The body of the hub, therefore, must be capable of handling the forces coming from the blades and transferred to the nacelle. Blade loads are transferred to the hub through a number of bolts holding the blade bearing in variable pitch machines (in fixed blades there is no bearing, and the blades are bolted directly to the hub). The bolts and the bearing between each blade and the hub, therefore, must take the loads. At the rear end of the hub there is another bearing between the rotor and the nacelle. Since the rotor (the assembly of hub and blades) rotates with respect to the nacelle, this bearing must take all the loads from the rotor.

In order to study the loads from the rotor to the rest of the turbine we examine the various loads on a blade. These are mainly the **aerodynamic force,** the weight, and the **centrifugal forces** when a turbine is running at a constant speed. During start-up and shutdown, **inertial force** comes into the picture as well. Some other forces may also come to existence, but their effects can be small or only for a short period of time. Aerodynamic force, as learned in chapter 3, is the force from wind on the turbine blades. Centrifugal force is what any rotating object experiences: it is a force pushing out an object that rotates about an axis or moves on a curved path. You can feel it in a vehicle when it turns. Inertial force is the force required to accelerate an object. To change your car's speed from stop to any speed or from 40 to 50 m/hr, for example, it is necessary to overcome the inertial force. Also, to do the reverse action of slowing down a moving object, the inertial force opposes the deceleration. In the example of your car, the brakes effort is used to cancel the inertial force.

aerodynamic force
Force exerted by moving air or any gas on an object in the airstream.

centrifugal force
The outward force in a rotating object that tends to break away the object.

inertial force
The force required to accelerate or decelerate an object.

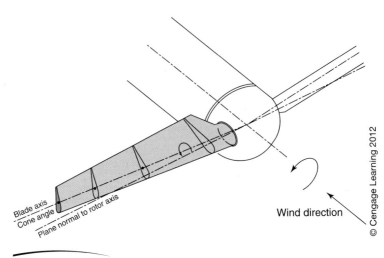

© Cengage Learning 2012

FIGURE 13.1 *Blade configuration with respect to rotor hub.*

cone angle
The angle that the
blades of a wind
turbine make with
a plane perpendicular
to the axis of rotation.

     Figure 13.1 depicts a blade and a hub. Note that the blade axis, about which the blade pivots, is normally not perpendicular to the rotor axis; there is a small angle between the blade axis and the perpendicular direction. This angle is called the **cone angle**. This angle changes when a blade bends due to aerodynamic forces. Some rotors are designed such that the cone angle is not initially zero and leans toward the front of the turbine; thus, when aerodynamic forces bend the blade back, this angle becomes smaller.

13.2.1 Aerodynamic loads on the blade

As you can recall from chapter 3, each blade of a turbine is subject to aero-dynamic forces from the wind. The airstream has a relatively fixed direction for the area swept by the blades, at an instant, but since blades are twisted the relative direction of wind with respect to a blade is not the same for various segments of the blade. Moreover, whereas the windstream has a constant speed (for a short period under consideration), the speed of air as a result of the blade rotation is smaller for the segments of the blade closer to the hub than the segments closer to the tip of the blade (see the text in chapter 7 on "Speed of a point on the blade of a turbine"). Consequently, the relative motion of air with respect to the blade as a result of wind and blade motion varies along the length of a blade. The resulting aerodynamic force, thus, varies in both magnitude and direction along the blade span. The typical force for a segment is shown in **Figure 13.2** (a blade can be assumed to be made of any arbitrary number of segments).

    The force on each blade segment consists of two components: one com-ponent along the direction of wind (the drag force, almost in the horizontal direction) and one perpendicular to wind (the lift component, in a near verti-cal plane). These forces are depicted for a two-blade turbine in **Figure 13.3**.

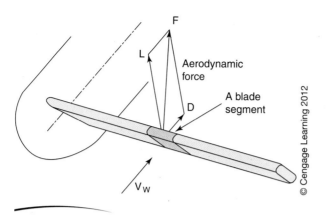

FIGURE 13.2 *Aerodynamic force on a blade segment.*

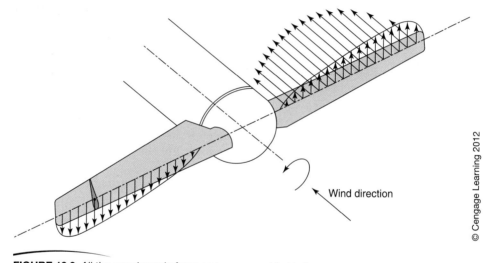

FIGURE 13.3 *All the aerodynamic force on two symmetric blades.*

A two blade-turbine is more appropriate to demonstrate the fact that all the components along the wind direction have the same force direction in the two blades, whereas the forces normal to the wind have opposite directions in the two blades, since the blades are symmetric to each other. The forces shown correspond to when the blade is feathered and does not catch much energy (the lift components are smaller than the drag components). This is just for the sake of clarity of the figure. This implies, also, the fact that the horizontal force on the blades is greater when a turbine is parked than when it is working.

> The horizontal force on the blades is greater when a turbine is parked than when it is working.

It is easy to verify that the resultant of the two sets of forces is a push on both blades in the wind direction and a torque about the turbine shaft axis. In other words, all those force components along the wind direction contribute to a backward push on the blades, and do not generate any rotational motion. However, those force components that are in the opposite direction in a (near) vertical plane are the only ones that generate a torque that makes the turbine rotate.

Note that the just mentioned aerodynamic forces are functions of wind speed, rotational speed, and pitch angle. Therefore, they are not the same for different operating conditions and for when the turbine is parked. A blade must be able to withstand these forces in the harshest condition; that is, when these forces are at their highest.

> All the forces on a turbine must ultimately be transferred to the ground through the tower.

13.2.2   Other blade loads

In addition to the preceding forces, a blade is subject to its own weight and when rotating at a constant speed a longitudinal force; that is, a force along the length of the blade. The latter is the centrifugal force. The magnitude of the blade weight force is always constant; its direction is also always the same, in the vertical direction and downward. Thus, during rotation this force changes with respect to the blade. The centrifugal force is, on the other hand, always along the blade. Its magnitude, however, is a function of the blade rpm. The faster the blade rotates the larger is this force. Its magnitude is normally more than the force of the weight, even for turbines with a low rotational speed of 12–14 rpm. For example, for a 6150 kg (13,576 lb) turbine blade, 44 m (144 ft) long, running at 12 rpm, the centrifugal force can be 145,000 N (32,955 lb).

During start-up and shutdown there is a speed change, and an additional force comes into effect: the inertial force. The inertial force is lateral to the blade direction. Its magnitude depends on how fast the turbine resumes its normal speed of rotation from start-up, or how fast it stops during a shutdown. The direction of the inertia force is in the opposite direction to the rotation during start-up, and in the same direction as rotation during shutdown. The inertial force has the effect of bending the blades in the plane of rotation, and it directly affects the bolts and the bearing at the root of the blades. Like other forces on the blade, this force is transferred to the hub. This force can be very high, especially during an emergency shutdown that the speed is abruptly dropped from operating speed to zero.

The centrifugal force is always in the radial direction and outward from the center of rotation. In this sense, when a blade is hanging downward the centrifugal force and the force of weight add together, and when the blade is in the upright position (tip up), the centrifugal force and the weight are in

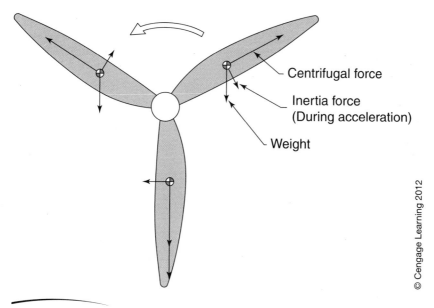

© Cengage Learning 2012

FIGURE 13.4 *Centrifugal, inertial, and weight force at three positions of a blade.*

different directions and subtract from each other. The inertia force, when it exists (that is, during speeding up and slowing down motion of rotor), is always lateral to the blade. The direction of the force of weight varies with respect to the blade direction, since it is always vertical. These forces are shown in **Figure 13.4.**

In those positions of the blade between horizontal and vertical, the weight force can be decomposed into two parts, one along the blade and one perpendicular to the blade, as shown in **Figure 13.5.**

As can be realized, the effects of the forces thus far studied are a lateral load (a component of weight force, and inertial force when present) and a longitudinal load (the other component of the weight and the centrifugal force). The blade is, therefore, subject to a combined stress: tensile stress as a result of longitudinal forces as well as tensile and compression stress due to lateral forces that tend to bend the blade. Note that the drag component of aerodynamic forces act at a 90° angle from the lateral and centrifugal forces. The blade is, therefore, bending in two directions. This is illustrated in **Figure 13.6.** Note also that the lateral forces just described do not cause the blade to always bend to the same side, as their direction can switch; for instance, the weight force effect switches direction by crossing a vertical position.

The loads on a turbine blade have a periodic nature. As the blade turns one revolution, the aerodynamic forces change from a maximum value to a minimum value.

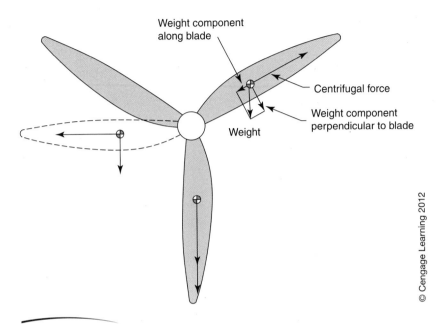

FIGURE 13.5 *Decomposition of force of blade weight when at an angle.*

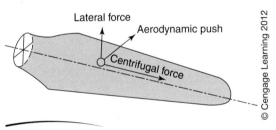

FIGURE 13.6 *A blade can bend in two directions 90°
from each other.*

mass center
The point in an object
where the whole mass
can be assumed to be
concentrated.

Another load on the blade is due to the fact that the **mass center** of the
blade is not necessarily on the blade axis. This causes additional load, such as
moments. Mass center is a point that all the mass can be assumed to be accumulated there. Consider when a blade is bent due to aerodynamic forces. Then the
mass center, even if initially on the blade axis, displaces from its original position. The centrifugal force consists of many smaller forces all over the blade, but
the resultant of all these forces can be shown as a sole force at the mass center
of the blade. The direction of the centrifugal force is perpendicular to the rotor
shaft axis. As a result of displacement of the mass center, the centrifugal force
that must be along the blade axis is now at a distance from it. It is also at an
angle to the blade axis.

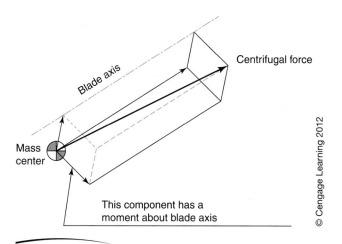

Mass center

Blade axis

Centrifugal force

This component has a moment about blade axis

© Cengage Learning 2012

FIGURE 13.7 *Position of the centrifugal force with respect to blade axis.*

Figure 13.7 shows a possible position of the centrifugal force with respect to the blade axis. As shown, the centrifugal force can be decomposed into three components, one parallel to the blade axis, one intersecting the axis, and the third one perpendicular to and at a distance from the axis. The only component that generates a moment is this third one. Whereas it would not create a moment if the mass center was exactly on the blade axis, now the centrifugal force generates a moment that tends to twist the blade about its axis. As a result, the blade is subject to torsion. Torsion introduces another stress, **shear stress,** which tends to tear an object (for *shear stress* see the text on "Force, pressure and stress" in chapter 9).

As we can see, any blade, no matter whether or not the turbine is working, is subject to various loads. The magnitude of these loads vary based on (a) the aerodynamic forces (which depend on wind speed, blade surface area, blade length, and the angle of attack—which corresponds to the pitch angle—and the rotational speed); (b) the position of the blade in its 360° course of rotation; (c) the rotational speed of the turbine; (d) the weight of the blade (which depends on its size and design and the material used); and finally (e) the inertial force during speed change, which can be significant in many cases. These loads are not constant and each one can continuously fluctuate between a maximum and minimum range of values.

shear stress
A type of stress due to a cutting force. The force from a pair of scissors causes shear stress in the object to be cut.

No matter whether or not the turbine is working, it is subject to various loads.

FORCE AND STRESS IN A BEAM

The loads on the blade of a turbine due to the aerodynamic forces and the loads on the tower are similar to the cases shown in this page for a (general) beam. The presentation here is, in fact, the fundamentals of finding the stresses for the design of a blade, tower, and the like (see discussions of force, pressure, and stress in chapter 9). While there is much to be discussed on the subject and is beyond the capacity of this book, the material here shows the minimum knowledge that a technical person must have.

Both the tower and each blade are supported at one end and are subject to lateral forces. They can be regarded as *beams* for their load analysis. A beam with such a configuration is called a *cantilever beam*. This is shown in the following figure. The lateral force can be either concentrated at one point or distributed along the length (or a combination of both).

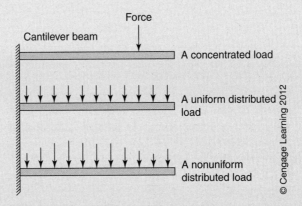

As can be understood, in both cases the effect of the load is to bend the beam. As a result, the beam is bent, although the amount of the deflection can be so small that it cannot be noticed by eye, unless the beam is sufficiently long or appropriate measurement devices are used. When a beam is bent, the grains of material in one part of the cross section at one side of the so-called neutral axis are stretched (the upper part in the figure shown). At the other side of the neutral axis, the grains are compressed together. Thus, when a beam is bent down, its upper half is subject to *tensile stress* and the lower half is subject to *compression stress*; this is shown in the following figure. If a beam is bent upward, the reverse happens and the upper half undergoes compression stress. The stress that results depends on the magnitude of the load, the type of the loading (concentrated, distributed), and the size and form of the cross section of the beam.

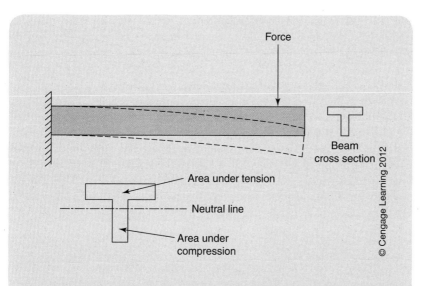

Other factors such as the environmental temperature and its variation come into the picture, and these change the overall scenario. The most important factor, however, is the nature of the load; that is, whether it is a constant load, or its value fluctuates (which can cause vibration), or it is in the form of sudden shocks. The easiest load to handle is a constant force.

Obviously, the design process is based on finding the stress and making sure that the stress is below the amount of allowable stress for the material, taking safety factors into account.

In the preceding discussion, the beam is shown as horizontal and with a T shape (for clarity of the two separated areas). However, the subject matter applies to any long mechanical component at any angle, fixed at one end and free at the other, subject to lateral forces that bend the piece. The shape of the cross section can be any profile, solid or hollow.

The discussion here applies very well to any tower, including the tower of a wind turbine. A beam can be subject to both lateral and longitudinal (along the direction of its length) loads at the same time. A wind turbine tower is under the load of the rotor and nacelle weight and the wind forces.

Vibrations 13.2.3

For any mechanical component, a constant (or nearly constant) force is better than a force whose magnitude, direction, or both vary. A component subject to variable force must be designed to be stronger than if it was subject to a constant force. In addition, vibration may be induced in mechanical systems and parts, which is a basis for rapid destruction and failure. A cyclic force—that is, a force whose magnitude changes, but repeats the same values with time—can lead to vibration. (See the figure in the text under "Mechanical Vibration" in this chapter.)

Consider the rotor of a wind turbine. When a blade is at its top position it is subject to stronger aerodynamic forces than it experiences at a lower position, because the wind speed is higher at a higher altitude (see chapter 2). For instance, in a turbine with 50-m (165-ft) blades, the difference between the highest and the lowest points through which a blades passes is 100 m. The speed of wind is not the same between these two points, and therefore, there are more aerodynamic forces on each blade when it sweeps the top part of its course during rotation. In addition, in the lower course of rotation, a blade crosses the shadow of the tower. A 10-ft-diameter tower has sufficient effect on the wind flow that it can significantly reduce the wind speed in front of it all along its length.

> A mechanical component subject to variable force must be designed to be stronger than if it was subject to a constant force.

With a constant speed of rotation the preceding phenomena lead to a cyclic variation of the forces to the blades and the rotor. This cyclic alteration of the force magnitudes leads to vibration of the blades as well as vibration of the rotor. When these vibrations have a small **amplitude** no harm is done, but if the vibrations have a frequency the same as or near the **natural frequency** of the blade, then undesirable results can follow. For a brief definition of these new terms, such as amplitude and natural frequency see the text for "Mechanical Vibration" in this chapter.

Vibration in machine parts is a "killer," since the vibrating element undergoes a repeated alteration of stress, say from tensile stress to compression stress. A mechanical component that normally can handle a fixed load for a long time fails in much less time (hundreds or even thousands of times smaller) if the load is switching direction in a repeated manner.

> Vibration is not desirable in any machine part.

In order to prevent vibration buildup in the rotor, the blades must be designed such that their natural frequency is sufficiently away (usually higher than) from the rotor rpm. Also, they must not be an integral multiplier (that is, 2 times, 3 times, etc.) of the rotor rpm. For example, if a rotor rotates at 14 rpm, the natural frequency of the blades must not be equal to 14, 28, 42, or any other multipliers of 14. This is for possible vibration of the blades about their point of connection to the hub. The rotor bearing is subject to another vibration of the rotor assembly. If the operating speed is 14 rpm, since there are three blades and each blade passes the same course 14 times per minute, the rotor is subject to a varying force with a frequency of $14 \times 3 = 42$ cycles per minute (which is 0.7 Hz).

amplitude
The instantaneous value of a cyclic variable that changes with time between a maximum and a minimum.

natural frequency
The frequency (or frequencies in a more complex system) inherent to any flexible object, part of a machine, or structure that can vibrate in any way. If the object receives a shock or an impulsive force, it vibrates with a frequency equal to its natural frequency.

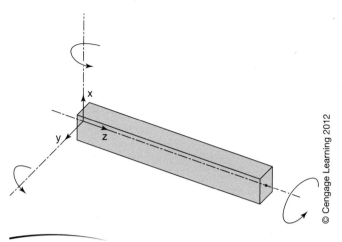

FIGURE 13.8 *Possible vibrations in a turbine blade.*

The variation of aerodynamic and other forces in the blades, thus, can induce severe vibrations both in each blade and in the rotor. Note that, in general, the vibration of a long piece like a blade can occur about three axes, as shown in **Figure 13.8.** Those about x- and y-axes are in conjunction with the bending of a blade; they affect the body of the blade and the bolts attaching it to the hub, as well as its bearing in a pitch-controlled turbine. Another possible vibration in blades is about the blade axis (the z-axis in figure 13.8); it is in conjunction with blade torsion due to the moments about this axis (see figure 13.7). This type of vibration is called **flutter.**

Vibrations in the entire rotor affect the bearings on the main shaft and are transferred to the gearbox. The effect of vibrations is never desirable, and if vibration is detected, its source must be found and corrected as soon as possible.

Another source of vibration of a rotor is an imbalance in the blades. The three blades must be completely uniform, similar in design, and have the same weight and their mass centers be the same distance from the rotor axis. Any deviation from these requirements causes an imbalance in the rotor and leads to vibration.

flutter
A vibration of a wind turbine blade about its own axis.

Blade fatigue

Fatigue is a term used for the stress effect of cyclic loads on a mechanical component. As we learned in this chapter, a part of a machine that is under a continuously varying load, either due to vibration or the nature of load, can fail much faster than expected. This failure is called **fatigue failure.** In fact, this is a way that we use to break a piece of metallic bar by hand, repeated bending in opposite directions.

All the rotating components in machinery, including wind turbine blades, are subject to fatigue even if there is no vibration. In the presence of vibration, a machine part may fail after a week. But even if there is no vibration, a blade

13.2.4

fatigue
Undergoing repeated alternative tension and compression stress.

fatigue failure
A failure due to fatigue.

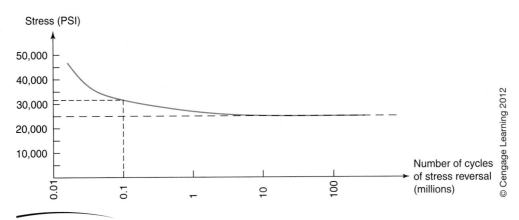

FIGURE 13.9 *A typical fatigue curve.*

of a turbine running at 14 rpm undergoes alternating stresses approximately 7 million times per year.

A machine part subject to fatigue must be designed based on fatigue test data. These data usually show the expected life for a material (more for metals) under various stress values. A typical curve can look like that in **Figure 13.9,** where the horizontal axis shows the number of cycles and the vertical axis shows the stress. These data are used extensively for the design of airplane propellers and blades in blowers, compressors, and so on. For wind turbines, because it is a relatively young and limited industry, such data do not exist. Compared to blades in other machines, the rotational speed of a rotor in a wind turbine is much lower and the loads are much smaller.

> Under fatigue stress a mechanical component in a machine can break much faster than it should.

The curve in figure 13.9 shows that if the stress is, say, 32,000 psi, the specimen tested failed after 100,000 cycles, but for a stress of 25,000 psi the specimen did not fail through 100 million cycles. Thus, based on these test results, the stress for the part must not exceed 25,000 psi (25 kips) if subject to stress reversals.

13.2.5   Blade construction

Wind turbine blades are hollow; otherwise their weight would be tremendous, even with a lightweight material, and their use would be impossible. They might not even start to move the rotor. It is, however, necessary that they have a sturdy structure to withstand all the stress from various loads that were discussed in this chapter, and work for the useful life of a turbine.

The blades of a turbine spinning at 14 rpm must go through 175 million cycles in 25 years.

Blade shells are made from composite material that is light and strong. Inside, they must have structural support for the shell. The internal structure can be made from wood or a similar lightweight material. (For a picture of the inside of a wind turbine blade see figures 5.7 and 5.8). The shape of blades, their construction, and the other related subjects are to be determined based on careful and extensive studies, and experiments if needed. This subject is beyond the scope of this book.

One of the important issues in new blades is the integration of a lightning rod in the blade structure. The lightning rod is a strip of copper along the blade. At some point at the tip of a blade it is exposed to the outside by a small disk. At the other end, it is connected to a metallic counterpart in the hub. Through a number of metal connectors, the strip in the blade ultimately makes a connection to the ground through the tower.

If a blade becomes damaged in operation, it must be repaired. Normally the broken parts must be cut, replaced by pieces of the same size, and glued together by the proper material. Depending on the severity of damage, this can be done while the blade is in the air, or it may have to be brought down to the ground. Any repair work on the blades is a costly job and involves loss of production.

Gearbox   13.3

You learned about wind turbine gearbox in chapter 9. Here we want to only add some other information that one has to be aware of. All the power collected from wind by the rotor goes to the generator through the gearbox. And, as you learned in chapter 9, all the power transferred by a gear is concentrated on each tooth of that gear. This explains how much load is on the teeth of the gears.

We also learned that during rotation the rotor is subject to a periodic force due to the nonuniformity of the aerodynamic forces at the top position of a blade and when it passes across the tower shadow at the bottom position. All this variation leads to a pulsating torque on the rotor shaft that is transferred to the gearbox. This pulsating vibration causes fatigue in the teeth of a gear.

In addition to the preceding, another phenomenon has an undesirable effect on a turbine gearbox. This is due to the nature of wind, which does not blow uniformly. Gusts of wind can generate sudden power fluctuation on the rotor, which ends up as shocks on the gearbox. In industrial machinery, a flywheel usually smoothes out the load variation and reduces the effect of uneven loads. In a wind turbine, the rotor itself plays the role of a flywheel. At the same time, however, it is the source of the uneven load. In other words, the smoothing effect of the rotor on the shaft is smaller than the disturbance that it injects to the shaft. With all these fluctuations carried to the gearbox, we can see that the gearbox in a wind turbine is subject to very hostile loading conditions.

Presently one of the major shortcomings of wind turbines is the uncertainty about the gearbox and its behavior. Some wind turbine gearboxes have given up and failed after 6–7 years of service. Changing or repairing a wind turbine gearbox is a major cost and it changes the whole scenario of the economics of a wind turbine. Six or seven years are much less than an expected life of 20-plus years. Proper gearbox maintenance is one of the key issues in prolonging its service life.

> One of the major shortcomings of present wind turbines is the uncertainty about the gearbox and its behavior during its life expectancy.

13.4 Nacelle

Nacelle is the room-size box on the top of the tower. Its main role is to house all the components that are to be on the top of the tower, such as the gearbox and the generator. In most wind turbines, the nacelle consists of a floor, called the **bedplate,** on which the equipment is installed, and the roof, which is a shell-type cover to enclose the equipment. In another type of nacelle, the floor is a part of the structure of the gearbox body.

bedplate
A main structure in a nacelle where all the components are mounted or attached.

The force of the weight of all the components, including the rotor, and the aerodynamic forces are transferred through the body of the nacelle to the tower. In addition to the force of weights of the rotor and the components inside the nacelle, the nacelle must transfer the torque received from the rotor to the tower. In this sense, the nacelle must be sufficiently strong for all of these loads.

MECHANICAL VIBRATION

> Vibrations can happen in all systems, including electrical and economical systems. However, we confine this introduction to mechanical vibrations, without getting deep into the subject.
>
> **Cyclic entity**
> If the magnitude of an entity such as force and displacement has a repeated pattern in time, as shown in the following figure, that entity can be called *cyclic* or *periodic*. The repeated pattern in the figure is shown three times, but it continues indefinitely as long as a device is in operation. Based on the figure, the magnitude changes from a positive value to a negative value and continuously fluctuates between a maximum to a minimum according to the pattern. This is the case in many practical cases. Nevertheless, the magnitude can change only in the positive or negative region, as shown in the second figure.

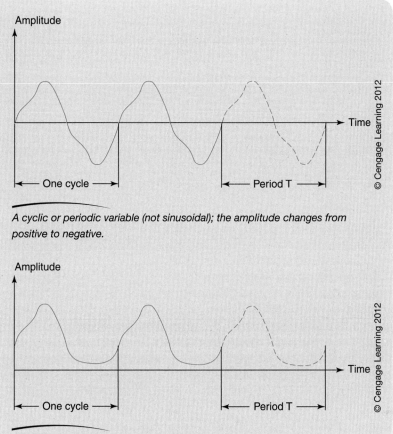

A cyclic or periodic variable (not sinusoidal); the amplitude changes from positive to negative.

A periodic variable whose value is always positive.

In a cyclic entity, the repeated pattern is called a *cycle*, and the duration of time for one cycle to take place is called **period,** measured in seconds. The number of cycles that happen in 1 second of time is called **frequency.** The magnitude of a periodic entity is referred to as *amplitude*. So, if the subject is force, for instance, the amplitude represents the magnitude of force at any instant of time. The amplitude oscillates between a minimum and a maximum value. The frequency of oscillations is measured in hertz (Hz), or cycles per second. The most common periodic pattern is *sinusoidal waveform*. The alternating current electricity has a sinusoidal waveform.

If the sum of all the forces applied on a mechanical component has a cyclic nature, then the effect on the component also has a cyclic nature. The effect of a force is (1) a displacement and/or (2) stress. None of these can necessarily be visually noticeable by a human eye, and sensory devices such as vibrometers and strain gauges can be used to measure them. A periodic displacement in a mechanical part is **vibration** and is

(continued)

period
The duration of completing one cycle in a vibrating action.

frequency
The number of repetitions, in one second, of a cyclic phenomenon.

vibration
Physical oscillations in a mechanical system. A continuous and cyclic variation around a nominal value in a system.

accompanied by the corresponding stress that oscillates between two values. Often the two values have opposite signs (positive and negative), and the stress in the part alters between tension and compression in each cycle, for instance.

As long as the amplitude of vibration is small (that is, below a specific value for a particular item), there is nothing to worry about. For different parts of machinery acceptable values of amplitudes for harmless vibrations are defined by manufacturers. These values depend on frequency (as the frequency increases the acceptable value is smaller). But if the vibrations are excited, no machine part can survive the rapid path to damage and failure. Vibration *excitation* occurs if the frequency of vibrations is near the *natural frequency* of a mechanical part; that is, if the periodic force acting on a mechanical component has the same or nearly the same frequency as the natural frequency of that component. This often happens in rotating machinery, such as blades of turbines, tires in cars, loads on motors, and so on.

All mechanical components in machinery, such as rods, beams, plates, gear teeth, including wind turbine blades, have some flexibility in their structure. Based on this flexibility and the mass of the component, another property comes into existence. This property, which is inherent and specific to each mechanical entity, is called *natural frequency*. If then, the piece under consideration is subject to periodic forces with the same frequency as the natural frequency, vibration with amplified amplitude results. In other words, the natural frequency of a piece is excited by external forces, leading to severe vibrations.

This excitation does not happen only when the two frequencies are equal. It also happens if the frequency of the external force is two times, three times, or multiple times the natural frequency. These are *harmonics* (second harmonic, third harmonic, etc.) of the natural frequency. The effect becomes, however, gradually smaller for the second, third, and higher harmonics.

It is the bedplate (or its equivalent) that takes the loads, and the roof must support only its own weight plus any other item that goes on the top of the nacelle, such as anemometers, oil coolers, and so on. (Each wind turbine has a flashing beacon that is normally installed on top of the nacelle.)

The magnitudes of the weights and rotor shaft torque depend on the size of a turbine. They are also dependent on the design and the material used. Just to have an idea about typical values, the following data are for Gamesa G87, a 2-MW turbine with 87-m (290-ft) blade diameter turbine:

Weight of blade:	6150 kg (13,576 lb)
Rotor weight:	37 ton (metric) (81,000 lb)
Total tower top mass:	107 ton (235,000 lb)
Four-section tower:	203 ton (446,000 lb)

Tower   13.5

The tower is like a vertical slender bar fixed at one end (cantilever beam) subject to

a. A compression load due to the weight of the rotor and nacelle, at its free end.
b. A lateral force due to the aerodynamic forces from wind on the rotor, at the free end.
c. A moment at its free end, as shown in **Figure 13.10.**
d. Aerodynamic forces on the nacelle.
e. Aerodynamic forces on the tower.
f. Weight of any other item attached to the tower.
g. Its own weight.

As discussed earlier, the forces from wind and the torque have variable values and have a periodic nature as well as being impulsive when gusts of wind hit a turbine. Depending on a turbine's geographic location, and its design, a tower

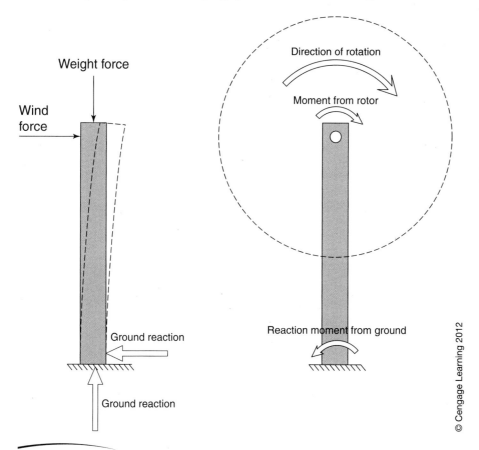

FIGURE 13.10 *Force and moment on the top of the tower.*

© Cengage Learning 2012

consists of three to four segments. A rule of thumb is that the height of the tower is the same as the diameter of the blades. But this is not always the case.

The aerodynamic force on the tower is only a drag force, pushing the tower in the direction of wind. Assuming a linear graduation from zero (at the ground level) to a maximum on the top of the tower, it can be substituted by a horizontal force at the 2/3 height of the tower from ground. The magnitude of the force of wind on the tower is given by (see chapter 3)

$$F = \tfrac{1}{2}\, c\, \rho\, v^2\, A$$

where ρ is the air density, v is the wind velocity, and A is the surface area blocking the wind flow. Thus, A is the projection of the shape of the object in the direction perpendicular to the wind flow. Also, c is the drag coefficient, a factor that depends on the shape of the object. For instance, for a rectangular prism c is larger than for a circular pipe. The values of c for different shapes have been experimentally measured and can be found in the form of tables (see Appendix B)

In order to have an idea about the magnitude of force on a typical tower, consider the following example.

EXAMPLE:

In this example, we assume that at the height of an 80-m tall tower, wind blows with a speed of 16 m/sec (35 mph). We further assume that the flow of wind is linearly changing and it is zero at the ground level.[2] The tower is conic and has a diameter of 4 m at the bottom and 3 m on the top. The drag coefficient (or shape factor) is 0.8 for a cylindrical surface. If the air density is 1.225 kg/m³, what is the force from wind on the tower?

Solution: The problem would have been simpler if there was no taper in the tower. But because of the change of diameter across the tower, there are two linearly changing parameters with opposite effects, the wind speed and the area subject to wind force. One can analytically find a mathematical expression for the force in question. However, since that falls beyond the scope of the book, the best way is to divide the tower into a number of segments, and find the force on each individual segment. Then the total force is determined by adding up these segmental forces.

The 80-m tower is divided into 16 segments, each 5 m long. All segments are assumed to be rectangular, but each lower segment is larger than the upper segment. The variation of wind speed is formulated such that at the lowest segment it is 1 m/sec (not zero), and on the highest segment (top of the tower) it blows at 16 m/sec speed.

Figure 13.11 illustrates when a tapered shape is divided into segments. For each segment the middle point defines the average width, so that the segment area remains unchanged. For the tower in question, Table 13.1 shows the various values.

---

[2] This assumption is made for simplification of the problem. The variation of wind speed with height is not linear.

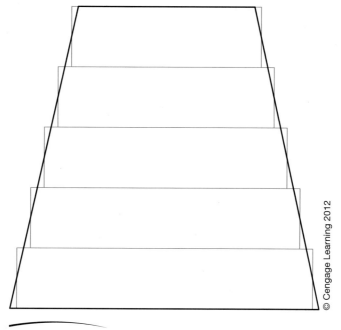

FIGURE 13.11 *Dividing a tapered shape into segments.*

TABLE 13.1

Forces on tower segments

Segment number	Average height (m)	Average width (m)	Area (m²)	Wind speed (m/sec)	Force (N)
1	2.5	4.0	19.8	0.5	2.4
2	7.5	3.9	19.5	1.5	21.5
3	12.5	3.8	19.2	2.5	58.9
4	17.5	3.8	18.9	3.5	113.5
5	22.5	3.7	18.6	4.5	184.5
6	27.5	3.7	18.3	5.5	271.0
7	32.5	3.6	18.0	6.5	372.0
8	37.5	3.5	17.7	7.5	486.7
9	42.5	3.5	17.3	8.5	614.0
10	47.5	3.4	17.0	9.5	753.2
11	52.5	3.3	16.7	10.5	903.2
12	57.5	3.3	16.4	11.5	1063.2
13	62.5	3.2	16.1	12.5	1232.2
14	67.5	3.2	15.8	13.5	1409.3
15	72.5	3.1	15.5	14.5	1593.6
16	77.5	3.0	15.2	15.5	1784.2

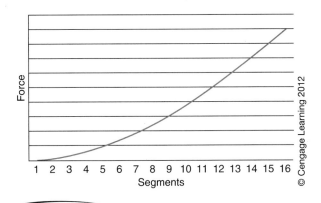

FIGURE 13.12 *Variation of wind force along the tower from bottom to top.*

The last column of this table shows the force on each segment. The total force is, thus, the sum of all the values in the column, that is, 10,863 N (about 2444 lb). As for the point of action of the force, **Figure 13.12** shows the variation of the magnitude along the tower. If this variation was linear, then the point of action of the resultant would be 1/3 from the top of the tower. Here with good approximation we can assume that the equivalent force acts at 1/3 from the tower top.

13.6 Chapter summary

- The loads on all parts of a turbine must be ultimately transferred to ground through the tower.
- Each component, such as a blade, must be strong enough to withstand its loads.
- The loads are of different forms and they vary with the operation of a turbine. Thus, they depend on the turbine rpm, as well as other parameters.
- Aerodynamic forces are due to wind; thus, they depend on the wind speed.
- The forces do not diminish when a turbine is at rest. Only their values change.
- The aerodynamic forces during a revolution of blades are more on the top and less at the bottom. The blades and rotor are, thus, subject to periodic forces.
- The frequency of these forces is three times the speed of rotation for a three-blade turbine.
- Natural frequency is an intrinsic property of any mechanical component with flexibility. Each blade, the hub structure, the main shaft, and the tower are the principal components whose natural frequencies are important to turbine operation.

- The natural frequencies of the turbine components must not be excited by the periodic forces. In particular, the natural frequencies of the blades, the main shaft, and the tower must not coincide with the multiples of the frequency of periodic forces on turbine elements.
- A blade, in particular, and any other turbine components under cyclic loads are subject to fatigue. They can fail due to fatigue stress.
- Vibration is very undesirable in any part of machinery; wind turbines are not an exception. A part of a machine experiencing vibration can fail much sooner than it is designed for.

ADVANCED LEARNING

Support reaction to a rotating load

In rotating machinery, it is usually a shaft that transfers a torque from the driving element to the driven element. The rotating shaft is supported by a number of bearings that MUST be fixed to ground, either directly or indirectly. Consider a pump or blower, as shown in the figure.

The load needs a torque (moment) to be driven and exhibits this torque as a resistance to rotation on its shaft. The motor provides a torque and overcomes the resistance of the load. When a constant speed is reached, this is a state of equilibrium and the torque provided by the motor is equal to the resisting torque of the load.

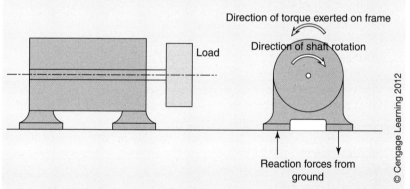

© Cengage Learning 2012

The torque exerted on the shaft by the motor is in the same direction as the rotation. Since the cause of motion is internal to the motor, in order for the motor to be able to exert this torque it is necessary that the motor body has supports on the ground; otherwise it is not able to turn the shaft without itself being turned in the opposite direction. It is similar to when you want to push an object; your feet and body must be supported by the ground. You are not able to do that on ice; your body moves in the opposite direction.

(continued)

The support from the ground can be in the form of two forces on the motor body legs. These two forces are equal in size but in the opposite direction to each other, as shown, and constitute a *couple*, which is equivalent to a torque.

In a wind turbine, since the cause of motion is external to the turbine body, the scenario is different from that of a motor. While the rotor turns it tries to turn its support (the nacelle) in the same direction; thus, exerting a torque on the nacelle in the same direction as that of the rotation. The support (nacelle), on the other hand, must resist this torque; otherwise the tower and nacelle bend. This resistance comes from the ground. The ground applies a torque (in the opposite direction of rotation) to the tower, which is transferred to the nacelle through the tower. The magnitude of this torque is the same as that created by the rotor. The tower is, thus, in equilibrium and does not topple. The tower body is subject to this torque, and must withstand it. The magnitude of this torque is not constant and depends on the power captured by the turbine at each instant of time (see the discussion about power and angular speed in chapter 7).

Review questions

1. What is aerodynamic force?
2. What is the direction of aerodynamic force?
3. What is centrifugal force?
4. What is the direction of centrifugal force?
5. What is inertial force?
6. What is the direction of inertial force?
7. What is the difference between inertial force when a turbine blade is speeding up and when it is slowing down?
8. How much is the inertial force when a turbine is turning at a constant angular speed?
9. Name all the forces on a turbine blade.
10. What is a cyclic load?
11. Define frequency and amplitude in a cyclic quantity.
12. What is the difference between cyclic and periodic?
13. What is period in a periodic motion?
14. Define fatigue.
15. What is fatigue failure?
16. Can you show an example of fatigue?
17. Can you guess what thermal fatigue is?
18. What is meant by mass center?
19. What can lead to vibration in a machine part?
20. Is vibration in a machine element good or bad?

21. Define natural frequency.
22. What happens if the rotating speed of a blade is equal to its natural frequency?
23. Of what is a turbine blade constructed?
24. Is it better or worse if a turbine blade is made up of heavy material? Why?

Projects

PROJECT 1

In this project you are to find the force from wind on a tower, in the same way that was done in the example in this chapter. Assume a tower 280 ft long, 10 ft in diameter at the top and 12 ft in diameter at the bottom. Wind speed is 40 mph, and the drag coefficient of the tower shape is 0.8. Air density is 0.075 lb/ft³.

   Make all the assumptions as given in the example. Draw a table showing the results.

PROJECT 2

Think about 25 problems that can happen in a car. Write them down in a column on a sheet of paper. For each problem, think of a wind turbine and see if the same problem can happen in the wind turbine. If the answer is "yes," briefly identify where and in what form. Write that in front of the problem. If the answer is "no," write "No." Discuss your findings in the class and compare them with the other students.

Economics
of Wind Energy

© Cengage Learning 2012

KEY TERMS

compound interest
corrective maintenance
cost of capital
discount rate
future value of money
initial cost
net present value
operating cost
present value of money
preventive maintenance
scheduled maintenance

OBJECTIVES

After studying this chapter, you should be able to:

- Explain various categories of costs of a wind energy project.
- Itemize the initial costs and the running costs.
- Realize that developing a wind farm must be economically viable.
- Do simple economics calculations.
- Realize that the cost of developing a wind farm very much depends on accessibility to the site and to the grid.
- Explain why the unit cost depends on the number of turbines in a farm.
- Analyze the economic state of a project and determine whether it is profitable.
- Realize why other types of power plants are more affordable to an investor.
- Explain how not having a good maintenance program can skyrocket the repair costs of wind turbines.

14.0   Introduction

initial cost
The money
necessary to spend
at the beginning
of a project to pay
for the equipment,
transportation, and
installation.

operating cost
Continuous expenses
of running a business,
manufacturing a
product, or producing
goods, such as
salaries, rent for
property, purchase
of raw material,
maintenance cost,
insurance, and so on.
This is as opposed
to a one-time cost
for purchase of land,
machinery, patents,
and the like.

It is necessary for someone involved in wind energy to know somewhat about the economics of wind turbines, and how this matter can affect a wind energy project. From the financial point of view, a wind turbine, or a number of wind turbines in a farm, must generate revenue. In other words, from an investment viewpoint, installing a wind turbine is an economic activity. If a wind turbine cannot generate revenue, it is not a profitable activity.

In this chapter, we study in brief the economic side of wind turbine operation. At the commercial level, usually a developer, that is, an investment organization, decides to make an investment in creating a wind farm. From the developer's viewpoint, it does not matter if the investment is made in some other type of business. The developer chooses the project with the highest return, putting all the numbers together.

For any wind turbine installation, no matter whether in a wind farm or as one unit, whether connected to a grid or as a stand-alone turbine, the associated cost falls under two categories: **initial cost** and **operating cost.** This is also true for any other large-size project. For wind turbines, the initial cost corresponds to those expenses pertinent to purchase and installation, and the operating cost corresponds to the operation and maintenance. We are going to study the costs involved in installation and running a wind turbine, mainly at the commercial level.

First in this chapter, we study the principle of investment and the way a project is seen from a business angle. This is very brief, and shows only the principle in broad terms. Then, various costs associated with a wind energy project and a comparison with other types of energy projects are discussed.

14.1   Fundamentals of economics

future value of money
A term in economics
representing the value
of an amount of money
at a given future time.

**present value
of money**
The value of an
amount of money,
expressed for a
future time (say in
10 yr from now),
in terms of today's
money; i.e., a smaller
amount obtained
by subtracting the
interests for the period.

The basis of economics in many industrial countries is that "money today has more value than money tomorrow." This concept, right or wrong, structures the economics of many countries. The difference between the values of money "today" and "tomorrow" is the "interest." The money you have today has more value tomorrow, since it gains interest.

Here, by "today," we mean the present time, and by "tomorrow," we mean the future time. In this sense, the **future value** of an amount **of money** x that one owns presently is more than x. Similarly, the **present value** of an amount **of money** y in the future is less than y. In this way, one can buy more of goods with a fixed cost, today, with the same amount of money (say with $1000) than in the future. The difference is due to the **cost of capital** or **discount rate,** which is the rate of return investors expect to receive for their money. This implies that it costs money to generate capital. For simplification purposes, we will assume that this rate is the same as the market interest rate. The interest rate is the yearly yield that a bank expects from borrowers.

Another aspect of economy is the mechanism for business. A businessperson borrows money from a bank in order to run a business and earn money. The money borrowed must be paid back to the bank, with interest. The net income from the business must pay for all the expenses as well as the interest that must be paid to the bank. If the profit made from the business is not sufficient to pay for the expenses and the interest, the business does not work and it must close down.

From a business point of view, developing a wind farm is a project that must be profitable to be attractive to investors and developers. That is, at the end of a period it must add to the value of the money invested. This addition to the value of money must be greater than or equal to the added value from other options (projects or businesses) that the investors have.

Large projects, including developing and operating a wind farm, are not simple to start and terminate easily, and must be based on a lot of preliminary studies and thorough analysis. Often the profitability does not even start until after a long time (years). This is also included in the analysis of the economy of a wind project. In order to evaluate a project (of any kind) all the expected profits are converted to their present values. If altogether the **net present value** (that is, the revenues minus the money invested and costs, based on today's money) of the execution of a project is positive, the project is acceptable.

Calculating the future value of money and the present value of money requires an interest rate. Interest is added to the principal (the initial money borrowed or invested) at the end of a specific period of time, say every 3 months, or yearly, after which a new period starts with a higher amount as the principal, containing the accumulated past interest. This is called **compound interest** and is a common practice in financial organizations. You need to study the text on "Value of Money" and "Payback Installment Calculation" in this chapter, in order to follow the examples.

cost of capital
The money one has to spend (in terms of percentage) to generate capital for an investment or to pay for a project.

discount rate
A term used in economics to imply the rate at which one can borrow money (practically it is the same as the interest rate.)

net present value
A term used in economics to represent all the revenues minus the money invested and all the costs during the life of a project, converted to today's money. Today's money has more value than future money. Thus, the numbers become smaller when expressed in terms of today's money.

compound interest
A term used in economics implying that the interest on capital is periodically added to it, thus increasing the capital.

EXAMPLE 1:

With an annual interest rate of 8%, determine the future value of $15M (15 million dollars) investment after 5 years. The interest is compounded every 6 months.

Solution: Note that the annual interest rate is 8%, but the compound interest is calculated every 6 months. After 5 years we have 10 intervals, and the interest rate is 4%. Thus, the values for use in the formula are (see the text on "Value of Money" in this chapter):

$$P = 15,000,000 \qquad r = 0.04 \qquad n = 10$$

And the amount of $15M today is equivalent to having (with FV meaning future value)

$$FV = (15,000,000)(1.04)^{10} = \$22,203,664$$

at that time (5 years from now).

EXAMPLE 2:

Suppose that $15M is spent for a project. It takes 1 year for the project completion, after which the project does not have any revenue for another 2 years, but after that its yearly revenue is $5M for 5 years. We want to see if investment in this project is profitable. The interest rate is 10% and it is compounded annually.

Solution: This problem is a bit more involved, meaning that it requires more calculations. One way to do it is to find the present values of all the revenues in 8 years and compare it with the money to be spent today. Note that there are no additional expenses associated with the 3 years during which the project has no revenue.

Since the revenues correspond to various years, the present value of each one must be calculated separately, as follows. We may, thus, write (using PV for present value)

$$PV = P_1 + P_2 + P_3 + P_4 + P_5 = (5,000,000) [1/(1 + 0.10)^8 + 1/(1 + 0.10)^7 + 1/(1 + 0.10)^6 + 1/(1 + 0.10)^5 + 1/(1 + 0.10)^4] = (5,000,000)(0.4665074 + 0.5131581 + 0.5644739 + 0.6209213 + 0.6830135)$$
$$= (5,000,000) (2.8480742) = \$14,240,371$$

We see that the present value of all the revenues in today's money is less than the money to be invested and, therefore, the project is not financially acceptable. Obviously, if the interest rate is less than 10% the numbers change, and the project can become attractive.

From this example you can also see the way to solve the problem in a more general form. That is, even if the revenues for various years were not the same and if the interest rate varies from year to year, one can insert the appropriate figure for each. Note also that since the revenues are accounted at the end of each year, we have considered the present values of money after 4, 5, 6, 7, and 8 years (no revenues after the first, second, and third years).

14.2   Initial costs of a wind energy project

As was mentioned in the introduction to this chapter, there are two types of costs for a wind energy project. The first category is the initial costs. Initial costs are associated with the purchase and installation of wind turbines up to the point that they are ready for production. It is normally a one-time expense that a developer must pay for at the beginning of a project. However, in addition to the price of a turbine and the installation cost, there are other costs that are not so evident, and that one may overlook. We may categorize the initial cost, thus, into direct and indirect costs, as discussed here. Direct costs are for turbine(s) and its ancillary components (controls, transformers, etc.), turbine foundation and installation, transportation, and connection to the grid. Indirect costs are for the purchase of land and/or the site, roads to access the site, extra cost if the site is far from the grid, legal fees, and so on.

In this section we consider the various initial costs of a wind energy development project.

Cost of a turbine   14.2.1

The cost of a turbine is what one must pay to the manufacturer for all the components of a turbine. Normally a turbine is sold as a whole, like an automobile. So, although the blades, the gearbox, the generator, the nacelle, and the tower can be manufactured by different companies, they are already matched together and are delivered to a customer. At delivery to the site, however, a turbine is not yet assembled because of the large size. Each major component is delivered separately, and the assembly takes place on site only after the foundation is ready.

The cost for a turbine, thus, is the price one must pay for:

tower (including what is inside the tower, such as ladder, cables, lights and so on);

gears and motors for yaw motion;

nacelle (containing all auxiliary parts as oil heat exchangers, space heaters, etc.);

generator;

gearbox;

rotor (hub, blades, all the controls inside them);

turbine controller;

pad mount transformer; and

all other small components for measurements, instrumentation, communication, and control.

Installation   14.2.2

Installation of a turbine consists of

1. Foundation construction
2. Erecting the turbine
3. Connection to the grid, testing, and commissioning

The foundation of turbine is a giant block of concrete that must be able to hold the turbine; that is, it must withstand and transfer to ground all the force of weight and the lateral forces that a turbine gets from wind (see chapter 13). The diameter of a turbine tower can be around 10–12 ft (3–3.6 m), whereas the diameter of the foundation can reach 50–55 ft (15–17 m). The depth of the foundation is accordingly proportional. In this regard, the bulk of the foundation is hidden in the ground, covered by soil. Only the part of the foundation a turbine is bolted to shows from outside, like an iceberg, only the tip of which is out of the water.

Considering the size, weight, and the forces exerted on a turbine, the weight of the foundation block must be comparative. The dimensions of the foundation, depending on the type of soil and the ground condition, are determined to withstand the forces. In this respect, the construction of the foundation implies digging the ground and replacing the soil with thousands of cubic yards of reinforced concrete. Obviously, the cost of this operation is not trivial. But it is a one-time expense that comprises a significant percentage of the initial cost of a turbine.

After the foundation is ready, erection of a turbine is usually carried out by setting up the base (lowest segment) of the tower. The tower arrives to the site in three or four segments, depending on the height and the design—the base, one or two middle segments, and the top. The base is fixed to the bolts mounted in the foundation. This follows by fixing one by one the other segments of tower on top of each other, and bolting each to the previous one.

After the tower is completed, the nacelle, the gearbox, the generator, and other components, depending on the way a turbine is designed, are lifted and fixed in their places. Finally, the rotor, which was assembled on the ground, is lifted and attached to the nacelle.[1] All the lifting must be done by strong and large cranes that are able to reach higher than the nacelle (see **Figure 14.1**). This installation of the turbine is also costly and the operation counts for a non-negligible percentage of the initial cost.

Courtesy of Clipper Windpower, Inc.

FIGURE 14.1 *Installation of wind turbines constitutes a large part of initial cost.*

---

[1] Sometimes, depending on the turbine design, manufacturer, and site conditions, the hub will be mounted on the nacelle without the blades and then each blade will be lifted and mounted on the hub individually.

Transportation   14.2.3

Another portion of the initial cost is the expense of transporting the turbine(s) to the site. That is, from the manufacturing site to the wind farm. Whereas the previous two cost items are almost the same for two identical turbines, this cost can be very different and depends on factors such as the distance from the manufacturer, how far the site is from major roads, and how difficult access is to the site.

All the pieces of a turbine are either heavy or long. For on-road transportation of a large turbine (over 1 MW), each blade is carried separately and each segment of the tower is also carried separately on a special truck for long loads. On large roads and highways, this is normally not a problem provided that the highway safety code is respected. On smaller roads and in mountainous regions, nevertheless, a long vehicle may have difficulties in turning, in addition to disturbing the other traffic, which can lead to travel time limitations and long delays. As well, if no appropriate road exists from a main route to the site, new temporary roads must be constructed. These are the extra costs that cannot be avoided. See **Figure 14.2.**

All of the preceding road factors can add significantly to the cost of a project. In other words, two exactly similar projects at two different locations may have a significant cost difference because one is near a major road and the other one is in a remote area.

Grid connection   14.2.4

The same thing that was mentioned for the transportation and road factor is true also for connection to a grid. That is to say, if a transmission line exists in the vicinity of a wind farm, connecting the farm to the grid is much less expensive than if a wind farm is in an area where either there is no transmission line or there is no more capacity for an existing transmission line. Installing transmission lines, when necessary, adds to the cost of a project.

UpWind Blade Solutions, Inc.

FIGURE 14.2 *Transportation of large blades requires special vehicles, special arrangements, and sometimes new roads.*

VALUE OF MONEY

Projects are evaluated based on the future value of money spent at the time of project start and the present value of the revenue gained in the future. The formulas for calculation of the future value of money and the present value of money are given here.

Interest is compounded at a fixed interval. This interval must be known. Also, the rate of interest must be known. If the interest rate changes, the problem becomes more complicated. Here we consider the simplest case that the interest rate does not change.

Interest rate is normally given as annual rate. If the interval of compound interest is shorter (it is never longer), for example, 3 or 6 months, then the proper rate for that period must be used in the formula. For example, the quarterly rate is ¼ of the annual rate.

Future value of money

If the *principal* (the initial money spent or invested) is P and the interest rate for the interval of compounding is r, then at the end of one interval the amount is $P(1 + r)$. This new amount is the principal for the second interval, after which the amount is

$$P(1 + r)(1 + r) = P(1 + r)^2$$

In the same manner, the formula to use to find the *future value* (FV), the amount of money, and the accumulated interests for the duration of n intervals is

$$FV = P(1 + r)^n$$

Present value of (future) money

In the same way and by the same reasoning, one can determine the present value (today's value) of a given amount in a given future from now. Present value is today's worth of a given future amount of money. For example, an amount of $1 million 10 years from now is equivalent to a lesser amount that is available now. In the simplest case, the same formula can be used.

If the interest rate is r, the *present value* (PV) of the *amount A* at a time n periods from now is

$$PV = A/(1 + r)^n$$

You see that the total of this amount and its accumulated interest after this n periods of time is A.

For a wind farm, all the generation, that is, the output electricity from turbines, is put together in a collector. The collector, which operates like a substation, is connected to a grid. The cost of the substation and the cables from turbines (usually buried underground cables) is included in the initial cost of a wind farm.

Legal and other costs 14.2.5

Among the indirect initial cost of a turbine there also are those expenses for legal issues such as right of way, agreements with land owners where the turbines are located, various contracts, insurance for operations, and similar items. This cost is not very high, but still has to be acounted for when a wind farm is to be developed.

Operating cost 14.3

The operating cost or running cost for any plant or business activity, including a wind farm, is the regular day-to-day expenditure for running the business. For a general plant, this is normally addressed as operations and maintenance. The breakdown of this cost depends on the type of the activity. The cost associated with a wind power plant can more specifically be categorized into:

Operations
Maintenance
Insurance
Lease and Royalty
Taxes

For a production or manufacturing plant, normally the cost of raw material must also be added to the list. For an activity that requires people, the building to accommodate them, heating, office supplies, and so on are to be also added. For a power plant, in addition to these there is a larger cost for fuel. Compared with gas, fossil, coal, and nuclear power generation plants, the fuel for a wind turbine is wind, which is free. With the others, one has to pay for gas, coal, oil, or uranium.

> Compared with gas, fossil, coal, and nuclear power generation plants, the fuel for a wind turbine is wind, which is free.

For wind energy, since a wind farm is considered as a unit of activity from investment point of view, no considerable office cost is involved and the list is short, as already mentioned. Relatively, the maintenance cost, nevertheless, is more outstanding and can constitute a larger percentage of the total operating cost. In this respect, for more discussion we may separate maintenance from the rest.

Running cost 14.3.1

We consider here the other wind turbine operating costs except the maintenance for further discussion.

When a set of wind turbines are in production they work in open air in the land where they are installed. In many cases, this land does not belong to the developer of the wind farm. It can belong to one or more owners. Based on a

PAYBACK INSTALLMENT CALCULATION

As you go further through this chapter you will see that you need to add up the total amounts of periodic payments and their interests after a given time. For example, one wants to know the income from the yearly sales of electricity in a wind farm after, say, 25 years. Another example is the monthly payments that a person makes to a bank, until the loan is fully paid back.

In both of these problems, normally the payment is constant (or nearly constant in the first example). Not all the installments carry the same interest after the period, since the duration is not the same for all. The very first payment or income remains in the operation for the longest time, whereas the last one does not carry any interest. The duration to be considered for interest calculation of each installment decreases by one after each payment period.

For a production project, the sum of all the installments and their interests is the total income in terms of a future value of money. For a loan repayment, the sum of all the installments and their interests must become equal to the sum of the money borrowed plus all the interest it accumulates for the maturity period. Here you learn the involved mathematics.

Suppose that A is *the amount of the income at the end of each period*, and there are n periods. If the interest rate is r, the *total* amount T after n periods is

$$T = A(1 + r)^n + A(1 + r)^{n-1} + A(1 + r)^{n-2} + A(1 + r)^{n-3} \\ + \cdots + A(1 + r)^2 + A(1 + r)^1 + A \tag{1}$$

In order to find this sum, one may multiply T by $1 + r$:

$$T(1 + r) = A(1 + r)^{n+1} + A(1 + r)^n + A(1 + r)^{n-1} + A(1 + r)^{n-2} \\ + \cdots + A(1 + r)^3 + A(1 + r)^2 + A(1 + r) \tag{2}$$

Now, subtracting all the elements in both sides of equation (1) from equation (2) gives (note the similar terms cancel each other by subtraction)

$$T(1 + r) - T = A(1 + r)^{n+1} - A$$

or

$$Tr = A[(1 + r)^{n+1} - 1]$$

T, the total amount after n periods, thus is equal to

$$T = A[(1 + r)^{n+1} - 1]/r \tag{3}$$

In the case of a loan, if we want to know what the monthly payment for a loan P from the bank is, we can combine the above formula with the one we learned for the future value of money. If T is the total amount of the money borrowed plus its compound interests after a given period, the total future value of all the monthly payments and their interests after the same period must also be equal to T. The *monthly payment* (installment) A, for the borrowed money P, for the period n, thus, can be found to be

$$A = Tr/[(1 + r)^{n+1} - 1] = Pr(1 + r)^n/[(1 + r)^{n+1} - 1] \qquad (4)$$

contract, a yearly amount is paid to the owner of the land in the form of rent or lease for the land. At today's rate this amount is \$5000 to \$6000 for each turbine.

A second item of this cost is the amount to be paid for the insurance. Insurance of an operating turbine is necessary. Again, this is based on the content of the insurance policy. It could be for the coverage on the equipment failure, damage to surrounding area, injury to people, fire, and so on. This could be much more than the aforementioned figure for rent, for each turbine.

A third item that comes into effect is the tax that a company must pay. Tax is usually calculated based on the income of a corporation. For a wind farm, however, certain tax breaks are granted by governments to help or persuade companies to invest in clean energy. The tax sometimes makes a big difference in the financial status of a wind energy project. Companies seriously take this into account, since the tax to pay determines a "go" or "no go" decision for a project.

The salary to be paid to technicians for the maintenance and upkeep of a turbine can be regarded as part of the maintenance cost. Nevertheless, supposing that there is no repair for a turbine, but technicians are paid and are available to do any repair job, it could be considered a part of this cost category.

Maintenance cost

14.3.2

Like any other machine, wind turbines can break every now and then and they need to be fixed. Also, regular maintenance such as inspection of parts and changing oil are necessary for proper work of a turbine. Maintenance, in general, falls into two types: fault correction maintenance (or corrective maintenance) and preventive maintenance. **Corrective maintenance** refers to bringing back a turbine to working order after a part has failed and has caused the machine to shut down; **preventive maintenance,** on the other hand, is beforehand scheduled inspection, detection, correction, and repair of machine parts in order to prevent future breakdowns and failures that could be more costly. In preventive maintenance, some parts that are prone to fail soon are replaced, even if they are still working. The cost of maintenance can become very high and prohibitive, and companies must reduce this cost to a minimum by paying careful attention to what the causes of problems are and eliminate them.

> The cost of maintenance can become very high and prohibitive, and companies must reduce this cost to a minimum by paying careful attention to what the causes of problems are and eliminate them.

corrective maintenance
A maintenance philosophy that a component is fixed after it breaks (if it works, do not fix it), as opposed to preventive maintenance.

preventive maintenance
The systematic repair and maintenance work to prevent a fault or failure before it happens.

The cost of maintenance, as we'll see here, is not just the price of the items that need to be replaced and the labor. Here we can again categorize this into direct and indirect classes. When a turbine shuts down because of a fault or for maintenance the following costs are involved:

a. Loss of revenue for the whole period of downtime.
b. The cost involved for the problem to be fixed.

Since turbines are normally remote from the offices and people to maintain them, it takes a long time for technicians to reach a turbine and diagnose the problem. Climbing a turbine with the proper safety gear on also adds to the required time. Moreover, the size and weight of a piece to be replaced, and the time of the year (consider below-freezing temperature in a windy region with a lot of snow on the ground) directly influence the downtime and the amount of revenue lost.

If a major component, such as a blade, gearbox, generator, and so on needs a repair, the cost is very high and in most cases requires a crane to be brought to the site. This is a noticeable cost for a wind turbine. By properly looking after the turbine and doing scheduled inspection and preventive maintenance, the maintenance cost can be minimized. **Figure 14.3** depicts how the maintenance cost for a turbine can be kept to a minimum, whereas it can skyrocket to a much higher amount. This figure illustrates the effect of preventive maintenance (**scheduled maintenance**) for a turbine in a number of years.

scheduled maintenance Regularly planned repair and maintenance work as part of preventive maintenance.

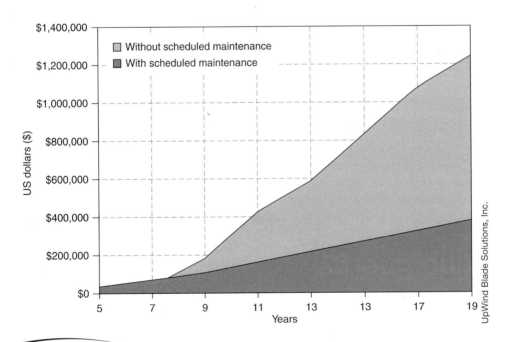

FIGURE 14.3 *Proper scheduled maintenance can eliminate many unnecessary costs and revenue losses.*

Comparison with other energy sources   14.4

A distinct difference between a wind power generation facility and a conventional fossil power plant of the same size is the initial cost. The initial cost for a wind farm is much higher than the initial cost for a fossil power plant. Since the initial cost has to be paid up-front, it makes it more difficult for investment. Nevertheless, since one has to pay for the fuel, but wind is free, the difference gradually changes the course.

Each year the cost of maintenance, other running costs, and the price for fuel adds up to the amount of expenditure from an economical standpoint; that is, the invested capital. The total cost of a project, thus, can be shown by adding all the cost in each year to the cost of the previous year. The result is an ascending curve. **Figure 14.4** illustrates the matter in broad terms (that is, without any variation of the cost from one year to the other) for a typical wind farm and a fossil plant.

In this figure the price of oil (gas, coal) is considered to remain constant over the 20 years that the graph shows. Also, no inflation of prices is taken into account for more clarity of the comparison. According to this graph, after 15 years the cost of the fossil plant exceeds that of the wind farm. In reality, inflation is not zero and the increase in the oil price cannot be ignored. These will affect the number of years one surpasses the other.

A plant is expected to be productive for 20 to 25 years. In practice, it may perform well beyond that, say, up to 30 years. Considering the total cost of a fossil plant after this time, it is obvious from figure 14.4 that the cost of a wind farm is much lower than the cost of a fossil plant.

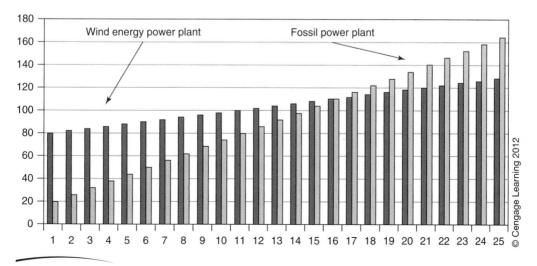

FIGURE 14.4 *Comparison of total long-term cost of a fossil power plant with that of a wind power plant.*

Any industrial plant (power generation or manufacturing) comes to the end of its useful life, and operations will be stopped when the cost of running the plant exceeds the revenue from the plant. In such a case, the plant will be abandoned or salvaged. The 20 to 25 years, mentioned above, is for many large projects. For an advancing industry such as wind turbines, a plant can become obsolete before that period. The rule of thumb for industrial machines, and a wind turbine is not an exception, is that the larger a unit, the more efficient it is.

For wind turbines, the new generations are around 2–3 MW, compared to those at 500–750 kilowatts. So, it will be no surprise if a wind turbine is *decommissioned* (put out of work) even before its life expectancy has ended.

> The rule of thumb for industrial machines, and a wind turbine is not an exception, is that the larger a unit, the more efficient it is.

14.5   Cost per unit

It is sometimes necessary to know the cost per unit for an item. For instance, what is the cost for a 2-MW turbine? Or, what is the cost per watt in a wind farm? In section 14.3 the initial cost of wind turbines was analyzed and the various components of the cost were explained. From that discussion it is easy to realize that the cost associated with a single running turbine depends on a number of factors, and it is not easy to just assign a number to the cost. In other words, the cost for each individual case can be specific for that case.

For two wind farms, the cost of purchase of a turbine may not change that much, but the cost of transportation and connection to the grid can be significantly different. Consider two wind farms, one with 150 turbines and one with only 10 turbines. The cost of constructing access roads and pavement to bring the turbines to the site, and the cost of a substation for grid connection, will be shared between 150 units in the first case, whereas it is for only 10 turbines in the latter.

On the other hand, when comparing the older turbines with a smaller capacity with today's larger turbines, a difference must be observed for advancement of technology, increase of efficiency due to the larger size, and decrease in cost of installation for less number of turbines (consider installation of 300 units of 400-kW turbines versus 60 units of 2-MW turbines, both totaling 120 MW of installed energy). Thus a figure for the average cost, based on the total cost of many wind farm projects, can be obtained. This figure can be used as an index or a rule of thumb in order to roughly determine the cost of a project, or compare past and present costs. At today's prices one can say that the cost of a wind turbine is 1.2–1.4 dollars per watt. Thus, for example, the cost of a 2-MW turbine is about $2.6 million.

A case study   14.6

In this study, we take the example of a wind farm with a number of turbines. We want to see if the project of developing this wind farm makes a profit, or if it should be rejected. Each turbine to be installed is 2 MW and the total cost of developing the farm is based on the simpler value of $1 per watt. In this sense, calculation can be made for one dollar, but here for the sake of simplicity it is carried out for one turbine at $2M. The following data are necessary (numbers are all assumptions):

- Years of operation = 25
- Interest rate = 7%
- Operating cost = 5% (100,000$/year)
- Selling price per kilowatt-hour = $0.04
- Capacity production per year = 40%

Note that the installed power is 2 MW. But, the production is less than the installed power. Production varies during 24 hours and also it varies based on the season. (See chapters 2 and 7.)

For this project there are three items that we must consider and find their values to compare: (a) the cost of investment (initial capital), (b) the cost of operations, and (c) the income from selling the product. If the total cost after 25 years is less than the total income, the project is not profitable; otherwise the difference shows the profit after 25 years. The first year is considered the development stage and neither the operating cost nor the income are considered.

a. Cost of investment after 25 years = $2,000,000 \times (1.07)^{25}$ = (2,000,000) (5.4274) = $10,854,865.

This is calculated directly based on compounded interest.

b. Cost of operation after 25 years = cost of operation of each year and its interest after number of years until 25 years.

For this item, the interest to be calculated starts from the end of the second year (construction work, no operations cost in the first year), 23 years for the second year, 22 years for the third year, and so on until the 25th year with no interest. Instead of adding the individual values for each year, the following formula can be used (See equation 3 in the text on "Payback Installment Calculation", in this chapter):

$$100,000 \times [(1.07)^{24} - 1]/0.07 = (100000)(58.17667) = \$5,817,667$$

c. Income calculation

$$\text{Income each year} = 0.4 \times 2,000 \text{ kW} \times 365 \times 24 \times 0.04$$
$$= \$280,320 \text{ (No tax considered)}$$

This amount is the income each year based on the average production of 40% of 2 MW each day. The interest for every year income must be accordingly

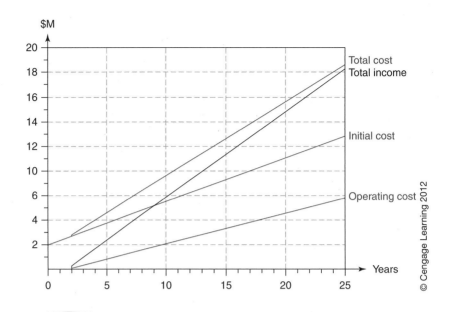

FIGURE 14.5a *Project cost and income in 25 years (electricity at 4 cents per kilowatt-hour).*

calculated and added to the amount. Similar to the cost, the interest starts from the end of the second year; thus, it must be 23 years for the second year, 22 years for the third year, and so on.

Using the same formula as above:

$$\text{Accumulated income after 24 years} = 280{,}320 \times [(1.07)^{24} - 1]/0.07$$
$$= (280{,}320)(58.17667) = \$16{,}308{,}084$$

$$\text{Net profit after 25 years} = \$16{,}308{,}084 - \$10{,}854{,}865 - \$5{,}817{,}667$$
$$= -\$364448 \text{ (\textbf{Negative})}$$

Figure 14.5a illustrates the above numbers and how the cost and income are related.

Note that instead, we can subtract the cost each year from the income, and use the same formula for the difference, which is the net profit each year. The reason it was done separately is that we may want to see if the electricity is sold at a different rate, then what the income would be. For example, consider that the sale price is 5 cents per kilowatt-hour, instead of 4. Then the income from sales will be (no tax or tax break is considered)

$$\text{Income in year} = 0.4 \times 2{,}000 \text{ kW} \times 365 \times 24 \times 0.05 = \$350{,}400$$

$$\text{Income after 25 years} = \$20{,}385{,}105$$

$$\text{Profit after 25 years} = 20{,}385{,}105 - 10{,}854{,}865 - 5{,}817{,}667$$
$$= \$3{,}712{,}573 \text{ (\textbf{Profitable})}$$

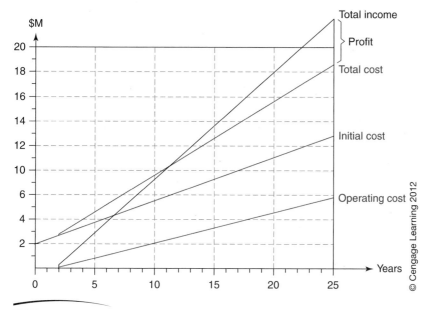

FIGURE 14.5b *Project financial details (electricity at 5 cents per kilowatt-hour).*

The corresponding graph is shown in **Figure 14.5b.**

Instead of finding the values of the investment and net income after 25 years, it is possible to base the calculations on the present values of them. In this way, the net present value can be found. It is another way of doing the calculations based on today's worth of money. The results lead to the same conclusion. That is, for a price of 4 cents per kilowatt-hour there is a loss, but for 5 cents per kilowatt-hour there is a profit.

Chapter summary   14.7

- Value of money is not constant; it changes with time. Money today is worth more than the same money in the future.
- In order to evaluate an investment project, the net income after the useful life of the project must be calculated. This is done by either reflecting all the future income and cost in terms of present value of money, or reflecting all the income, cost, and investment in terms of a future value of money, usually at the time the project is completed and must be salvaged.
- Future value of money depends on interest rate.
- A developer decides to invest in a wind farm if it is profitable.
- The cost of developing a wind farm for generating energy, like any other large project, consists of initial cost and operating cost.
- The initial cost is the expenses that must be paid up-front for the purchase, transportation, erection, and connection of turbines to electric grid.

- The operating cost is the annual expenses that are required for maintenance and running the generation and delivery operations.
- Compared to other types of power plants, the initial cost of a wind turbine operation is relatively much higher.
- The operating cost, on the other hand, is relatively much smaller, since one does not have to pay for the fuel.
- Proper preventive maintenance can significantly reduce the cost of maintaining wind turbines.
- Preventive maintenance is scheduled inspection and repair work, which can enable one to avoid or reduce long delays, downtime, and unnecessary additional costs.

ADVANCED LEARNING

Cost modeling and scaling

For large industrial projects and large machinery it is often required to estimate the cost before decisions are made. This will be a tedious, time-consuming, and expensive process, since a lot of the components in the project have to be designed for the required purpose and with the desired capacity. For example, a company that is expert at manufacturing 600-kW wind turbines is asked to bid for construction of 20 wind turbines of 2 MW power. The company wants to know, based on the cost of 600-kW turbines, how much the approximate cost of a 2-MW turbine will be. The company can do the calculations for the major components only and come up with the estimate.

In this example and similar cases, companies use a formula to correlate the larger size component to the smaller size and use the price of one to obtain a number for price estimation.

The following formula can be used for estimating the cost of a component based on that of another one with a different scale: The machine or project for which all the past data are already available is called the "baseline."

$$C(x) = C_B \left(\mu \frac{m(x)}{m_B} + (1 - \mu) \right)$$

$C(x)$ is the estimated price for the new component/machine, C_B is the baseline (previous and already known) price, $m(x)$ is the mass when the design parameter takes the value x, m_B is the baseline mass values, and μ is the proportion of cost that varies with mass.

For example, consider the case of the turbine, as mentioned above. In order to estimate the cost of the generator, suppose that 40% of the generator cost is associated with the material and the rest is the labor. Thus, a larger generator requires more material for manufacturing. Therefore, the value of μ is 0.40. The value of x in this case is simple, thus the ratio $m(x)/m_B$ is 2,000,000/600,000 = 20/6 = 3.34. Applying these values in the equation leads to

$$\begin{aligned} \text{Cost for new generator} &= \text{previous cost} \\ \times \, (0.4 \times 3.34 + 0.6) &= 1.936 \times \text{previous cost} \end{aligned}$$

So, although the size is 3.34 times larger, the cost is only about twice as much.

Review questions

1. Why is economics of wind turbines important?
2. What is the future value of money?
3. What is the difference between the present value of money and future value of money?
4. What happens to the future value of money if the interest rate is zero?
5. Describe the two categories of cost for a wind turbine project.
6. Name the various items of expenditure associated with the initial cost of a turbine.
7. What makes a project financially attractive?
8. What is the main difference in financing a wind energy project and other types of power-generating plants?
9. What makes a wind energy project competitive if the initial cost is much higher than that of other types of energy?
10. How can the maintenance cost be reduced?
11. What is the effect on the cost if turbines get poor maintenance work?
12. What is meant by scheduled maintenance?
13. What is preventive maintenance?

Problems

1. If the annual interest rate is 6%, what is the total interest and capital of $1 million after 1 year, and after 2 years if the interest is compounded on a yearly basis?
2. Find the same values in question 1 if the interest is compounded every 6 months.
3. With a constant annual interest rate of 6% in 5 years, if someone pays you $5,000 5 years from now, how much is that money worth now?
4. Show that having $10,000 now is better than if you have $14,000 6 years from now, if the interest rate remains constant at 6%.
5. If you borrow $20,000 dollars at a rate of 8% and want to pay it back with only one payment after 4 years, how much do you have to pay?
6. An insurance company must pay you 5 yearly payments of $10,000 starting now. They offer you to take instead a lump sum of $40,000 in the first

year. If the current annual interest rate to borrow money is 7.5%, but the interest on saving is 4%, will you accept the offer?

7. The initial cost of a project is $20 million. The expected income for 10 years is $2.2 million per year, starting from the second year. The annual interest rate to borrow money for this project is 5%. The project can be sold for $5 million after this period. Determine if the project is profitable.

PROJECT

The cost of a small wind farm is $27 million and it is projected for 20 years. The present interest rate is 7.5%. The interest rate is expected to decrease to 5% after 5 years, but then after 6 years increase to 8% and remain constant for the rest of the project life. The installed power is 16 MW and the monthly variation of production in the region has a pattern as shown in figure 7.8. Determine the selling price of electricity per kilowatt-hour so that the project makes at least 15% of net return on the investment.

Environmental Concerns

© Cengage Learning 2012

KEY TERMS

flashing
flickering
high-voltage direct
   current (HVDC)
infrasound
offshore
onshore
prevailing wind
shadow flicker
spar platform
tension leg platform (TLP)

OBJECTIVES

After studying this chapter, you should be able to:

- Explain the effects that wind turbines may have on the environment.
- Become more alert about the environment.
- Describe the main difference between onshore and offshore wind farms.
- Know why the older wind turbines were noisier.
- Understand that noise can be at various frequencies, some of which we cannot hear.
- Explain that if a turbine is far enough that one cannot hear it, it does not have adverse sound effects on the neighborhood.
- Realize that for wind farm development, one must consider community approval.
- Know that not everybody will be happy about having a new wind farm in the view.
- Describe why the newer turbines are less annoying to people than the old ones.
- Understand the meaning of flashing and flickering.
- Realize that birds and bats can be killed by wind turbines.
- Explain that on all environmental matters, research has been carried out for scientific investigation.

(continued)

- Realize that many more birds are killed by many other man-made structures and machines.
- Realize that wind turbines can also interfere with communications.
- Know that for offshore wind turbines high-voltage DC is preferable.
- Realize that offshore wind turbines can have a fixed tower or a floating tower.
- Describe why, despite the cost, we have offshore wind farms.

15.0   Introduction

One of the contradictory aspects of a wind turbine is its effect on the environment. People with different views see things differently, and sometimes statements are made based on insufficient experimental measurements and scientific backup. It is necessary that a reader become familiar with various effects of wind turbines on the environment. In this chapter, these effects are studied without making any comparison with any other ways that electrical energy is generated.

On some of the issues a great number of studies have been carried out in the past, and if no satisfactory results have been obtained, necessary investigations are still in progress to the present time.

Wind farms primarily fall into two categories: onshore and offshore. As the name implies, **offshore** wind farms are those installed in the middle of water, away from land. Offshore turbines cost much more. The reasons for offshore wind farms can be manifold. One reason is the limitation of available land in certain regions. Other reasons such as noise and view in the landscape could also play a role, particularly in the older turbines where the noise level is higher and the turbine color has more contrast with the sky. Also, at sea there is more wind, and wind is steadier. **Onshore** turbines are installed on ground. The environmental aspects of the two classes are a little different, as we will see in this chapter.

offshore
On a lake, sea, or ocean, away from land.

onshore
On the land.

> Our responsibility dictates that we keep the environment clean. This implies that we use the land in a responsible manner, and not destroy nature by overexploiting it, even if it is for installing wind turbines.

Onshore wind farms   15.1

Offshore wind farms are far into the sea, sometimes up to 50 km (31 mi). Therefore, contrary to the onshore turbines, they are not seen by people, some of whom may feel disturbed by a large number of rotating blades in their view. Other than that, in today's turbines, the other differences are the land that onshore turbines occupy and the underground cables from each turbine to the collector. Offshore turbines are also made much larger. Obviously, from the points of view of cost and maintenance, onshore turbines have preferences over the offshore turbines.

Noise   15.2

Noise, contrary to voice and music, can be considered a type of contamination, no matter what the source. It is, in fact, an unwanted sound that is not pleasant to the human ear; it could also be music or voice, when it is not desirable or when it reaches a disturbing level. Elevated levels of noise can be found in and around airports and near train tracks and highways. Nevertheless, in the modern world and in industrialized regions (the factories), a large number of people are exposed to this contamination. Noise can also be at a low level, but disturbance comes with its continuity, such as the noise from air conditioners, refrigerators, and the like. When one is irritated by hearing other people talking to each other or on their cellular phones, playing music, or listening to music are other examples.

One of the issues associated with wind energy that has gained weight is the noise from wind turbines. Even if it is not realistic today, this began in the time of the earlier turbines. Older wind turbines had a faster speed of rotation and the blades did not have a refined design. Thus, they were noisier than today's turbines, which feature a smaller angular speed and particularly a better aerodynamic design for blades.[1] The older wind turbines were also smaller and the nacelle was closer to ground; thus, the sound from the gearbox could be heard in their vicinity. This noise can well travel to the bottom of a tower, due to its metallic structure. On the contrary, the newer turbines are taller and the nacelle and the noise of the gearbox, as a result, are farther from the ground. In modern wind turbines, a listener on the ground can hardly hear the noise from the gearbox.

Sound is generated at different frequencies, and a human ear is sensitive (not all humans are the same) to only the frequencies in approximately the range between 100 Hz and 10,000 Hz. Nevertheless, sound at lower frequencies can also exist, though not audible by a human. Sound below the audible level is called **infrasound.** The effect of very low frequency sound (say, 1–3 Hz) is in the form of vibration, since sound is indeed generated by vibration, and it can travel through any medium, including air. This low-frequency sound is, in fact, structural vibration that travels through structures and the ground. In this

infrasound
A sound with a frequency below the threshold of human hearing so that it cannot be heard by a human.

---

[1] Note that although the rpm is lower in newer turbines, the relative velocity of parts of the turbine blade can be higher due to the larger blade size.

respect, some people have raised the issue that wind turbines generate low-frequency sound that leads to headaches and some other physical sicknesses.

These complaints may once have been made by the people living in the neighborhood of a wind turbine, but later they were used as a weapon against the installation of wind farms. Obviously, for some people in the vicinity of wind farms, a wind turbine would be an annoyance before being a health hazard. However, the issue of wind turbine noise is not public knowledge and thus can mislead many people.

Basic (single-frequency) sounds are characterized by two factors: the frequency and the intensity. The intensity (or loudness to a listener) is the obvious factor for irritation or annoyance when it exceeds the bearable level, no matter how desirable the source may be. The intensity of sound depends on the distance from the source and it decreases proportional to the square of that distance. Also, the direction of wind in the case of wind turbines plays a significant role, by bringing the audible sound toward a person, or taking it away. The loudness of a sound when it is heard depends on other conditions as well, such as reflections. The noise from modern turbines in a wind farm can, indeed, be less than the noise from wind through fields and trees.

Table 15.1 at the end of this chapter shows a comparison of noise levels of various sources. Nowadays there are governmental regulations for allowable noise level during the day and at night for various districts (residential, industrial) in a city. Accordingly, there are regulations about how far from or near to residential areas a wind farm is allowed to be built. This is based on significant amounts of studies and research on the subject. The conclusions of the scientific studies are as follows:

1. So far as a wind turbine noise cannot be heard, there are no hidden sound-related aspects that can be harmful to humans. If the noise from a wind turbine can be heard, but it is not loud enough to be irritating, then there is no infrasound at a level harmful to the human body.
2. If a wind turbine is far enough from residential areas so that it cannot be heard, although it can be seen, there is no sound-related cause for irritation and annoyance.

15.3   View

It is of no doubt that wind farms introduce a change of view to the landscape. In many regions, wind turbines are relatively new entities added to the neighborhood. Since "change" is often difficult for human beings to accept, for some people this change in view can cause some irritation, particularly in the beginning. Nonetheless, as is the case for many other changes that occur, such as new buildings going up in an area, the fact is that one gets used to it. For those people who are far from any wind turbine or wind farm the subject is immaterial.

For an individual looking out of his or her window and seeing a wind turbine or a lot of them instead of the previous natural view, there can be a disturbing feeling, at least for some time. The fact is that (a) this is a feeling that can gradually fade away, and (b) in many cases, there is little that an individual person can do about it.[2] For wind farms in remote and rural areas, the view is not a major problem, but when wind turbines are installed close to residential areas, the rights of the people must be respected and wind farms should not disturb people's lives.

Looking at this matter more logically, we know that wind is a source of plentiful clean energy and with the ever-increasing demand for energy the advantages outweigh the disadvantages. Without doubt our responsibility dictates that we keep the environment clean, meaning that we use the land in a responsible manner and not destroy nature by overexploiting it, even if it is to install wind turbines.

Newer turbines are painted in a color that has less contrast with the sky, compared with the older turbines. In this sense, when in operation they can be less disturbing to a viewer. Also, research has shown that three-blade turbines have a less negative effect than two- or one-blade turbines on people who may be sensitive to wind turbines in their view and those who find them in their view every day.

In general, noise is more annoying than an unwanted view since one can hear the noise when it exists, whereas one has to look in a certain direction in order to see the view.

> Noise is more annoying than an obstructed or unwanted view, since one can hear the noise when it exists, whereas one has to look in a certain direction in order to see the view.

Flashing and flickering

15.3.1

When wind turbines are installed very close to residential areas, in addition to just being in the view, they can have other disturbing effects. When a wind turbine is working, it is possible that in a specific condition (time of the day and sunshine directions and at a particular angle of view) the blades show a reflection of sunshine. This is called **flashing** and can last for quite a while until the angle of the sun changes sufficiently. With the modern turbines there is less possibility of flashing because of the color. Older turbines sometimes had blades with shiny colors.

Flickering occurs when the shadow of the rotating blades repeatedly and in a cyclic manner falls on the surrounding land or a building. Turbine blades are not small and their shadows can cover a large area, depending on the time of the day. The effect of **shadow flicker** is more pronounced at locations closer to a turbine. The regulations set today by municipalities and authorities are much broader than those in the past and take into account many more concerns, including turbine shadow flicker.

The shadow of a turbine blade's size and position vary depending on the time of the day and the time of the year. At each instance, the shadow depends on the sun direction. **Figure 15.1** shows the shadow in a possible case. The

flashing
A repeated reflection of sunshine on a wind turbine blade that can annoy a viewer.

flickering
The repeated shadowing effect of a rotating blade (see also *shadow flicker*).

shadow flicker
The moving shadow cast of a rotating blade.

---

[2] Before a wind farm is constructed, acceptance and/or support is sought from those people who live in the neighborhood. This process may take a long time and the result can be decided by court rulings. In the case where a project is passed by a yes/no decision or by a majority vote or a court ruling, an individual cannot do much about it.

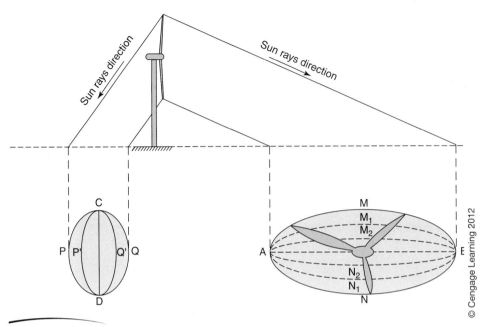

FIGURE 15.1 *Possible shadow of a wind turbine at a specific time.*

shadow can be larger or smaller than the real size of the turbine blades. Blade shadows always move inside an ellipse. Examples of these ellipses are shown in the figure. The blade tips move on the circumference of the ellipse. For instance, in the case represented on the right side of the figure, the shadow can sweep the ellipse denoted by AMBN, AM_1BN_1, AM_2BN_2, etc. As can be seen, the ellipse can be very narrow. At a particular instant, it can only trace the line AB.

In the case shown on the left in figure 15.1, CPDQ and CP′DQ′ illustrate two of the many possible ellipses swept by the shadow. At a particular moment the shadow can only sweep the line CD. Shadow flicker and flashing can be very disturbing to residents of houses very near to wind turbines. The minimum distance of a wind farm from residential areas must bring this into account.

> Shadow flicker and flashing can be very disturbing to residents of houses very near to wind turbines. The minimum distance of a wind farm from residential areas must bring this into account.

15.4 Landscape

A wind farm, in general, consists of a number of wind turbines. This number can be as small as five or six or it can be over 100. Nevertheless, sometimes one or two stand-alone turbines can be seen in some regions. One or two turbines in the view may not have the same impact on the landscape as when they are in a group. Concern about the landscape and the impact of wind turbines, like most of the other matters, started in Europe where the land is scarcer compared to North America.

But even if most of the current wind farms in North America can be far and not seen from residential areas, eventually with the increase in their number, one will see more of them, at least when traveling. Wind turbines then become part of the reality in the landscape, in the same way as the trees and meadows are. As is the case, one can see wind turbines in the fields together with cows, sheep, or other animals grazing, or in the fields with crops of wheat, corn, and the like.

Those who have not seen these scenes yet, will do so later. Wind turbines are there for energy that we need. This is the way to get the energy from wind. And it will take a long time, if it even happens, before our entire electric energy requirement is generated somewhere windy although very far away.

The placement of turbines in a wind farm depends on the topography of the land and variation in the wind direction. Turbines can form a row with respect to wind direction, or they can form a column. A more common case is that they are placed in an array (or matrix form). These are shown in **Figures 15.2** and **15.3.**

If in a region, the direction of wind does not change much, then the relative form of turbines with respect to this **prevailing wind** remains unchanged. That is, a row will always be a row. This can be helpful for determining the distances

prevailing wind
A most-often-blowing wind.

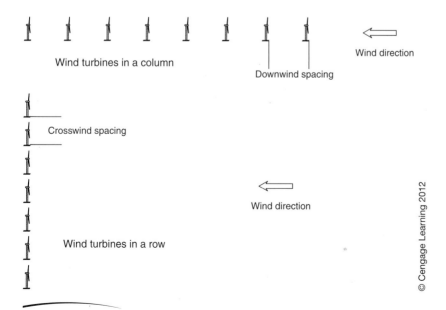

© Cengage Learning 2012

FIGURE 15.2 *Row and column arrangement of turbines in a wind farm.*

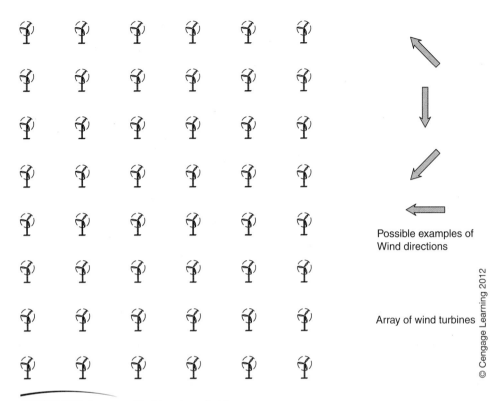

Possible examples of
Wind directions

Array of wind turbines

© Cengage Learning 2012

FIGURE 15.3 *An array of turbines in a wind farm.*

between turbines. A minimum distance between two neighboring turbines is very important in order to avoid the adverse effect of turbines on each other. Wind turbines create a wake of rotating air around and behind them. If another turbine is placed in this wake, there is certain loss of the wind power in the second turbine. For this reason, in the calculation of the total power in a wind farm a factor for this loss must always be brought into effect. This factor depends on how far the turbines are from each other, plus other parameters.

A rule of thumb for placing turbines in a wind farm is as follows: For a single row of turbines the minimum cross wind distance is 1.5 times the rotor diameter; for an array of turbines this minimum spacing must be increased to 2.5 times the diameter. In the case of a column arrangement, the minimum downwind distance is 10 times the rotor diameter. These numbers are the very minimum values and are valid if there is a prevailing wind. **Figure 15.4** shows a wind farm with two rows of wind turbines with a prevailing wind.

U.S. Department of Energy

FIGURE 15.4 *Wind farm with prevailing wind.*

Effects on birds   15.5

One of the major critical issues that have received the most attention is the complaints made by environmentalists about wind turbines killing birds, bats, and other flying creatures that can be hit by wind turbine blades. A noticeable amount of research has been devoted to study the matter. Separate research has been carried out for birds and bats that are randomly killed by hitting or being hit by the turbine blades, and migrating birds that can run into a wind farm and be affected by turning blades.

In the first case, study has shown that the number of birds killed by wind turbines is much smaller than the numbers killed by hitting buildings, cars, and the like. It is true that the numbers of vehicles and buildings are presently much greater than the number of wind turbines. Nevertheless, even with their increasing number, there are other facts that cannot be ignored, as the research shows.[3]

1. The largest causes of mortality among birds include loss of habitat due to human infringement, environmental exploitation, and collisions with man-made objects. Other man-made items in addition to buildings and vehicles leading to avian mortality include:
   – Utility transmission and distribution lines.
   – Communication towers and their guy wires.

---

[3] A good report on the matter can be found in http://www.awea.org/faq/sagrillo/swbirds.html.

  – Bridges and smoke stacks.
  – Agricultural pesticides.
  – Other human-caused contaminations, such as oil spill and so on.
Additional natural causes are other animals (cats and raptors) and trees.

2. Older turbines were mostly mounted on lattice towers. Lattice towers provided ideal nesting spots for birds. This increased the presence of birds in the vicinity of wind turbines, and thus the chance of being hit by turbine blades.

In the case of migrating birds, although study has shown that flocks of birds readjust their flying path and go around the wind farms, one of the considerations in wind turbine siting is that wind farms should not be developed in places that endanger the ecosystem, including interference with the routes of migrating birds.

The studies so far have shown that in the United States, human infrastructure and industrial activities are responsible for 1 to 4 million bird deaths per day! Out of this, commercial wind turbines cause the direct deaths of only 0.01% to 0.02% of all of the birds killed by collisions with man-made structures and activities.

> Studies have shown that in the United States *human infrastructure and industrial activities are responsible for 1 to 4 million bird deaths per day!* Commercial wind turbines are the cause of 0.01% to 0.02% of this.

15.6   RF interference and effect on communications

In all RF (radio frequency) broadcasting and communications such as TV, radio, and aviation and weather radars, solid objects between a radio transmitter and a receiver will cause signal loss. In addition, signals emitted by a transmitter are reflected by any object in an area around the path between the transmitter and a receiver (called the Fresnel zone). This reflection causes multiple signals to be received at the receiver, which will result in a variety of negative outcomes such as shadows of the main signal for analog broadcasting and loss or disturbance of the signal in digital communications.

Obstruction and reflection of communication signals is also made by all the man-made and natural objects, such as buildings, bridges, transmission line towers, trees, mountains, and so on. The interference depends on the material. Metals have a greater effect. The resulting distortions are not desirable, but cannot be avoided. The difference between wind turbines and the above objects is that while those are stationary, wind turbine blades are not. The adverse result can be in the form of inconsistency in the distortion. Whereas remedial solutions can be found for the nonvarying distortions and obstruction by high-rise buildings, stacks, and the like, it is technically much harder to deal with inconsistent effects.

Although wind turbine blades are made of composite materials, the metallic strip for lightning protection can have a considerable effect. It is common for the latest wind turbine blades to reach heights of 150 m (492 ft) and sweep areas as large as 7854 m^2 (84,496 ft^2). Thus, the effect on any nearby communication signal can be significant. Such a turbine can entirely block a point-to-point microwave path. The effect is multiplied for a wind farm that contains many turbines.

Whether or not any individual wind turbine or wind farm has the afore-mentioned adverse effects on any broadcasting and communication activity depends on the particular case and how far from or near to the wind farms communication stations are located. If they are near, or the wind farm is in the direct path of an antenna, the effect is inevitable. It is also useful to know that if a wind turbine is within 1.5 mi of a high-power AM broadcast station, it can cause a shock hazard to the maintenance/construction personnel in the area.

As can be seen, the adverse effect of a wind turbine or a wind farm, if it is not directly in the path of the signal, is like any other man-made construction. It can block a signal from reaching an area or distort a particular signal, but this cannot be generalized for all areas and all signals.

> A wind farm can have distortion effects on broadcasting and communication signals. This depends on the relative geographic location in each particular case. The effect cannot be generalized for all communications around a wind farm.

Offshore wind farms   15.7

Offshore wind farms can be many miles out to sea. Although not yet practiced in North America, in Europe it is common to have wind farms of over 30 mi from shore. Obviously, they cost much more to install and the cost of their maintenance is also higher. The initial motivation for moving wind farms out to sea is a lack of sufficient land and avoiding problems such as view, noise, and the like. Although they cost more and the mechanical problems can increase due to humidity and salt, they benefit from steadier winds and higher wind speeds. The generated electricity, nonetheless, must be transported to land by cables. This is another additional cost compared with onshore farms, and can lead to problems that stem from high voltage. Imagine a high-voltage line sitting on the uneven sea floor, subject to salty water and all microscopic and macroscopic creatures.

Depending on the sea depth, offshore wind turbines are either mounted on a foundation fixed to the seabed, or they have a floating foundation. The floating foundation must be anchored to the seabed with cables so that it does not drift. Where water is not deep (say, around 20 ft or 6 m) a fixed foundation is used. For deeper water, floating foundations are more economical.

tension leg platform (TLP)
A floating platform that is kept in place by cables attached to the seabed. The cables are under tension in order to prevent the platform from moving with water.

spar platform
A platform in the form of a vertical cylinder at the top of which a wind turbine can be installed. The cylinder is hollow, but partly filled with sand and stone to have a heavy weight. It must stay afloat and remain vertical. Some cables limit its horizontal displacement on the water.

Offshore wind turbines use the same technologies utilized for offshore oil and gas floating platforms for deep water. There are many types of floating platforms for oil and gas, but those practiced for wind turbines are either of the type **tension leg platform (TLP)** or spar platform. In both cases, the weight of the platform and the turbine must be such that they do not sink. Also, the assembly must stay stable and the forces from wind and wave loadings must not overturn the turbine. In a TLP, the floating platform is fixed in position by cables that are under tension and attached to the seabed. A **spar platform** consists of one or more cylindrical bodies that can float. Part of the cylinder is filled with sand and stone (and the rest is air). The mass center of each spar is as low as it can be, so that it can stay vertical. Each spar is held by cables attached to the sea bed limiting its horizontal motion. **Figure 15.5** shows the three arrangements of offshore wind turbine bases.

Unlike onshore wind turbines for which the transmission lines are mainly overhead, offshore wind turbines use cables that rest on the seabed to bring electricity to shore. Usually one cable with three conductors is used. Thus, the conductors for three phases of AC are very near each other. All AC lines along their lengths behave as capacitors, since each two conductors separated by an insulator form a capacitor. For this reason, there is always some loss in the AC lines due to capacitive effect. The capacitive effect of cables is more than that of the overhead lines.

Because of the capacitive loss, AC cables have length limitations. Also, in AC there is skin effect in the conductors, which although small for low frequency cannot be ignored for long conductors. Because of all these reasons, it is preferable that offshore wind farm production is transmitted in DC rather than AC. DC suffers less electric loss, as well as the fact that for DC there

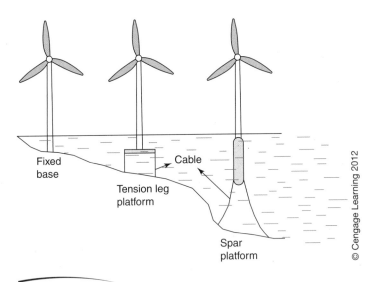

FIGURE 15.5 *Offshore wind turbine structures.*

are only two conductors in the cable instead of three; thus, it is less costly. One can say that for the same size of cable, DC can transfer more power than AC.

For DC electricity transfer over long distances, high voltage must necessarily be employed to reduce the conductor loss. The term **high-voltage direct current (HVDC)** is used for DC electricity at very elevated voltage, for example 250,000 V. This requires that electricity be raised to that level of voltage before the transmission and then be converted back to AC after the transmission, so that it can be integrated with the grid.

high-voltage direct current (HVDC) A method for transmission of electricity that uses DC at very high voltage (e.g., 300 kV) instead of AC.

It is not easy or cheap to perform these two conversions. Despite the aforementioned advantages of high-voltage direct current, the cost is prohibitive. The cost effectiveness of using HVDC in a project must be carefully studied. For example, the cost of overhead high-voltage AC lines is about $1 million for a mile. The cost of a buried (in-ground) high-voltage cable is between $1 and $4 million per mile, and the cost of HVDC is around $10 million for each mile, considering the cost for convertors at the two ends of the line.

Although offshore turbines are out of sight, some other matters such as throwing ice (if there is a potential for it) still exist. This has to be of concern for fishing boats if they get close to the wind farms. Also, the effect of salt on the body and the components of the turbine cannot be ignored.

Effects on birds and marine animals   15.8

Any adverse effects of offshore wind turbines lie in their effect on birds, notably migrating birds, and marine animals. The effect can be divided into short term and long term. Short term refers to the construction period; long term refers to the lifetime of an offshore wind farm that could be 20 or more years. The effect during construction is definitely a disturbance to the natural structure and, thus, the mammal life in the sea. In addition, the noise from ships, helicopters, and construction work is high. Mammals are very dependent on their sense of hearing, used for several functions, including communication. But, this is for a relatively short period of 6 months to 1 year.

There has been relatively little study of the ecological long-term effects of offshore turbines on marine species passing by a cable carrying high-voltage AC or DC electricity. There is also noise from turbines that is transferred to the sea. This adds to the noise from ships and other man-made machines.

Investigation of migrating birds has shown that they adapt their path of flight to avoid the disturbance, and change their way around the wind farm. In bad visibility conditions, the risk of collision with wind turbines is higher. More specifically, for both onshore and offshore towers (communication towers, as well), continuous lights at night cause more casualties than the intermittent flashing lights.

On the other hand, recent research has shown that wind turbine tower foundations at the bottom of shallow water, together with the rock boulders that are placed around them, create hard surfaces that act as reefs on the seabed. This makes it more favorable for fish and other species for shelter and habitation.

15.9   Chapter summary

- It is important to know the effects that wind turbines may have on the environment.
- The main reasons for offshore wind farms are environmental concerns and the lack of sufficient land in some regions.
- The older wind turbines were noisier, mainly because of their design. They were faster, too; and they were closer to the ground.
- Sound is characterized by its frequency. It is a mixture of many frequencies, some of which we cannot hear.
- Noise is unwanted sound.
- Infrasound is the sound or noise with a very low frequency, not audible by humans.
- If a turbine is far enough from us such that we cannot hear it, it does not have adverse sound effects on us.
- Not everybody likes wind turbines. Some people do not like having wind turbines in their view.
- Newer turbines are more pleasant to people because of the tower and color. Also, three-blade wind turbines are less annoying than those with two or one blades (still some exist).
- Flashing is the cyclic reflection of sunshine on the turbine blades during rotation.
- Flickering is the cyclic moving shadow cast of turbine blades on buildings or on ground, called shadow flicker.
- Because of the tower and huge structure some birds and bats can be killed by hitting wind turbines.
- Everyday a significant number of birds are killed by hitting buildings, bridges, towers, and so on.
- A wind farm must not be developed where the ecosystem of natural habitants can be disturbed.
- Wind turbines can also interfere with communications. This depends on the location of wind turbines and transmitters.
- Offshore wind turbines can be 30–40 miles out to sea.
- Their electricity production must be brought onshore by cables installed under the seabed.
- HVDC stands for high-voltage direct current. HVDC is preferable for offshore wind farms, but it is much more expensive than high-voltage AC.
- Wind turbines in shallow water have fixed towers rising from the seabed; those in deep water can have floating towers.

ADVANCED LEARNING

Sound level measurement

It is often necessary to measure or compare sound or noise levels. Here we want to briefly explain how this is done without going deep into the science and characteristics of sound. Since sound is a variation in pressure (in the air or other medium carrying the sound wave), it is measured by a microphone or a similar device that works in the same way as the ear senses sound.

We normally compare and judge sound levels by loudness. But, human perception of sound loudness is not a mere measurement of sound level. What we perceive as a loud or quiet sound depends not only on factors such as reflections from the surrounding environment, it also depends on the frequency and duration. This is due to the way the sensing system in our body functions. We cannot say a sound is loud unless it persists for a minimum of between 0.6 to 1 sec, depending on the frequency.

Forgetting about the extra effects of reflections from walls and so on, in the open air, the level of sound depends on the distance from the source. As one moves away from the source the sound level decreases, and as we move toward a sound source it increases. We need to represent this by an appropriate number and a unit. Sound and noise level is measured in *decibels* (dB). Sounds can exist from the lowest level of 0 dB to a very annoying level of 130 dB, corresponding to military aircraft taking off when heard from a 50-ft distance. Zero dB corresponds to a sound that can barely be heard by a sensitive human ear; that is, it is the lower threshold of human hearing.

As can be readily understood, sound (and noise) level and loudness depend on how far the distance is from the sound source. At some distance, the noise from the military aircraft can be barely heard. For this, the allocation of a value for sound level is based on relative association with a known sound level value. For example, knowing that at 50 ft from the aircraft the sound level is 130 dB can be used to find the sound level at other distances, such as 100 ft. Here we focus on this problem first.

Decibel measures a ratio of two values, and is defined as

$$dB = 10 \log\left(\frac{Value\ 2}{Value\ 1}\right) \tag{1}$$

where, log is the decimal logarithm. Thus, it represents a change in magnitude. Therefore,

$$Change\ in\ sound\ level\ (in\ dB) = 10 \log\left(\frac{Sound\ intensity\ 2}{Sound\ intensity\ 1}\right) \tag{2}$$

(*continued*)

The sound level of a given sound at a different distance is found based on the following facts:

1. Sound is generated by a pressure wave in the air.
2. Sound intensity is proportional to the power (strength) of sound.
3. The change in sound power due to distance change is proportional to the square of sound pressure ratio.
4. The change in pressure of a sound wave is inversely proportional to the distance ratio.

Thus, we have

$$\log (P_2/P_1) = \log (p_2/p_1)^2 = \log (d_1/d_2)^2 = 2 \log (d_1/d_2) \tag{3}$$

where P denotes power, p represents pressure, and d stands for distance. For example, the sound level of a fighter jet, which is 130 dB at a 50-ft distance, at 100 ft is

$$130 + 20 \log (50/100) = 130 + 20 (-0.03) = 130 - 6 = 124 \text{ dB}$$

In fact, -6 dB is the change for any sound at a distance twice as much. In other words, if the distance from a sound source is doubled, the sound level is decreased by 6 dB. Conversely, if the distance is halved, the sound level is increased by 6 dB.

In equation (1) if "value 1" is chosen for a reference sound for which the sound level is 0 dB, then the sound level for any other sound can be determined, if the intensity ratio is known or measured. **Table 15.1** shows the level for certain common noises.

Sound loudness is more complicated to evaluate than the sound level because, as mentioned earlier, it depends on a number of other factors. To make it simpler, a rule of thumb is "If the level of a sound is increased by 10 dB it is perceived twice as loud." This statement is more accurate for the sounds having the same frequency, but since it is not precise anyway, can be used for all cases. Based on equation (1) this rule says that ten violins sound twice as loud as one violin.

TABLE 15.1   **Sound levels and relative loudness**

dB	Loudness	Examples:
0	Threshold of hearing	
10	Just audible	
30	Faint	Quiet library, quiet bedroom (35 dB)
40	Lowest level of urban ambient	Living room, bird calls (44 dB)
50	Quiet	Quiet office, light traffic

dB	Loudness	Examples:
60	Noisy	Conversation (60 dB), busy office (60 dB), dishwasher (Rinse) at 10 ft, sewing machine
70	Moderately loud	Radio or TV-audio. vacuum cleaner (70 dB), living room music (76 dB), passenger car 65 mph at 25 ft (77 dB)
80	Loud	Noisy office, garbage disposal (80 dB), diesel train 45 mph at 100 ft (83 dB), food blender (88 dB), propeller plane flyover at 1000 ft (88 dB)
90	Quite loud	Motorcycle at 25 ft (90 dB), city traffic (95 dB), Boeing 737 or DC-9 aircraft at one nautical mile (6080 ft) before landing (97 dB)
100	Very loud	Pneumatic drill, Bell J-2A helicopter at 100 ft (100 dB), jet flyover at 1000 feet (103 dB), Boeing 707 or DC-8 aircraft at one nautical mile (6080 ft) before landing (106 dB)
110	Extremely loud	Riveting machine (110 dB), turbo-fan aircraft at takeoff power at 200 ft (118 dB)
120	Uncomfortably loud	Jet takeoff at 200 foot, oxygen torch (121 dB), thunderclap, military jet aircraft takeoff from aircraft carrier with afterburner at 50 ft (130 dB)

Review questions

1. Why is it important to know about the effects that wind turbines have on the environment?
2. Why is it very important that wind turbines do not have any adverse effect on the environment?
3. What are onshore and offshore wind farms?
4. Are offshore wind farms cheaper to develop?
5. Why do we need to have offshore wind farms?
6. Are the older wind turbines noisier, or the new larger wind turbines? Why?
7. What is the difference between sound and noise?
8. What is the name for sound with frequencies below our threshold of hearing?
9. If a turbine is far enough away that you cannot hear it, what bad effect can it have on you?
10. Why is it important for wind farm development to consider community approval?
11. Do you like to see wind turbines around where you live? Why or why not?
12. Describe a few differences between newer turbines and the old ones.
13. What is meant by flashing?
14. What is shadow flicker?

15. What happens if birds and bats hit wind turbines?
16. Whose responsibility is it to take care of environmental concerns of wind turbines?
17. Are more birds killed by wind turbines than other man-made structures?
18. Why do wind turbines interfere with communications?
19. What is HVDC?
20. Count a few advantages of HVDC versus high-voltage AC buried cables.
21. What are the two ways that offshore wind turbine foundations are constructed?
22. What are the advantages of offshore wind turbines?
23. What are the disadvantages of offshore wind turbines?

Safety and Other Issues

© Cengage Learning 2012

KEY TERMS

arc flash
lightning bolt
OSHA
skew wind

OBJECTIVES

After studying this chapter, you should be able to:

- Describe the importance of safety in operation of wind turbines.
- Explain what hazards one can face with respect to wind turbines.
- Give examples of safety issues.
- Define the conditions of operation of wind turbines.
- Understand the effects of nature on a turbine.
- Realize the effects of a turbine on the surrounding area.
- Explain possible damage from lightning.
- Realize the potential dangers in cold climate.
- Describe the potential use of wind turbines in residential areas.

16.0   Introduction

Like any other device or machine, a wind turbine must operate in a safe manner; that is, safe for an operator or operators; safe for the surrounding people, buildings, and environment; and safe for itself. No unsafe operation of a wind turbine is acceptable. The first category, the safety of the operator, in the case of wind turbines is different from many other machines, since unlike manufacturing machines there is no operator working on the wind turbine; thus, operator safety comes up only during repair and maintenance. This subject is studied in more detail in chapter 17.

In this chapter, a number of matters associated with safe operation of wind turbines are discussed. These matters have not been touched on in the previous chapters, but a person knowledgeable of or working with wind turbines must be aware of these issues. These are related to safety of a turbine and the surrounding areas. Also, the subject of wind turbines in urban areas, a concept that occurs to many people, is discussed.

16.1   Safety and working with wind turbines

OSHA
Occupational Safety and Health Authority.

In working with wind turbines, safety is rule number 1. Technicians must know and follow a number of rules that are set by the Occupational Safety and Health Administration (**OSHA**) standards in the United States, and by similar institutions in other countries. A wind turbine technician deals with electricity, height, limited space, and mechanical devices and tools. There are potential dangers with all of these that, even if the probability of occurrence may be very low, one should not take a chance and risk the consequences.

arc flash
A flow of electricity outside a conductor (in the air), caused by a short circuit, which generates a high pressure and temperature in the air.

Risk of electric shocks, electrocution by elevated voltage, **arc flash,**[1] fire, and high-pressure hydraulics are some of the dangers that a technician must be aware of, know the precautions for, and learn how to deal with in case they happen. Also, he or she must know about the proper clothing and how to wear it and the appropriate tools and how to use them.

It is also essential for a wind turbine technician to go through training for rescue operations in case called on to rescue a colleague hurt or at risk in an accident. All of the above are among the safety rules that are of the highest priority for working with wind turbines. It is necessary that technician safety training be updated at suitable intervals, for instance every year, or every other year (read more about safety in chapter 17).

> In working with wind turbines, safety is rule number 1.

---

[1] Arc flash is a flash of light that occurs when a switch is turned on or off. At higher voltage and currents this flash is large and can burn one's face and hands. Proper protection is necessary.

Effect of temperature and environmental changes

16.2

Wind turbines are installed in various geographically diverse regions with significant differences in ambient conditions, from cold and humid places (offshore) to hot and dry locations. As well, since they are outdoor machines, they are subject to change in climate, which can be entirely dissimilar from summer to winter.

Temperature change has numerous effects on the operation of a wind turbine. For example, the selection of gearbox oil and greases for bearings and so on are issues that come into the picture. After a seasonal change, the same turbine is subject to new conditions. Normally, for a turbine to be installed in a particular region, the manufacturer recommends the proper grade of oil and grease. In this respect, a manufacturer must be aware of the variation in the ambient conditions.

The nacelle of a wind turbine is normally equipped with fans, blowers, and heaters for cooling during summer and warming during winter. During summer, all the heat from the generator and gearbox must be taken away. Blowers circulate outside air into the nacelle for this purpose.

During winter, if the heat generated by the generator and gearbox is not sufficient, the heaters kick in and inject more heat to keep the nacelle at a minimum required temperature. Nonetheless, during a long shutdown, the temperature of the gearbox oil can fall to ambient temperature.

In some regions, a temperature alteration from summer to winter, for instance, from 38°C (100° F) to −25°C (−13°F) is not unusual. Although it is not necessary to change the gearbox oil from summer to winter and vice versa, one should make sure that the selected oil has the appropriate property for the two conditions. Otherwise, a turbine might face some difficulty, in the same way that a car gets starting problems in the winter.

The effect of temperature change is not only on the lubricating materials. Snow falls and freezing rain make any maintenance job much slower and more time consuming. Another nonnegligible effect is on anemometers. Cap anemometers that have a rotating shaft, although equipped with a small inside heater, can stop operating when freezing rain hits. In such a case, the reading from the anemometer to the turbine controller is a zero velocity wind. The turbine controller, accordingly, sets all the automated actions for no wind condition. Days of production will be lost in such a case until the anemometer starts to work again.

For offshore and coastal installations, the effect of sea salt on various components must be taken into consideration, whereas the onshore wind farms are immune from these effects. On the other hand, in dusty regions a turbine eventually gets dirty. A cleaning of the blades and tower, although relatively expensive, may prolong the useful life as well as the efficiency of operation of a turbine. As a matter of fact, the inside of a turbine nacelle, hub, and tower must always be kept clean from oil spills, grease, and other materials used by a turbine or for maintenance works. It is the responsibility of an operator to make sure that after maintenance work, everything is back to normal and no pieces of tools or extra stuff are left behind.

> It is the responsibility of an operator to make sure that after maintenance work, the turbine environment is back to normal and no tools, leftover material, parts, equipment, and so on are left behind.

16.3   Protection from lightning

lightning bolt
Very high voltage and bright arc of lightning.

Lightning is a discharge of millions of volts of static electricity between clouds and ground. All the electrons in the **lightning bolt** must travel to ground in a fraction of a second. This is why it is accompanied by intense light, thunder, and a lot of destruction power.

As wind turbines become larger, the tip of a blade tracks a higher altitude in the sky. It has been observed that every year a number of wind turbine blades have been damaged by lightning. Lightning can strike any metallic or nonmetallic object standing out at the instant it happens. All tall buildings have a lightning rod on the top that catches the high-voltage discharge (bolt) and shorts it to the ground. In this way, the discharge voltage of lightning does not damage the building; otherwise parts of the building would be torn off. The same thing is true for the blades of a wind turbine. **Figure 16.1** shows the extent of damage to an unprotected turbine blade.

UpWind Blade Solutions, Inc.

FIGURE 16.1 *Damage from lightning to a turbine blade with no protection.*

Today, all wind turbine blades have a lightning rod embedded inside them along the blade. It is a copper wire or strip that through connections in the hub and nacelle it finally makes a connection to the body of the tower. The rod is exposed to the outside somewhere near the blade tip so that if lightning hits, the rod catches the discharge. The tower is metallic and transfers any electrical charge to the ground.

A lightning rod inside a blade is a preventive device. In case of lightning, a blade without a rod is more likely to be damaged. On the contrary, if lightning strikes a blade with a lightning rod, it is not 100% sure that it gets to the metallic rod at the point of exposure to outside; thus although the chances are smaller, the lightning may still damage the blade, but to a much less extent.

> Today, all wind turbine blades have a lightning rod embedded inside them along the blade.

Ice forming on the blades   16.4

One of the not-so-evident phenomena in wind turbines in cold regions that get freezing rain is the formation of ice on the blades. The formation of ice on blades can be of different scales. At lower ambient temperatures above but not far from the freezing point, the temperature of the air passing through the blade edge can fall below freezing. In such a case, drops of rain, or the humidity in the air, can freeze and make a layer of ice on the blade edge. This causes the profile of the airfoil to change, which alters the drag and lift coefficients. As a result, the power capture capacity of the blade is reduced and less power is harnessed from wind. The ice layer in this case is not very thick. In addition to the drop in power coefficient, the weight of the blades is increased by the weight of ice on the affected area. Thus, more energy is consumed by the blades themselves.

During freezing rain, a thick layer of ice is formed on the entire body of a turbine blade. The effect in this case is much harsher than in the previous one. In addition to the significant change in the weight due to ice, the lift and drag coefficients of the airfoil profile substantially deteriorate and the turbine power capture reduces to a large extent. Moreover, there is another important issue with this scenario. When the ice layer on the turbine blade starts to melt, or if sudden shakes occur to them (for instance, if a turbine is shut down), pieces of ice fly away. This is a dangerous situation and the areas at the reach of these large chunks of ice are not safe for humans, animals, buildings, and so on.

For any repair works on a turbine, if a prior formation of ice is suspected extra care must be taken when shutting the turbine down. Nobody must be around in the zones that can be hit by a possible flying piece of ice.

In order to get an idea about the weight increase of a blade due to ice formation, consider the following example. Note that ice forms on both sides of a blade.

As a result of ice on blades, the power capture capacity of a turbine is reduced and less power is harnessed from wind.

EXAMPLE:

Each blade of a Vestas V82 wind turbine has a length of 40 m (131 ft) and the maximum chord is 3 m (10 ft). It weighs 7.5 tons (16,556 lb). The projected area of the blade is 86 m² (925.7 ft²). The twist angle is 20°. What is the weight of ice if a layer of 6.4 mm (1/4 in.) is formed on the blade? The density of ice is 920 kg/m³ (57.36 lb/ft³).

Solution: This problem is solved in both the metric and the U.S. system of measurements. Since the blade has a twist, the actual surface area of the blade is more than the given data. The twist is not, however, along the entire length; that is, not all the surface has a 20° incline. The twist is zero for the blade root and is 20° for the tip. We assume that the twist is linearly distributed along the blade. Therefore, we consider an average twist of 10° for our calculation of the actual surface area. The calculation is, thus, approximate. Nevertheless, it serves the purpose of this example. Note that because of round-off errors and approximate conversions between the metric and the U.S. system of units, one cannot expect exactly the same values for the answer.

From geometry we get

$$\text{Actual surface area} = \text{Projected surface area} \div \cos (\text{Twist angle})$$

1. Metric system:

$$\text{Actual surface area} = 86 \div \cos(10) = 86/0.985 = 87.33 \text{ m}^2$$

$$\text{Volume of ice} = (2)(87.33)(0.0064) = 1.12 \text{ m}^3$$

$$\text{Mass of ice} = (1.12)(920) = 1028 \text{ kg}$$

The weight of each blade is thus increased by 1028/7500 or 13.7%.

2. U.S. customary system:

$$\text{Actual surface area} = 925.7/0.985 = 940 \text{ ft}^2$$

$$\text{Ice volume} = (2)(940)(0.25/12) = 39.16 \text{ ft}^3$$

$$\text{Ice weight} = (39.16)(57.36) = 2246 \text{ lb}$$

This gives a percentage increase of 13.57% to each blade.

Notice that converting from one system to the other is not exact. Thus, the final numbers from calculation are not exactly the same.

Wind energy in urban areas   16.5

Wind blows and it is free energy; why not use it where we can? This is a question of many people. It is true. It is clean and free energy. Here we are going to study whether such an idea works, or better, how it can work. We are going to put together all the information you have so far to see how the wind energy can be used in urban areas. In this way, indeed, various options that initially may look favorable are to be ruled out and only a few possibilities remain.

There are a number of issues that must be put into perspective; these are:

1. Safety
2. Environmental aspects (noise, view, etc.)
3. Economy
4. Technicality and practicality issues

Before continuing, it is better to remember that there are other types of wind turbines (see chapter 4) than what we have studied mostly so far. Furthermore, connecting wind turbine output electricity to the grid is not the only way that energy from wind can be used. This implies that the output of a turbine can be employed in other ways such as water and space heating.

Urban areas consist of houses, apartment buildings, office buildings, high-rise buildings, and the like. Generally, there is no free land or undeveloped space. In this sense, the only spots to mount a wind turbine can be on the buildings roofs. In fact, in windy regions there is a lot of wind on the top of high-rise buildings. Also, any roof-mount turbine is necessarily much smaller than those in a wind farm. In cities, because of the structures and buildings that have a braking effect on the wind, the speed of wind is relatively less than in an open space (see chapter 1). Thus a wind turbine for this purpose must be able to catch and use low-speed winds. This implies having a high starting torque, which is not the case with all types of wind turbines. Moreover, at the edge of high-rise building roofs, upward wind is also possible. This wind, which does not have a stream parallel to the roof, called **skew wind,** is not suitable for propeller type turbines, but can be used with other types of turbines.

skew wind
Wind whose direction
if very off from
horizontal.

> Energy from wind can be used for heating water or for space heating in buildings.

Safety   16.5.1

As we have learned, safety is the number 1 issue to take into consideration. This implies safety to the surrounding area and its inhabitants, and safety to the building on which the turbine is installed, plus other safety issues associated with electric lines and electric safety code.

From the discussion in section 16.3 the use of propeller-type turbines cannot be permitted for the following reasons:

a. Relatively high speed of operation and possible danger of throwing away ice or broken pieces, particularly in cold regions.
b. Exerting large loads to the building structure in the form of moments at the tower base. This can easily damage a building. The taller a tower is, the greater the moment of the forces from wind. Note that a propeller turbine cannot be without a supporting pole. It is possible for a small turbine to have the generator and the gearbox at the bottom of the tower base, but still the moment from the wind force can be significant.
c. Other criteria such as noise and view also discourage the installation of propeller wind turbines on top of the buildings.

Random cases of installing propeller-type wind turbines on top of low- and high-rise buildings have been practiced. The results have been negative and discouraging.

Among the various types of wind turbines only the low-speed turbines without a pole can possibly be utilized for rooftop installation. This reduces the choice to the Savonius rotor only (see chapter 4). This is only from the safety viewpoint, so far. **Figure 16.2** shows a possible design for a roof-top wind turbine.

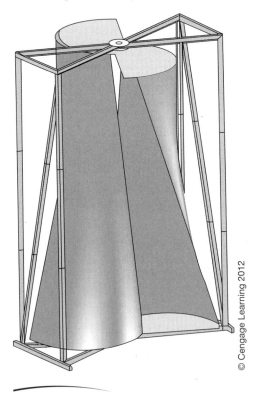

© Cengage Learning 2012

FIGURE 16.2 *A modified Savonius turbine.*
A possible design for a roof-top wind turbine.

> Safe operation of a turbine implies safety of people, safety for the surrounding area, and safety of buildings if a turbine is installed in a residential area.

Environmental aspects

From the environmental concerns, noise and view become very important issues. A turbine for urban areas must not generate any noise, or the noise must be so low that it does not bother anybody in the neighborhood. Also, it should not disturb people by being in the direct views and/or blocking their views. Roofs of high-rise buildings are appropriate from this point of view, since normally nobody can see what goes on the roof, or a wind turbine will be not much different from the other equipment that is already there, for example, the heating and cooling and similar installations.

A Savonius rotor or a similar design usually has a slow speed, thus it is possible that its noise can be within a tolerable level, particularly if it is located on top of taller buildings. Moreover, a Savonius rotor can use the skew wind blowing from below. Also, because of the design, which can be low and near to the roof, and low speed, less shaking of the building can be expected from its operation. If the edge of a building has a prevailing skew wind, then a Savonius turbine may be installed with its axis horizontal instead of vertical.

We recall from chapter 4 that a Savonius rotor is bulky and less efficient than propeller and Darrieus turbines, but for rooftop mounting, it is the only choice. Refinement in its design and use of light material can mildly compensate for the shortcomings, if rooftop wind turbines become realistic.

> Environmental concerns include noise from a turbine and view in the neighborhood.

Economy

As we learned in chapter 14 for a commercial wind turbine the economy of a project must be reasonable from an investment viewpoint. Unless there are some other issues to outweigh the economy of a wind project, such as health, clean environment, and urgency, economic concern always plays a role in decision making. No matter whether it is for a large building owned by a large corporation, or a small project for a residential house that the owner must pay for, it is always a matter of investment and providing the initial cost first, to be gradually paid back by the generated electricity from wind.

Like many other devices, such as gas turbines and steam turbines, the smaller a wind turbine is, the higher is cost per unit production (cost per kilowatt-hour of electricity). In this sense, the time to break even (recovery of the initial cost) is longer. From an economic viewpoint, although a wind energy project may imply clean and free energy, it may end up being more costly than if electricity were purchased from the grid. This, however, depends on the cost of electric power, and as expected, it depends also on the wind speed in a region and the total cost for the equipment. Thus, any individual project must be separately evaluated, and a general statement cannot be made for all cases.

> No matter what the size of a turbine is, it must be economically viable for the money to be spent on it.

16.5.4   Technicality and practicality issues

As we have learned so far, a wind turbine is either used as an isolated machine or is connected to a grid. This is true for any unit no matter what the size. Assume now that we are dealing with a wind turbine on top of a building catching a lot of wind energy without having any problem, and thus acceptable from an economic viewpoint for the turbine cost. How can we use the harnessed energy?

We assume that this is a relatively small turbine and the generator is a single-phase alternator or it is a DC generator. Taking into account the fact that wind is not necessarily blowing at the time energy is needed and that the rate of energy generation is neither constant nor uniform, the harnessed energy can be used in the following ways:

a. Connect the turbine output to the building electricity;
b. Use an independent circuit for the generated electricity;
c. Use the harnessed energy for heating water or for space heating in the winter;
d. Store electricity for using when required, for instance for lighting at night;
e. Combination of (b) and (c);
f. Combination of (b) and (d).

a. Connecting the turbine output to the building electricity

This is the best and the most preferred choice, but also more likely the most expensive. In such an arrangement, the voltage and frequency of the generated electricity must match those of the main (electric supply to the building). Equally important, the electricity from the turbine must be in perfect synchronization with the main. These conditions are similar to those described for the larger systems in a wind farm. The equipment to take care of and satisfy these conditions adds to the price of a turbine.

b. Using an independent circuit for the generated electricity

If a separate circuit is used for the generated electricity from a turbine, mainly for lighting purposes only, then the three aforementioned conditions no longer need to be tightly respected. Lights are not sensitive to frequency and voltage variation, and more importantly the synchronization is not necessary at all.

Consider, for instance, in a hotel that some of the lighting in hallways and in common areas are powered by wind turbines on top of the roof. Not all the lights, however, can be powered this way, since when there is no wind there is no light. In such a scenario for a small building, the voltage for this independent circuit can be completely different from the main voltage; 12 or 24 V, for example, corresponding to that of the halogen lightbulbs.

c. Using the harnessed energy for heating water or for space heating in the winter

This scenario is good for ordinary dwellings. Since the energy is used by a heating element, there is less restriction on voltage and frequency, thus making it the simplest/least expensive case. Either AC or DC can be used and electricity is generated at a variable rate, fluctuating with wind blow. A switching system can alternate the target between space heater(s) or water heater.

d. Storing electricity for using when required, for instance for lighting at night

This scenario requires a storage system. Nevertheless, it can be only an add-on to scenarios (a), (b), or (c). Storing electricity necessarily implies generating in DC or converting from AC to DC. The stored energy then might have to be converted back to AC, depending on the case. This application then calls for a separate circuit as in (b) and (c), or connecting to building electricity, as in (a), but then why store it (unless the connection from main is turned off)? This case is, thus, more costly than the previous scenarios.

e. Combination of (b) and (c)

Combining the two scenarios (b) and (c) gives more versatility as the generated power can be used for lighting and heating based on the demand. A manual or clock-operated switch can determine which application should be powered at each time and based on the production and consumption, as shown in **Figure 16.3**.

f. Combination of (b) and (d)

This combination was discussed under (d) (see above).

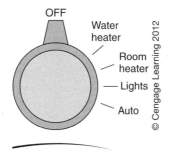

FIGURE 16.3 *A selector switch for directing the turbine output.*

Effect of ice forming on rotor

You learned that the formation of ice on wind turbine blades can change the aerodynamic property of an airfoil in addition to adding to the blade weight. Here you learn more about the effect of the increase of weight due to ice layers on the operation of the turbine from the mechanical and power capture viewpoints. As you learned in chapter 2, the (kinetic) energy in a moving object is

$$\text{Kinetic energy of a moving object} = \tfrac{1}{2}mv^2$$

where m is the mass and v is the speed of the object.

Similarly for a concentrated mass, such as a golf ball if in a rotational motion, one can write

$$\text{Kinetic energy of a rotating object} = \tfrac{1}{2}mv^2 = \tfrac{1}{2}m(r\omega)^2 = \tfrac{1}{2}mr^2\omega^2$$

For a golf ball one can assume that all the mass is concentrated in a point at the distance r from the center of rotation; ω is the angular velocity (see chapter 7).

For a wind turbine blade, contrary to a small ball, the mass is spread along the length of the blade. For this reason, we can divide the blade into a number of strips, each one having a different mass and at a different distance from the axis of rotation (see the figure). If each strip is identified by an index i, and its mass and radius (distance from axis) are denoted by m_i and r_i, respectively, the kinetic energy of the blade can be written as (note that ω is the same for all of the strips)

$$KE = \tfrac{1}{2}(m_1 r_1^2 + m_2 r_2^2 + m_3 r_3^2 + \cdots)\,\omega^2$$

What we have inside the parentheses is called the "mass moment of inertia" of the blade, and is represented by I. Thus, the kinetic energy of a rotating blade can be written as

$$KE = \tfrac{1}{2}I\omega^2$$

If the mass and shape of a blade (this is true for any rotating object) does not change, I remains constant. But, if the mass increases because of ice, then the magnitude of I changes.

Before a turbine starts to turn, its kinetic energy is zero. Then during speeding up, its kinetic energy gradually increases. In this process, the energy from wind is consumed to raise the kinetic energy of the rotor. After the rotor reaches its operating speed its kinetic energy will remain constant. No more wind energy is spent to increase ω (and increase KE). But still a part of wind energy is consumed by the rotor to compensate for overcoming friction in the bearings and maintaining the speed. The rest of wind energy captured by the rotor is delivered to the gearbox, to be transferred to the generator.

When the weight of a rotor increases due to ice on the blades, the moment of inertia of each blade increases accordingly. More energy is required to bring the rotor speed to the operating speed. Also, more energy is required to maintain the speed constant, since the bearings must support a heavier load and the forces on each element (ball or roller) have augmented. You can better comprehend this if you imagine the energy you must spend for turning a light wheel and a heavy wheel, and keep them running.

Now imagine the same wind that turns a turbine in the regular condition must turn it with the required speed while the weight has swelled. The turbine may hesitate to start in the first place, and the rotation is very reluctant. Assuming it starts and runs, then a larger portion of the wind energy is to be used for keeping the turbine running, before the rest is transferred to the gearbox. The output of the turbine is, thus, reduced.

Adding the fact that the blades lose part of their aerodynamic property of capturing wind energy, the turbine output can become much smaller. This all depends on the thickness of the ice layer.

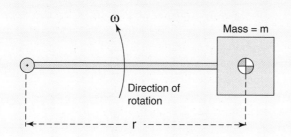

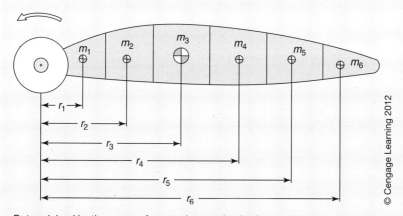

Determining kinetic energy of a complex rotating body.

© Cengage Learning 2012

16.6   Chapter summary

- Like any machine, a turbine must always operate in a safe state. That includes safety for the people, safety for the surrounding area and buildings, and safety for the machine itself.
- Safe working includes cleaning of any dirt, oil, and leftovers of any kind after a maintenance work.
- A wind turbine is subject to getting struck by lightning. A blade struck by lightning may undergo severe structural damage.
- Today's wind turbine blades are equipped with a lightning rod, which is a strip of copper embedded inside the blade structure, from tip to the root, where it will connect to the tower with provided contacts in the hub and nacelle.
- The effect of temperature change from summer to winter, and from warm regions to cold regions must be considered for a turbine and its operation. This will affect the performance of a turbine, its maintenance, and its cost. The first component to be influenced by temperature change is the gearbox and its lubrication oil.
- Among the secondary effects due to cold weather is the formation of ice on the blades. This not only alters the characteristic of a blade, it brings up a hazard when the ice melts and pieces of ice can fly around from a blade.
- The likelihood of ice flying away is greater when a turbine is to be stopped or started, since both of these maneuvers introduce a sudden change in motion during acceleration (deceleration) of a blade.
- Wind turbines have the potential of being used in urban and residential areas. This can be more logically on top of buildings instead of towers.
- A propeller turbine is not appropriate for urban areas.
- A turbine for urban areas must be safe, quiet, and not obstructing people's views.
- In the first place, such a turbine must be safe for surrounding people, the neighboring building, and the building on which it is mounted.
- Like a large turbine a turbine for residential areas must be economically viable and be worth the cost and endeavor.
- There are a few ways that use can be made of an isolated wind turbine integrated in or near a residential building.

Review questions

1. Under normal operating conditions, is a wind turbine considered dangerous?
2. What are the hazards of an operating wind turbine?
3. What are the hazards of a parked wind turbine?
4. What makes working in a nacelle involve more risk?
5. What are the most important effects of temperature change on a turbine?
6. Can hot temperature have a harmful effect on wind turbines?
7. In what ways can freezing rain disturb a wind turbine?
8. In what ways does a wind turbine affect its surrounding area?

9. What are the potential damages from lightning?
10. Has snow any bad effect on a wind turbine?
11. Can large wind turbines be installed in residential areas?
12. Why can't a propeller turbine be installed on top of a house?
13. What are the points to be considered for employing wind turbines in residential areas?
14. What are the points to be considered for employing wind turbines on rooftops?
15. Why can't we just put a wind turbine on our rooftop and get free electricity?
16. What can be wrong if we put up a wind turbine and connect its generator output directly to our electric line in the house?

Problems

1. A turbine blade is 85 ft long and its widest chord line (see chapter 3) is 6 ft. Assuming that the blade is formed by (two) triangles (see the figure) and has a twist angle of 25°, find the weight of the ice on the blade if the ice thickness is 3/8 in. The density of ice is 57.5 lb/ft³.
2. Convert all the numbers from problem 1 to the metric system.
3. Solve problem 1 again, but using the metric system numbers.
4. Based on equation 2.1 in chapter 2, calculate the energy, in terms of ft-lb and joule, in a piece of ice whose weight is 2 lb and is thrown with a speed of 196 ft/sec (use the conversion table at the beginning of the book).
5. The speed of the tip of a turbine is 196 ft/sec. Determine how much it is in miles per hour.
6. In order to have a better feel for the energy in the piece of ice in problem 4, suppose that a motorcycle with its passenger has a total weight of 900 lb. What is its speed in mph if the kinetic energy is the same as obtained in 4?
7. The turbine in problem 4 has two speeds. At the lower speed if a piece of ice (with the same size) is thrown from the middle of a blade, its speed is 67 ft/sec. Compare its energy with what you found in problem 4.

PROJECT

This is a group project with the collaboration of a few students. It can be regarded as a final project toward the completion of the course. It thus takes more time and effort compared to previous projects.

In this project, you make a wind turbine, running a generator. The output of the generator is connected to a heating element through an on-off switch. The wind turbine blades' shape and size are your design. The generator can be DC or AC. A DC machine is easier and simpler, since any DC motor can serve as a generator, whereas for an AC machine

(*continued*)

starting it as a generator can be more difficult. Since you are connecting the output to a resistive element, it does not really matter if it is DC or AC. Instead of the heating element you may use lightbulbs. The switch helps you to see the difference in operation and speed when a load is added. You first let the turbine start running with blow of wind with the switch open. Then you close the switch. You will see a noticeable change.

With this project you may also see the effect of ice forming on blades. For this purpose you may stick cardboard or some other heavy material to the blade surfaces representing the ice. Then observe the performance and compare with and without the additional weight on the blades.

Working on Wind Turbines and Operator Safety

© Cengage Learning 2012

KEY TERMS

cable grip
carabiner
fall arrest
harness
lanyard
lockout
personal protective
   equipment (PPE)
tagout

OBJECTIVES

After studying this chapter, you should be able to:

- Explain various safety concerns of working on wind turbines.
- Categorize various potential hazards of working with wind turbines.
- Describe why it is important to have safety rules.
- Define common factors of most of the accidents with wind turbines.
- Describe commonsense safety rules to be followed for wind turbine workers.
- Know that rules are standardized in each country by authorities.
- Know that in the United States every industrial work environment must meet OSHA requirements.
- Identify hazards and risks of working in a nacelle.
- Describe what PPE means.
- Realize that PPE is not the same for different jobs.
- Explain PPE for wind turbine climbing and working.
- Become familiar with terms concerning gear for climbing wind turbines.

(continued)

- Know what lockout and tagout mean.
- Describe the two fundamental rules of safety and rescue.
- Become more concerned about safety and realize everything that can malfunction.

17.0 Introduction

In all professions dealing with equipment and machinery, an operator's safety at work is to be given the highest priority. It is the responsibility of employers to provide a safe working environment for the workers and, equally, it is the employees' responsibility to know and respect all the hazard and safety matters associated with a job.

Based on years of experience, observations, and analysis, authorities have come up with certain safety rules that must be respected by the employers and employees in any working environment. This is to minimize the risks that might endanger the workers' lives and health. Despite the rules, mistakes or special conditions lead to many accidents every year. It is everybody's responsibility to always respect the safety rules.

Work on wind turbines is subject to a number of hazards. A knowledgeable technician may save his own or a colleague's life, whereas a careless technician can endanger his or his colleagues' lives and health. The material in this chapter is a general basic overview of safety matters that a technician must know. Normally, an employer must make sure that workers have a sufficient level of awareness of hazards, and the essential knowledge of appropriate actions and rescue work in the case of accidents, before they are engaged in their duties. For this reason, conducting safety lessons is a vital part of training of workers and technicians who are supposed to do the maintenance work on wind turbines. This is carried out by most of the companies in wind energy and by wind turbine manufacturers.

This chapter provides you with some preliminary common knowledge for safety. For someone willing to work in the wind energy industry and on wind turbines there is much more to learn. Everyone must realize that he or she is the person most responsible for his or her own safety. Respecting safety rules can save your life.

It is everybody's responsibility to always respect the safety rules.

Working environment   17.1

In working with a wind turbine a technician must normally carry out the work inside or outside the nacelle, and in the hub. In addition to the conditions dictated by weather, the working environment is (a) at a height, (b) limited in space and maneuverability, (c) with possible oily and greasy floors and equipment, (d) surrounded by electric devices and possible electric shocks, and finally (e) containing mechanical devices that are prone to move. Corresponding to each of these conditions there are hazards. For instance, working at heights has the danger of falling.

Falls can occur from various heights. One can fall from a high area such as a roof or from a ladder, or the ladder itself may fall. Other heavy and/or sharp objects may also fall during the course of one's work at heights. The person working at heights as well as people on the ground nearby are at risk when a fall occurs. Depending on the height involved when one of these events happens, the result may be minor or major injuries or even a fatality. For example, in 2008 the number of work-related fatal occupational injuries in the United States due to falls from heights was 700.[1]

Another problem with working at heights is when, due to other hazards, an accident occurs. For instance, if a fire breaks out in the tower, or a person slips from the turbine top and hangs in the air (held by safety gear), then escaping from the scene, rescuing the injured person, or receiving help are not simple matters (compared to if the accident happened on the ground).

Limitation of space introduces difficulties and prolongs the time to carry out jobs. Limited space can also lead to panic when an accident happens. Electricity has its own hazards. Being in the vicinity of electricity and devices that may have electric charges dictates that a technician be knowledgeable about electric hazards and the procedures to comply with, such as putting up warning signs and using tags to lockout equipment and forbid access.

All these hazards are more pronounced for wind turbines, since they combine together. This fact does not mean that working on wind turbines is a dangerous task and to be avoided. In reality, there are other working environments that can be more dangerous, because of the presence of inflammable gases, or poisonous gases, lack of oxygen, and so on. This all emphasizes that the safety regulations must be more strictly respected and followed.

The most responsible person for someone's safety is himself or herself. Respecting safety rules can save your life.

---

[1] Source: U.S. Department of Labor.

Common factors associated with most accidents include failure to:

- Recognize that a problem exists.
- Ensure that the work environment provides a safe system to work with.
- Comply with regulations.
- Use appropriate equipment.
- Have sufficient instruction or training.
- Have or provide a safety procedure to follow if something goes out of control.

17.2   Possible hazards

Hazards in working with wind turbines, as far as an operator or worker is concerned (that is, not something that can happen to a turbine such as speeding, blade failure and breakage, tower collapse, turbine catching fire, while in operation, and the like) can be categorized as caused by:

a. Height
b. Electricity
c. Mechanical subjects
d. Equipment
e. Environment
f. People and human interaction

The above can be put in a table, as shown in **Table 17.1.**

For someone to work on wind turbines it is absolutely necessary to know the devices and the working environment to be dealt with and the various possible hazards. The gained knowledge about electricity, gearbox, and so on from the previous chapters helps one to understand better the seriousness of hazards and the importance of following safety rules. **Table 17.2** shows the effect of electric current on the human body. The current depends on the voltage and the body resistance. The body resistance depends on the points of contact; the skin

TABLE 17.1

Potential hazards in a wind turbine

Height	Electricity	Mechanical subjects and equipment	Environment	Human interaction
Falling from roof	Electric contact	Equipment failure	Temperature	Wrong tool
Objects falling	Static charges/ residues	Tool failure	Confined space	Wrong mood/stress
Falling from ladder	Arc flash potential Improper wiring	Physical barriers	Oil spills Uneven floor Lighting conditions Weather	Wrong instructions Wrong assessment Wrong information Taking risk/failure to follow safety rules

© Cengage Learning 2012

Effect of electric current through human body		TABLE 17.2
Current (mA)	**Effect**	
Less than 1	No sensation	
Less than 3	Mild sensation and possible sudden shock	
3–10	Painful shock, let go current for most of people	
10–15	Local muscle contraction, hands freeze to the conductor for some people	
15–50	Loss of muscle control, freezing to the conductor, burns	
50–100	Difficulty breathing, collapse, and unconsciousness. Death for prolonged contact	
100–200	Heart problem (ventricular fibrillation), and death if more than ¼ sec	
Over 200	Clamping action of the heart, respiratory paralysis, death if current is not stopped	

© Cengage Learning 2012

and how damp it is; and the clothes at the point of contact, including the shoes if one of the contact points is the ground. If the current does not pass through the heart, local burns and irreparable body damage may result; a current can kill a person if its path involves the heart.

Commonsense safety rules 17.3

Of most importance in any work that requires safety measures is to use common sense rules if no written or prescribed procedure is available. If there is a potential for hazard, one has to make an assessment of the situation and evaluate the possible risks. Something may be, or look likely to be, a simple matter, but can cause accidents (for example, fires have been started by carelessly discarded cigarette butts). As a first safety rule, if there is a potential for a hazard or failure of equipment never take a risk. Have in mind that "Whatever can go wrong will go wrong." For instance, in using a ladder, if it is not secure and stable, fix that first. A second safety rule is that "If you are not sure about certain risks and what you must do, then ask."

It is imperative that the following common rules be respected when working on wind turbines:

- Never climb or work on a wind turbine alone.
- Never should two people be climbing or descending on the same sections of a tower (between closing hatches).
- Always close the hatch between two tower sections.
- Do not work on an energized circuit, unless it is absolutely necessary.
- Take the time to do things properly (do not rush through completing a job).
- Check the tools first (before climbing) to be sure they are correct and in working order.

- Check your climbing gear (see next section) before putting it on.
- Check your climbing gear regularly after every use.
- Have a radio and/or cell phone with you for communication.
- Do not take keys to a service vehicle up the tower with you.
- Have proper clothing for the job, according to the weather and temperature.
- Do not wear metal jewelry, especially when working on electrical systems.
- Discharge capacitors and residual charges in power lines before working on equipment. Although electricity could be turned off, capacitors carry a charge. Also, transmission lines behave as capacitors; thus, even if electricity is switched off, they have a potential that must be neutralized by connecting a wire between them and the ground.

> As a first safety rule, if there is a potential for a hazard or failure of equipment, never take a risk

17.4   Standardized safety rules and official regulations

Safety rules for operations nowadays are regulated, standardized, and governed by an authority body in each country. In the United States safety standards are set and monitored by the *Occupational Safety and Health Administration* (OSHA). In Canada a branch of the *Canadian Standards Association* (CSA) takes care of occupational health and safety (OHS) matters. Both OSHA and CSA develop occupational health and safety guidelines and procedures. These are well documented in printed or electronic form and can be purchased.

Safety standards by OSHA, CSA, and their equivalent organizations in other countries describe in detail the equipment to use, the steps and/or procedures for testing equipment and places for defects and hazards, signs and symbols for warnings, and so on. All the rules are categorized and identified by rule numbers (code). For instance one can find the following categories:

- Construction safety.
- Electrical safety.
- Equipment and machinery safety.
- Ergonomics and human factor safety.
- General workplace safety.
- Emergency preparedness.
- Occupational health and safety management system.

These categories by themselves can be divided into a number of subcategories. Not all of the regulations necessarily apply to all occupations. For example, safety rules for a shipyard can contain matters that are not relevant to offshore wind turbines (and vice versa), or working in a manhole is different from working in a nacelle.

OSHA is one of the regulating institutions for safety and health at work in the United States. There are a number of other bodies to standardize and define regulations for procedures, tests, calibration of equipment, minimum strengths of mechanical parts, maximum tolerances and values of electrical components, and many pertinent matters and cases that come along in practice. The regulations by OSHA may refer to other standards and other organizations such as:

ANSI: American National Standards Institute.
CSA: Canadian Standards Association.
ISO: International Standards Organization.
IEEE: Institute of Electrical and Electronics Engineers.
ASME: American Society of Mechanical Engineers.

Climbing gear

17.5

In order to protect a person against the potential hazards in a working environment, in addition to the precautions for hazard control, wearing of **personal protective equipment (PPE)** is essential and compulsory by the safety standards regulations. Hazard control implies minimizing the potential risk by appropriate means. For instance, if there is a hole in the ground or a situation where one can fall, similar to the case for falling from a roof, it is necessary that protective guardrails be put around the hole, or on the roof. Warning signs and flashing amber light at night may also become required.

Personal protective equipment means the clothing and equipment that protect a person or reduce the risk of injury if involved in an accident. Since there are many different hazards, the appropriate clothing and equipment are not the same for everyone. For example, working in a chemical lab requires a PPE different from that for working on wind turbines, or working in a vessel with contaminated air.

Comparing the hazards associated with working in the nacelle or in the hub of a wind turbine, falling from a height has the greatest likelihood. It can happen during climbing, descending, or during the work on a device from the nacelle roof, and moving on the nacelle roof to reach the hub in many turbines. For this reason, it is absolutely necessary that when working on a wind turbine a worker has the necessary gear (clothing and equipment). The proper clothing is a long-sleeve shirt and pants, hard hat, safety goggles, and non-slip steel-toe shoes. The climbing gear to be used by a wind turbine worker is **fall arrest** equipment. A pair of gloves is helpful when using the ladder, though it is not part of the fall arresting PPE. If other hazards are involved, then one needs to accordingly adapt to the situation. For example, if there is a possibility for arc flash, then in addition to the above, clothing and face protection for arc flash must be included with the rest. In any place where there is a risk of fire, a worker is required to wear flame resistant cotton clothing.

An arc flash is an electrical current through the air instead of through a conductor. This can happen in case of short circuits that cause a high current and sparks. A high level of heat and pressure can accompany the uncontrolled

personal protective equipment (PPE) Equipment for protection of a person in a hazardous working environment. PPE is not the same for all situations and depends on a job and its environment.

fall arrest Stopping from falling down from a height.

energy released by the short circuit in the form of a blast. The short circuit can be created by human contact when working near electrically energized lines or equipment, or it could be because of tools incidentally touching different parts of a circuit or hot lines. An arc flash is very dangerous for a worker if body parts are exposed to it. It can become more hazardous if the person's clothing is not appropriate, since the wrong clothing can catch fire. The appropriate personal protective equipment, such as a face shield and rubber gloves, is necessary when working on live electricity lines and equipment when there is a potential for an arc flash to occur.

> In any working place where there is a risk of fire, a worker is required to wear flame resistant cotton clothing.

harness
Part of the *personal protective equipment* that must be worn to climb a turbine.

lanyard
Part of the *personal protective equipment* required to climb a turbine. A lanyard is used to hook a person to a safe point for fall arrest. It contains a spring-like piece that reduces the impact of a sudden stop on a person's body in case of falling.

cable grip
A piece of equipment for climbing that attaches to a cable or rope. It allows sliding along the cable in one direction, but blocks the motion in the reverse direction (preventing fall).

carabiner
A metallic device as part of equipment in climbing gear that acts as a closed hook, allowing safe connection of two rings. For wind turbine climbing only those with a safety lock must be used.

Fall arrest equipment includes a full-body **harness,** a **lanyard,** a **cable grip,** and a locking **carabiner.** These are available in many different forms and by different manufacturers. It is important that they are made according to standards acceptable by OSHA in the United States, CSA in Canada, and similar organizations in other countries. Fall arrest equipment must be such that it arrests a falling person without putting a sudden force on the person at the end of the arrest distance. Otherwise this can injure the person.[2] For this purpose, it must behave like a spring. A lanyard fulfills this need by functioning like a spring. OSHA requires that the force exerted by the equipment (the lanyard) on the person does not exceed 1800 lb (8000 N) and the free fall distance of a falling person be limited to 6 ft (1.5 m). The deceleration part of total fall must not exceed 42 in. (1.07 m). Once a lanyard has actively stopped a falling person, it must be discarded. It cannot be used again.

Figures 17.1 and **17.2** show a common cross-over style body harness and lanyard. The body harness consists of a waist belt, two leg straps, and a shoulder strap, all of which have quick buckles for opening and closing. A visual inspection before you put that on is necessary, to detect any sign of wear or tear. If you find any signs of wear or any tears, you must discard the equipment and no longer use it. All the straps must be adjusted on the body to be sufficiently tight, but comfortable. The chest strap and front D-ring must be located around midbreastbone and the back D-ring should be located between the bottom of the neck and midshoulder.

A lanyard has a small snap hook for attachment to the harness, and two larger snap hooks that are to be attached to a secure point on the platform, ladder, and so on during work. When not attached to any external point (when walking, for instance), they are attached to side rings on the harness, as shown in figure 17.2. **Figure 17.3** shows the snap hooks of a typical lanyard. Pressing

---

[2] Consider being attached to a rope and falling. At the reach of the rope end, the body receives a very large shock. This can break bones and significantly injure a person.

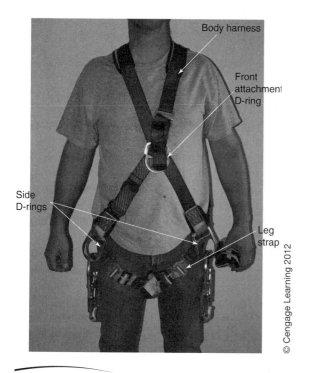

FIGURE 17.1 *Front side of a body harness.*

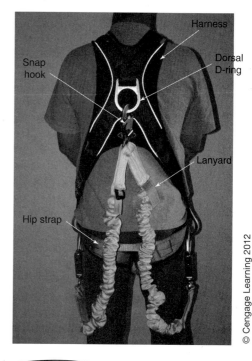

FIGURE 17.2 *Back side view of a body harness and lanyard.*

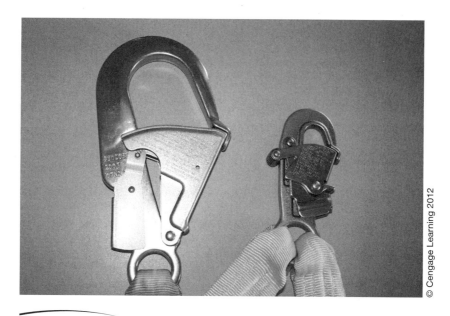

FIGURE 17.3 *Snap hooks of a typical lanyard.*

All climbing gear and PPE should be checked prior to use to make sure their condition has not changed since the last use.

Every time before wearing climbing gear it must be visually inspected to make sure it is in perfect condition. You must take care of the climbing gear. Your life can depend on it.

the gates of the snap hooks does not open the gate, since they are locked by a locking mechanism. Thus, these snap hooks cannot be accidentally opened if their gates hit against external subjects.

Figure 17.4 depicts a cable grip (also called cable sleeve, cable grab, or rope grab) and a carabiner. For climbing and descending a wind turbine, the carabiner is hooked to the front D-ring on the harness and the cable grip is engaged with the cable along the tower ladder (see the cable behind the ladder in figure 5.4). The cable grip has the mechanism that in case a climber suddenly loses control and falls off the ladder, it locks to the cable and does not let him/her slide down, thus holding the person in place. So long as the climber maintains control, the cable grip slides on the cable so that the climber can move up or down the ladder. The acceptable carabiner is of the locking type, so that if by accident its gate hits something it does not open. **Figure 17.5** shows how a cable grip is engaged with a cable.

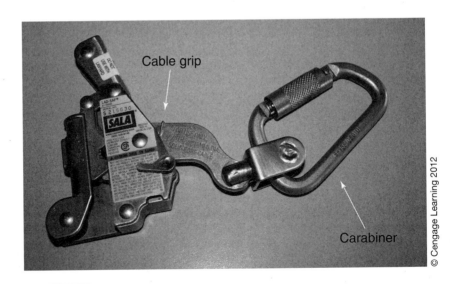

© Cengage Learning 2012

FIGURE 17.4 *A cable grip and a carabiner.*

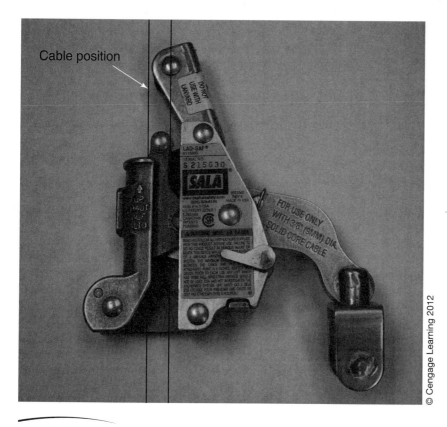

FIGURE 17.5 *A cable grip and the position of a cable when engaged.*

Minimizing the risk

17.6

Accidents happen mostly when more than one cause contributes to a hazard. And they happen, despite our attempts to prevent them. All one can do is to minimize the chance of an accident happening, and if it happens minimize the damage.

One of the very important issues in working with machinery and equipment that have hazard potential is to prevent other workers from unknowingly entering a dangerous zone or energizing a machine or a circuit when it is under repair or someone is working on it. It is quite possible that not everybody is aware of what goes on in a workplace, and someone tries to turn on a machine, or a switch for lights, or energize a line, while it has been turned off by another worker for some repair work. **Lockout** and **tagout** are two actions that must be practiced to prevent accidental energizing of machines and

lockout
The action of locking a dangerous area or device to restrict the access.

tagout
The action of putting a note on a device or to the entrance of a dangerous location to warn people.

electric circuits. Lockout implies that unsafe equipment or a hazardous device, a place containing such a device, and a door to a passage or location of the device or equipment, be physically locked. An accompanied note must clearly indicate who locked it and who the authorized person to contact is. An authorized employee can be the person who performs service or maintenance, or a supervisor of such a person. Tagout means putting a clear note on a device or an equipment, indicating the fact that there is danger involved and mentioning the name and contact information for the authorized person who put up the note.

In the case of an accident, one must deal with it in the best possible manner in order to minimize the damage. Working on wind turbines necessitates that, in addition to respecting and following the safety rules, one must learn how to escape from a dangerous zone and how to rescue a colleague caught in an accident, or faced with a hazard. Self-rescue and rescuing others requires special training that all workers on wind turbines must receive. These practices involve additional equipment that an employer must provide.

A wind turbine worker must learn how to use the equipment to get to the ground from the nacelle, not using the ladder (or the lift in the most recent towers), because ladders and lifts may not be usable because of fire or fumes in the tower. There is a special hatch somewhere on the floor in the nacelle that opens to the outside. This opening is used to escape from the nacelle from the outside of the tower.

A wind turbine worker also must learn how to rescue a colleague who has an accident or suddenly has faced a health problem and needs help. The first safety rule in such a case is *"Not to risk your own safety."* This can add to the problem instead of providing a solution, and instead of one injured person, more will then need help. When one is involved in such a situation, the first thing to do is to ask for additional help plus medical help. The injured person must be brought down the tower. Depending on the situation, this can be from the inside or outside of the tower, using the appropriate special equipment. After calling for help, one should assess the situation to decide on the right action.

It is outside the scope of this book to discuss more on the safety matters and rescue operations. Specialized training on the subject is essential. In summary, one should always have in mind that "Do not leave it to chance. Minimize the risks of any hazard."

The first safety rule in rescuing other people is "Not to endanger your own safety."

ADVANCED LEARNING

It was mentioned in the discussion in section 17.5 that the lanyard must behave like a spring. In fact, lanyards contain a flexible part that behaves as a spring in order to smooth out the stopping action of a fall arrest. The numbers given there for acceptable values stem from calculations.

Consider a rope (not a spring) with one end attached to a fixed point and the other end attached to a heavy block. The block is put on a shelf, somewhere that the rope is loose (not stretched) and is not supporting the weight of the block. If the block is pushed to fall off, the rope keeps it from falling on the ground. In this process, the rope will be stretched to its full length at one point in time. At this instant both the block and the rope experience a high force. The inertia force of the block pulls the rope down and causes a tension in the rope. Simultaneously, the rope pulls the block upward to stop it. These two equal forces (the block's inertial force and tension in the rope) depend on the length of the rope, or the fall of the block. The more length the rope has, the larger is the force.

Suppose now that a spring is added to the rope and for comparison the rope is shortened so that the total length does not change. If the block falls this time, at some point the inertia force is applied to the spring, which will be stretched. At the same time the spring will exert an upward force to the block depending on how much it is stretched (which gradually increases).

There are two differences here between the two cases: one is that the upward force to the block starts from zero and gradually increases to a maximum. Thus, the block is not subject to a shock (a large force exerted in a fraction of a second). The second difference is that the magnitude of the maximum force is less than that in the previous case.

The choice of the spring is very important. If it is too hard, then the force is large. In the extreme, it is like not having a spring. If it is too soft then the spring extension (which is the part of the fall with deceleration) is too much.

The two numbers in section 17.5 (1800 lb and 42 in.) work against each other. With mathematical formulation of the problem one can set the values and find the flexibility (or stiffness) of the spring. The numbers given are for a maximum weight of 300 lb (136 kg) for the person using the lanyard.

17.7   Chapter summary

- When working on wind turbines, technicians and workers are subject to a number of hazards. These hazards are mainly due to the height, the confined environment, the electrical equipment, and the possibility of turbine motion.
- Falling from a height is the most serious hazard. The number of work-related fatal occupational injuries in 2008 in the United States due to falls from height was 700.
- Human error or lack of knowledge and ignoring safety rules are among the other causes of accidents.
- It is necessary that technicians learn about safety rules and follow them carefully. Training for safety and rescue is provided to workers by employers.
- It is the responsibility of an employer to provide a safe environment for the workers. Because of the hazardous nature of work on wind turbines, moreover, it is a worker's responsibility to care for his or her safety as well as the safety of other workers.
- Safety regulations are set and monitored by authorities in each country. In the United States, these regulations are set by the Occupational Safety and Health Administration (OSHA).
- A mindset for a good technician is that "safety is rule number 1, always."
- For any job that involves hazards, *personal protection equipment* (PPE) is to be used in order to reduce the risk of having accidents. PPE is not the same for all jobs and depends on the nature of a hazard(s) involved in a job and the work environment.
- For working on wind turbines, PPE comprises climbing gear, a hard hat, and fall arrest equipment. In the simplest version, climbing gear includes a harness to wear, a cable grip, and a locking carabiner. The fall arrest gear is a lanyard with which a worker can hook himself or herself to a secure point during work.
- Long-sleeve shirts and pants, with no jewelry or unnecessary items on the body (or in the pockets) are part of the proper clothing that a wind turbine worker must bear in mind. A pair of gloves is often very useful when climbing a ladder.
- PPE must always be checked before each use to make sure it is in perfect order. A periodic professional inspection, for instance, a yearly inspection, may also become necessary, depending on the frequency of usage.
- A good safe action is to respect tagout and lockout practice at work. Tagout implies putting a note on a device or equipment, notifying others about a hazard or danger. Lockout implies that equipment or a device, or a place containing such a device, be physically locked to prevent it from being accessed.
- In addition to training for safety, a technician working on wind turbines receives training for self-rescue in case it becomes necessary to escape from a turbine without using the ladder inside the tower.
- It is also essential that a worker learns how to rescue a colleague who might get involved in an accident. A first rule to follow is that one must never put his or her safety at risk in such a rescue mission.

Review questions

1. Why is safety so important for working on wind turbines?
2. What are the potential hazards of working with wind turbines?
3. What is the risk associated with working at heights?
4. List the common factors of most accidents.
5. What is meant by commonsense safety rules?
6. Why can't two persons climb at the same time in a section of a turbine tower?
7. What is the risk if the hatch between tower segments in a wind turbine is left open?
8. What does OSHA stand for?
9. What does PPE stand for?
10. What is the most important risk when working with wind turbines?
11. What are the appropriate PPE for climbing a wind turbine?
12. What is the importance of a lanyard?
13. What does lockout involve?
14. What is the difference between lockout and tagout?
15. Who is most responsible for your safety?
16. What is the most important safety rule?
17. What is the first lesson (or rule) in rescue work?w

Density of Air at Different Temperatures

Air density, like any other gas, depends on its temperature and pressure. **Tables A.1** through **A.4** show the air density for a number of conditions associated with wind turbines. Temperature variations can represent transition from cold winter to hot summer weather. Since the pressure of the air changes with height (ground elevation), calculations are made in terms of increasing height. Tables A.1 and A.2 correspond to elevation change within the working area of a turbine around sea level. Tables A.3 and A.4 show numbers for elevated grounds.

These tables are for dry air. Moisture in the air changes the density. The corresponding numbers depend on the humidity.

TABLE A.1 Air density variation in lb/ft³ for various temperatures and pressures about sea level (Imperial system)

Height (ft)	Press (Pa)	Press (psi)	−50°F	−40°F	−30°F	−20°F	−10°F	0°F	10°F	20°F	30°F	32°F
Sea level	101325.0	14.69213	0.096720	0.094415	0.092218	0.090120	0.088116	0.086199	0.084364	0.082605	0.080918	0.080589
30	101215.2	14.67621	0.096615	0.094313	0.092118	0.090023	0.088021	0.086106	0.084272	0.082516	0.080830	0.080502
60	101105.5	14.66030	0.096510	0.094211	0.092018	0.089925	0.087925	0.086013	0.084181	0.082426	0.080743	0.080414
90	100996.0	14.64442	0.096406	0.094109	0.091918	0.089828	0.087830	0.085919	0.084090	0.082337	0.080655	0.080327
120	100886.6	14.62855	0.096301	0.094007	0.091819	0.089730	0.087735	0.085826	0.083999	0.082248	0.080568	0.080240
150	100777.2	14.61270	0.096197	0.093905	0.091719	0.089633	0.087640	0.085733	0.083908	0.082159	0.080481	0.080153
180	100668.0	14.59687	0.096093	0.093803	0.091620	0.089536	0.087545	0.085640	0.083817	0.082070	0.080394	0.080066
210	100559.0	14.58105	0.095989	0.093701	0.091521	0.089439	0.087450	0.085548	0.083726	0.081981	0.080306	0.079980
240	100450.0	14.56525	0.095885	0.093600	0.091421	0.089342	0.087355	0.085455	0.083635	0.081892	0.080219	0.079893
270	100341.2	14.54947	0.095781	0.093498	0.091322	0.089245	0.087261	0.085362	0.083545	0.081803	0.080132	0.079806
300	100232.4	14.53370	0.095677	0.093397	0.091223	0.089149	0.087166	0.085270	0.083454	0.081714	0.080046	0.079720
330	100123.8	14.51795	0.095573	0.093296	0.091125	0.089052	0.087072	0.085177	0.083364	0.081626	0.079959	0.079634
360	100015.3	14.50222	0.095470	0.093195	0.091026	0.088955	0.086977	0.085085	0.083273	0.081537	0.079872	0.079547
390	99907.0	14.48651	0.095366	0.093094	0.090927	0.088859	0.086883	0.084993	0.083183	0.081449	0.079786	0.079461
420	99798.7	14.47081	0.095263	0.092993	0.090829	0.088763	0.086789	0.084901	0.083093	0.081361	0.079699	0.079375
450	99690.6	14.45513	0.095160	0.092892	0.090730	0.088667	0.086695	0.084809	0.083003	0.081273	0.079613	0.079289

TABLE A.1 (Continued)

Height (ft)	Press (Pa)	Press (psi)	40°F	50°F	60°F	70°F	80°F	90°F	100°F	Height (ft)
Sea level	101325.0	14.69213	0.079299	0.077743	0.076247	0.074807	0.073421	0.072085	0.070797	0
30	101215.2	14.67621	0.079213	0.077659	0.076164	0.074726	0.073342	0.072007	0.070721	30
60	101105.5	14.66030	0.079127	0.077574	0.076082	0.074645	0.073262	0.071929	0.070644	60
90	100996.0	14.64442	0.079041	0.077490	0.075999	0.074564	0.073183	0.071851	0.070568	90
120	100886.6	14.62855	0.078956	0.077406	0.075917	0.074484	0.073103	0.071773	0.070491	120
150	100777.2	14.61270	0.078870	0.077323	0.075835	0.074403	0.073024	0.071696	0.070415	150
180	100668.0	14.59687	0.078785	0.077239	0.075752	0.074322	0.072945	0.071618	0.070338	180
210	100559.0	14.58105	0.078699	0.077155	0.075670	0.074242	0.072866	0.071540	0.070262	210
240	100450.0	14.56525	0.078614	0.077071	0.075588	0.074161	0.072787	0.071463	0.070186	240
270	100341.2	14.54947	0.078529	0.076988	0.075506	0.074081	0.072708	0.071385	0.070110	270
300	100232.4	14.53370	0.078444	0.076905	0.075425	0.074001	0.072629	0.071308	0.070034	300
330	100123.8	14.51795	0.078359	0.076821	0.075343	0.073921	0.072551	0.071231	0.069958	330
360	100015.3	14.50222	0.078274	0.076738	0.075261	0.073840	0.072472	0.071154	0.069882	360
390	99907.0	14.48651	0.078189	0.076655	0.075180	0.073760	0.072394	0.071077	0.069807	390
420	99798.7	14.47081	0.078104	0.076572	0.075098	0.073680	0.072315	0.071000	0.069731	420
450	99690.6	14.45513	0.078020	0.076489	0.075017	0.073601	0.072237	0.070923	0.069655	450

TABLE A.2 Air density variation in kg/m³ for various pressures and temperatures about sea level (metric system)

Height (m)	Press (Pa)	−50°C	−40°C	−30°C	−20°C	−10°C	0°C	10°C	15°C	20°C	30°C	40°C	50°C	Height (m)
Sea level	101325.0	1.58180	1.51396	1.45169	1.39435	1.34136	1.29225	1.24661	1.22498	1.20409	1.16437	1.12719	1.09231	0
10	101204.9	1.57993	1.51216	1.44997	1.39269	1.33977	1.29072	1.24514	1.22353	1.20266	1.16299	1.12585	1.09101	10
20	101085.0	1.57805	1.51037	1.44825	1.39104	1.33818	1.28919	1.24366	1.22208	1.20124	1.16161	1.12452	1.08972	20
30	100965.2	1.57618	1.50858	1.44654	1.38940	1.33660	1.28766	1.24219	1.22063	1.19981	1.16024	1.12319	1.08843	30
40	100845.6	1.57432	1.50679	1.44482	1.38775	1.33501	1.28614	1.24072	1.21919	1.19839	1.15886	1.12185	1.08714	40
50	100726.1	1.57245	1.50501	1.44311	1.38610	1.33343	1.28461	1.23925	1.21774	1.19697	1.15749	1.12052	1.08585	50
60	100606.8	1.57059	1.50322	1.44140	1.38446	1.33185	1.28309	1.23778	1.21630	1.19555	1.15612	1.11920	1.08456	60
70	100487.5	1.56873	1.50144	1.43969	1.38282	1.33027	1.28157	1.23631	1.21486	1.19414	1.15475	1.11787	1.08328	70
80	100368.5	1.56687	1.49966	1.43799	1.38118	1.32870	1.28005	1.23485	1.21342	1.19272	1.15338	1.11655	1.08199	80
90	100249.5	1.56501	1.49789	1.43628	1.37955	1.32712	1.27854	1.23338	1.21198	1.19131	1.15201	1.11522	1.08071	90
100	100130.8	1.56316	1.49611	1.43458	1.37791	1.32555	1.27702	1.23192	1.21054	1.18990	1.15065	1.11390	1.07943	100
110	100012.1	1.56130	1.49434	1.43288	1.37628	1.32398	1.27551	1.23046	1.20911	1.18849	1.14928	1.11258	1.07815	110
120	99893.6	1.55945	1.49257	1.43118	1.37465	1.32241	1.27400	1.22900	1.20768	1.18708	1.14792	1.11126	1.07688	120
130	99775.2	1.55761	1.49080	1.42949	1.37302	1.32084	1.27249	1.22755	1.20625	1.18567	1.14656	1.10995	1.07560	130
140	99657.0	1.55576	1.48903	1.42779	1.37139	1.31928	1.27098	1.22609	1.20482	1.18427	1.14520	1.10863	1.07432	140
150	99538.9	1.55392	1.48727	1.42610	1.36977	1.31771	1.26947	1.22464	1.20339	1.18286	1.14385	1.10732	1.07305	150
160	99421.0	1.55208	1.48551	1.42441	1.36814	1.31615	1.26797	1.22319	1.20196	1.18146	1.14249	1.10601	1.07178	160
170	99303.2	1.55024	1.48375	1.42272	1.36652	1.31459	1.26647	1.22174	1.20054	1.18006	1.14114	1.10470	1.07051	170
180	99185.5	1.54840	1.48199	1.42104	1.36490	1.31304	1.26497	1.22029	1.19912	1.17866	1.13978	1.10339	1.06924	180
190	99068.0	1.54657	1.48023	1.41935	1.36329	1.31148	1.26347	1.21885	1.19770	1.17727	1.13843	1.10208	1.06797	190
200	98950.6	1.54473	1.47848	1.41767	1.36167	1.30993	1.26197	1.21740	1.19628	1.17587	1.13708	1.10077	1.06671	200

TABLE A.3 Air density variation in lb/ft³ for various temperatures near ground at different elevations (Imperial system)

Height (ft)	Press (Pa)	Press (psi)	−50°F	−40°F	−30°F	−20°F	−10°F	0°F	10°F	20°F	30°F	32°F
0	101325.0	14.69213	0.096720	0.094415	0.092218	0.090120	0.088116	0.086199	0.084364	0.082605	0.080918	0.080589
1000	97728.7	14.17066	0.093287	0.091064	0.088945	0.086922	0.084989	0.083140	0.081370	0.079673	0.078046	0.077729
2000	94260.0	13.66771	0.089976	0.087832	0.085788	0.083837	0.081972	0.080189	0.078482	0.076845	0.075276	0.074970
3000	90914.5	13.18260	0.086782	0.084715	0.082743	0.080861	0.079063	0.077343	0.075696	0.074118	0.072604	0.072309
4000	87687.7	12.71472	0.083702	0.081708	0.079806	0.077991	0.076257	0.074598	0.073009	0.071487	0.070027	0.069743
5000	84575.4	12.26344	0.080731	0.078808	0.076974	0.075223	0.073550	0.071950	0.070418	0.068950	0.067542	0.067267
6000	81573.6	11.82818	0.077866	0.076011	0.074242	0.072553	0.070940	0.069396	0.067919	0.066503	0.065145	0.064880
7000	78678.4	11.40836	0.075102	0.073313	0.071607	0.069978	0.068422	0.066933	0.065508	0.064142	0.062833	0.062577
8000	75885.9	11.00345	0.072437	0.070711	0.069065	0.067494	0.065993	0.064558	0.063183	0.061866	0.060602	0.060356
9000	73192.5	10.61291	0.069866	0.068201	0.066614	0.065099	0.063651	0.062266	0.060941	0.059670	0.058452	0.058214
10000	70594.7	10.23623	0.067386	0.065780	0.064249	0.062788	0.061392	0.060056	0.058778	0.057552	0.056377	0.056148
11000	68089.1	9.87291	0.064994	0.063446	0.061969	0.060560	0.059213	0.057925	0.056691	0.055510	0.054376	0.054155
12000	65672.4	9.52250	0.062688	0.061194	0.059770	0.058410	0.057111	0.055869	0.054679	0.053539	0.052446	0.052233
13000	63341.5	9.18452	0.060463	0.059022	0.057648	0.056337	0.055084	0.053886	0.052739	0.051639	0.050585	0.050379
14000	61093.4	8.85854	0.058317	0.056927	0.055602	0.054338	0.053129	0.051973	0.050867	0.049806	0.048789	0.048591
15000	58925.0	8.54412	0.056247	0.054907	0.053629	0.052409	0.051243	0.050129	0.049061	0.048039	0.047057	0.046866

TABLE A.3 (Continued)

Height (ft)	Press (Pa)	Press (psi)	40°F	50°F	60°F	70°F	80°F	90°F	100°F	Height (ft)
0	101325.0	14.69213	0.079299	0.077743	0.076247	0.074807	0.073421	0.072085	0.070797	0
1000	97728.7	14.17066	0.076484	0.074984	0.073541	0.072152	0.070815	0.069527	0.068285	1000
2000	94260.0	13.66771	0.073770	0.072322	0.070930	0.069591	0.068302	0.067059	0.065861	2000
3000	90914.5	13.18260	0.071151	0.069755	0.068413	0.067121	0.065878	0.064679	0.063523	3000
4000	87687.7	12.71472	0.068626	0.067279	0.065985	0.064739	0.063539	0.062383	0.061269	4000
5000	84575.4	12.26344	0.066190	0.064892	0.063643	0.062441	0.061284	0.060169	0.059094	5000
6000	81573.6	11.82818	0.063841	0.062588	0.061384	0.060225	0.059109	0.058034	0.056997	6000
7000	78678.4	11.40836	0.061575	0.060367	0.059205	0.058088	0.057011	0.055974	0.054974	7000
8000	75885.9	11.00345	0.059390	0.058224	0.057104	0.056026	0.054988	0.053987	0.053023	8000
9000	73192.5	10.61291	0.057282	0.056158	0.055077	0.054037	0.053036	0.052071	0.051141	9000
10000	70594.7	10.23623	0.055249	0.054165	0.053122	0.052119	0.051154	0.050223	0.049326	10000
11000	68089.1	9.87291	0.053288	0.052242	0.051237	0.050270	0.049338	0.048440	0.047575	11000
12000	65672.4	9.52250	0.051396	0.050388	0.049418	0.048485	0.047587	0.046721	0.045886	12000
13000	63341.5	9.18452	0.049572	0.048600	0.047664	0.046764	0.045898	0.045063	0.044258	13000
14000	61093.4	8.85854	0.047813	0.046875	0.045973	0.045105	0.044269	0.043464	0.042687	14000
15000	58925.0	8.54412	0.046116	0.045211	0.044341	0.043504	0.042698	0.041921	0.041172	15000

TABLE A.4 Air density variation in kg/m³ for various temperatures near ground at different elevations (metric system)

Height (m)	Press (Pa)	-50°C	-40°C	-30°C	-20°C	-10°C	0°C	10°C	15°C	20°C	30°C	40°C	50°C	Height (m)
Sea level	101325.0	1.58180	1.51396	1.45169	1.39435	1.34136	1.29225	1.24661	1.22498	1.20409	1.16437	1.12719	1.09231	0
250	98365.7	1.53560	1.46974	1.40929	1.35362	1.30218	1.25451	1.21021	1.18921	1.16892	1.13036	1.09427	1.06040	250
500	95492.9	1.49075	1.42681	1.36813	1.31409	1.26415	1.21787	1.17486	1.15447	1.13478	1.09735	1.06231	1.02943	500
750	92704.0	1.44722	1.38514	1.32818	1.27571	1.22723	1.18230	1.14055	1.12076	1.10164	1.06530	1.03128	0.99937	750
1000	89996.5	1.40495	1.34469	1.28939	1.23845	1.19139	1.14777	1.10724	1.08802	1.06947	1.03419	1.00116	0.97018	1000
1250	87368.1	1.36392	1.30542	1.25173	1.20228	1.15659	1.11425	1.07490	1.05625	1.03823	1.00398	0.97192	0.94185	1250
1500	84816.4	1.32408	1.26729	1.21517	1.16717	1.12282	1.08171	1.04351	1.02540	1.00791	0.97466	0.94354	0.91434	1500
1750	82339.3	1.28541	1.23028	1.17968	1.13308	1.09002	1.05012	1.01303	0.99545	0.97847	0.94620	0.91598	0.88764	1750
2000	79934.5	1.24787	1.19435	1.14523	1.09999	1.05819	1.01945	0.98344	0.96638	0.94990	0.91856	0.88923	0.86171	2000
2250	77600.0	1.21143	1.15947	1.11178	1.06786	1.02728	0.98967	0.95472	0.93816	0.92215	0.89174	0.86326	0.83655	2250
2500	75333.6	1.17604	1.12560	1.07931	1.03668	0.99728	0.96077	0.92684	0.91076	0.89522	0.86569	0.83805	0.81211	2500
2750	73133.5	1.14170	1.09273	1.04779	1.00640	0.96815	0.93271	0.89977	0.88416	0.86908	0.84041	0.81357	0.78840	2750
3000	70997.6	1.10835	1.06082	1.01719	0.97701	0.93988	0.90547	0.87349	0.85833	0.84369	0.81586	0.78981	0.76537	3000
3250	68924.0	1.07598	1.02983	0.98748	0.94847	0.91243	0.87903	0.84798	0.83327	0.81905	0.79204	0.76674	0.74302	3250
3500	66911.1	1.04456	0.99976	0.95864	0.92077	0.88578	0.85335	0.82321	0.80893	0.79513	0.76890	0.74435	0.72132	3500
3750	64956.9	1.01405	0.97056	0.93064	0.89388	0.85991	0.82843	0.79917	0.78531	0.77191	0.74645	0.72261	0.70025	3750
4000	63059.8	0.98444	0.94221	0.90346	0.86777	0.83480	0.80424	0.77583	0.76237	0.74937	0.72465	0.70151	0.67980	4000
4250	61218.1	0.95568	0.91469	0.87708	0.84243	0.81042	0.78075	0.75317	0.74010	0.72748	0.70348	0.68102	0.65994	4250
4500	59430.2	0.92777	0.88798	0.85146	0.81783	0.78675	0.75794	0.73118	0.71849	0.70623	0.68294	0.66113	0.64067	4500
4750	57694.5	0.90068	0.86205	0.82659	0.79394	0.76377	0.73581	0.70982	0.69750	0.68561	0.66299	0.64182	0.62196	4750
5000	56009.4	0.87437	0.83687	0.80245	0.77075	0.74146	0.71432	0.68909	0.67713	0.66558	0.64363	0.62308	0.60379	5000

Drag Coefficient
for Different Shapes

The drag coefficients for various shapes are shown below. For relative drag force, all the objects must have the same frontal area.

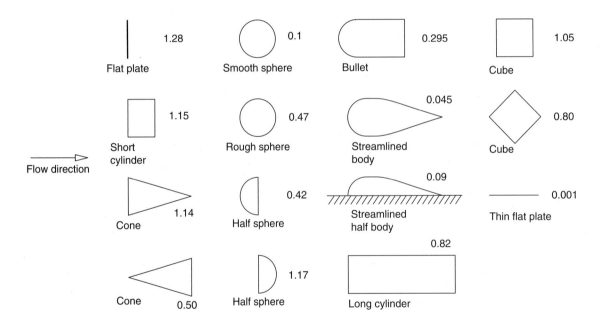

Flat plate	1.28
Smooth sphere	0.1
Bullet	0.295
Cube	1.05
Short cylinder	1.15
Rough sphere	0.47
Streamlined body	0.045
Cube	0.80
Cone	1.14
Half sphere	0.42
Streamlined half body	0.09
Thin flat plate	0.001
Cone	0.50
Half sphere	1.17
Long cylinder	0.82

Flow direction

Installed Power Capacity of the United States and Canada by Spring 2010

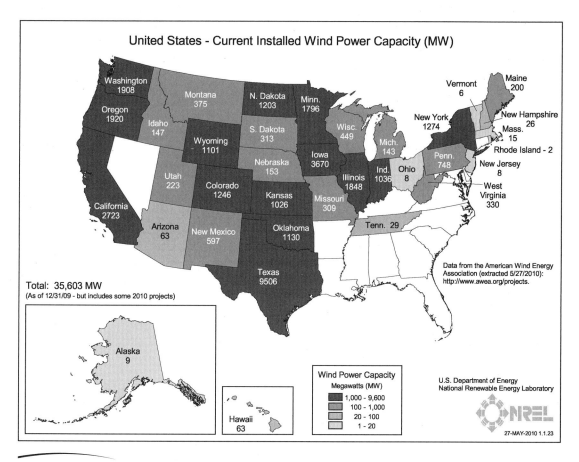

FIGURE C.1 *United States Installed Energy, May 2010*
(National Renewable Energy Laboratory/Department of Energy NREL/DOE)

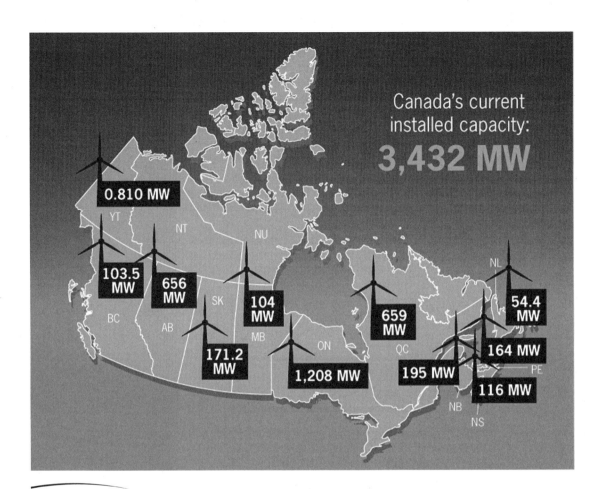

FIGURE C.2 *Canada Installed Energy, May 2010*
(Canadian Wind Energy Association CANWEA)

Glossary

A

abrasion   Wear in machinery components where two matching parts can grind each other.

active yaw   Having a controlled yaw motion with the aid of yaw motors to orient a turbine in the wind direction, as opposed to having a yaw motion regulated by the force from wind.

aerodynamic force   Force exerted by moving air or any gas on an object in the airstream.

airfoil   A profile for the outline of an airplane wing or a turbine blade with the property of having a large lift coefficient and a very small drag coefficient.

alignment   Having the axes (of two shafts, for instance) along a straight line.

alternating current (AC)   A type of electricity in which the flow of electrons is not in the same direction. This flow, called *current*, continuously alternates direction.

alternator   A machine generating alternating current electricity.

amplitude   The instantaneous value of a cyclic variable that changes with time between a maximum and a minim.

anemometer   A device to measure the speed of wind.

angle of attack   The angle an airfoil makes with the direction of airflow.

angular speed   The rotational speed of an object that rotates about an axis, measured in rpm, number of degrees of rotation per second, and in radians per second.

ANSI   American National Standards Institute

arc flash   A flow of electricity outside a conductor (in the air), caused by a short circuit, which generates a high pressure and temperature in the air.

arm   Part of a planetary gear that holds the planet gears together. Also called *carrier*.

armature   The rotating part in a motor, if composed of one or more windings.

ASME   American Society of Mechanical Engineers

asynchronous generator   Another name for induction generator; a generator that is not and does not work based on the same principle as a synchronous generator.

asynchronous machine   A general name covering both an asynchronous motor and generator.

autotransformer   A transformer where the primary and secondary windings are parts of the same winding. The primary and secondary windings are not electrically isolated.

AWEA   American Wind Energy Association

B

backlash   The free motion between two meshing gears due to noncontact free space between their teeth.

balanced load   When the three loads in a three-phase electric system have equal values.

bedplate   A main structure in a nacelle where all the components are mounted or attached.

Betz limit   The maximum value for a wind turbine power coefficient. It is 16/27 or 0.59.

bevel gear   A type where a pair of gears have their axes at 90°.

blade pitch control   The capability of turning a turbine blade about its axis with respect to the hub, thus changing the blade pitch angle.

blade root   The end of a blade where attached to the hub.

blade tip   The free end of a blade that makes the largest circle when moving.

bridge rectifier   A type of full rectifier where two sets of diodes alternatively conduct. The full voltage of the AC electricity is converted to DC.

brush   A part made of carbon in an electric machine that slides on a conductor and allows electricity transfer from a moving component to the stationary parts.

C

cable grip   A metallic device used in turbine climbing that attaches to a cable or rope. It allows sliding along the cable in one direction, but blocks the motion in the reverse direction (preventing fall).

capacitive reactance   The apparent resistance toward electric flow (thus, limiting current) that a capacitor exhibits in an alternating current circuit.

carabiner   A metallic device as part of equipment in climbing gear that acts as a closed hook, allowing safe connection of two rings. For wind turbine climbing only those with a safety lock must be used.

carrier   Part of a planetary gear that holds the planet gears together. Also called *arm*.

cavitation   An interaction between a metallic rotating piece and a liquid, such as in a gearbox or a pump, which causes the metal to be eroded and cavities to form.

centrifugal force   The outward force in a rotating object that tends to break away the object.

characteristic curve   A curve that exhibits some relationship corresponding to the main features of a device, a machine, or equipment, in terms of the variation of a major parameter.

characteristic diagram   An alternative term for characteristic curve.

chord   The distance between the leading edge and the trailing edge in an airfoil.

chord line   A line connecting the leading edge and the trailing edge in an airfoil.

collector   A point where all the turbine outputs in a wind farm are connected to be further connected to a grid by a transformer.

compound interest   A term used in economics implying that the interest on capital is periodically added to it, thus increasing the capital.

cone angle   The angle that the blades of a wind turbine make with a plane perpendicular to the axis of rotation.

corrective maintenance   A maintenance philosophy that a component is fixed after it breaks (if it works, do not fix it), as opposed to preventive maintenance.

cost of capital   The money one has to spend (in terms of percentage) to generate capital for an investment or to pay for a project

crowbar   A set of electrical resistors that come into operation only when a generator is disconnected from a grid and its load has vanished. These resistors form a temporary load for the generator so that the turbine does not overspeed.

CSA   Canadian Standards Association

current   The intensity of the flow of electrons in an electrical circuit, measured in amperes (A, or Amp). See the following:

> **alternating current**   A type of electricity in which the flow of electrons continuously alternates direction.
>
> **direct current**   A type of electricity in which electrons always flow in the same direction.
>
> **line current**   The current in the supply lines of a three-phase electrical system.
>
> **phase current**   The current passing through a load connected between each two lines of a 3-phase system (see also *phase current*.)

current direction   Direction of electron flow. Physically, outside of a battery this direction is from the negative terminal toward the positive terminal. However, conventionally, and in many books, this direction is considered to be from the positive toward the negative terminal.

cut-in speed   Same as Cut-In Wind Speed.

cut-in wind speed   A speed of wind below which a turbine is not designed to generate electricity and is cut off from a grid. A wind turbine stays in free-wheel until the wind speed passes this speed for a certain minimum period of time.

cut-out speed   Same as Cut-Out Wind Speed.

cut-out wind speed   A speed of wind beyond which a turbine is not designed to work, or is not safe to work. A turbine is shut down and disconnected from a grid; the blades turn to feather position and the brake is applied.

D

Darrieus turbine   A vertical-axis wind turbine that looks like an eggbeater.

delta connection   A method of connecting the three wires of three-phase electricity to a load in which the three individual loads form a triangle.

DFIG   Doubly fed induction generator.

diode   A semiconductor with two terminals, normally allowing the current in only one direction.

direct current   A type of electricity in which the flow of electrons is always in the same direction, as opposed to *alternating current*.

direct drive mode   Connecting a turbine directly to a generator, without a gearbox in between.

discount rate   A term used in economics to imply the rate at which one can borrow money (practically it is the same as the interest rate.)

dispatch system   A communication system in which all the participants are informed about the operation. A common example is that used by the city cab companies.

distributed generation system   An electricity generation system in which more than one power plant is involved.

downwind turbine   A turbine in which the rotor is such that the wind hits the tower before reaching the blades.

drive gear   Also called *driving gear*, a gear that transfers its energy to the mating gear.

doubly fed induction generator (DFIG)   An induction generator in which the rotor is made of windings and requires slip-rings to connect to electricity.

drag   The component in the direction of wind of an aerodynamic force.

drag coefficient   The ratio of a drag force to the aerodynamic force causing it. This depends on the profile of an object and not its size. See appendix B for the drag coefficients of some known shapes.

drag-type turbine   A turbine that works based on the drag force on its blade(s).

drive train   A set of gears arranged together to obtain a desired gear ratio.

dynamo   A direct current generator.

E

efficiency   The measure of how much of the energy given to a device or machine is consumed by it in order to function, when modifying a specification of that energy; efficiency is the ratio of the output energy to the input energy.

electric circuit   A setup of electrical components powered by an electric source.

electric current   The intensity of electricity flow, measured in amps.

electric generator   A machine that generates electricity.

electric load   A consumer of electricity, that is, a device that works with electricity, when placed in an electrical circuit. In general, it can be a combination of:

resistive load   A load that exhibits only resistance to the electric current.

capacitive load   A load that consists of capacitors.

inductive load   A load that contains motors and electric windings.

electric source   A battery or electric generator that provides electric energy in a circuit.

electromagnet   A magnet made of a coil (winding) and a core of ferromagnetic material inside it. This magnet can be turned on or off by connecting it to electricity.

electromechanical system   Any device or equipment that has moving mechanical components and works with electricity, such as an electric motor.

energy   Potential to do work.

epicyclic gear   Another name for planetary gear.

EWEA   European Wind Energy Association

F

fall arrest   Stopping from falling down from a height.

fatigue   Undergoing repeated alternative tension and compression stress.

fatigue failure   A failure due to fatigue.

fatigue stress   A stress that repeatedly alternates between positive and negative values.

feathered   A position of wind turbine blades (with respect to wind direction) with the smallest lift force and largest drag force, or the smallest aerodynamic force.

ferrous   Of or related to the iron family of metals.

fixed speed mode   A way of operating a wind turbine where the turbine speed must be kept constant.

flashing   A repeated reflection of sunshine on a wind turbine blade that can annoy a viewer.

flicker   A very short-time voltage variation in an electric power line due to connection or disconnection of a large load.

flickering   The repeated shadowing effect of a rotating blade (see also *shadow flicker*).

flutter   A vibration of a wind turbine blade about its own axis.

foundation   A massive block of concrete at the bottom of a turbine, covered by soil, containing the bolts that hold the tower.

free wheel   A standby state of a wind turbine when insufficient wind is available to turn the turbine.

frequency   The number of repetitions, in one second, of a cyclic phenomenon.

frequency converter   A device that takes AC electricity with a given frequency as input and delivers AC electricity with a (different) desired frequency.

full-wave rectifier   A type of AC-to-DC rectifier with two rectifier diodes that alternatively conduct. Each diode converts one cycle of AC to DC. Thus, all of the waveform is converted to DC.

fuse   A protection device in any electrical circuit that melts due to excess heat if an overcurrent occurs, thus opening the circuit.

future value   A term in economics representing the value of an amount of money at a given future time.

G

gearbox   A mechanical device, consisting of various gears in an enclosure, to convert the rotational speed of an input shaft to a different speed on the output shaft.

gear (mechanical)   A single part of a gear system or a gearbox component.

   arm   Part of a planetary gear that holds the planet gears together.

   bevel gear   A type for gears where the pairing gears have their axes at 90° angles from each other.

   driven gear   A gear that receives energy from its mating gear.

   driving gear   Drive gear; a gear that gives energy to its mating gear.

   epicyclic gear   Planetary gear.

   helical gear   A type of gear in which the teeth are angled with respect to the shaft axis in order to reduce the *backlash* compared to spur gears.

   idler gear   A gear standing between two main gears whose role is to change direction of rotation as well as filling the gap.

   internal gear   A gear in the form of a ring with its teeth on the inside rather than outside.

   planetary gear   See *planetary gear.*

   pinion   The smaller gear in a pair of gears or the output gear in a drive train.

   ring gear   The outermost gear with internal teeth in a planetary gear.

   spur gear   A gear in which the teeth are parallel to the axis of rotation.

   sun gear   The innermost gear in a planetary gear around which the planet gears rotate.

   worm gear   A type of gear set with a high gear ratio used only to reduce speed. The axes are at 90°. A worm gear set has the property that it is impossible for the driven gear to force the drive gear to rotate; thus, if the drive gear stops it also brakes the driven gear.

gear ratio   The ratio between the output and input rotational speeds of a pair or set of gears, or in a gearbox.

gear train   A set of gears arranged together to obtain a desired gear ratio.

generator   A machine that produces electricity, usually when turned; it converts mechanical energy to electrical energy.

   armature   The windings on the rotating part in a generator or motor. The current-carrying windings on a moving part.

   brush   Electrical connector made of carbon or a carbon composite in a motor or generator when electrical connection between the rotating and stationary parts is essential. A brush is stationary and slides on a metallic rotating ring.

   rotor   The rotating part in a generator or motor.

   stator   The stationary part in a generator or motor.

grid   An electric network.

H

half-wave rectifier   The simplest type of AC-to-DC rectifier, which consists of inserting one diode in the AC circuit.

harmonics   AC signals with twice, three times, four times, and so on the frequency that can become mixed with the base AC (with fundamental frequency) electricity in a circuit.

harness (1)   Part of the *personal protective equipment* that must be worn to climb a turbine.

harness (2)   Catching the wind energy by a turbine

HAWT   Horizontal-axis wind turbine.

helical gear   A type of gear in which the teeth are angled with respect to the shaft axis in order to reduce the *backlash* compared to spur gears.

hertz (Hz)   The unit for measuring frequency.

high-speed shaft   The output shaft in a wind turbine gearbox that must be aligned and coupled with the generator shaft.

high-voltage direct current (HVDC)   A method for transmission of electricity that uses DC at very high voltage (e.g., 300 kV) instead of AC.

horizontal-axis   Having its axis of rotation almost horizontal.

horizontal-axis wind turbine (HAWT)   A wind turbine whose axis of rotation is horizontal or nearly horizontal.

hub   The part of a propeller-type wind turbine to which the blades are connected.

HVDC   High-voltage direct current.

I

ideal transformer   An assumed transformer with zero loss or 100% efficiency.

IEEE   Institute of Electrical and Electronics Engineers

induction generator   An alternating current generator in which the rotor current is generated by induction rather than connection to electricity, as is the case with a synchronous generator.

induction machine   A term that embraces both an induction generator and an induction motor.

inductive reactance   The apparent resistance toward electric flow (thus, limiting current) that an inductor exhibits in an alternating current circuit.

inertial force   The force required to accelerate or decelerate an object.

infrasound   A sound with a frequency below the threshold of human hearing so that it cannot be heard by a human.

initial cost   The money necessary to spend at the beginning of a project to pay for the equipment, transportation, and installation.

input   Something given to a machine or a system, in the form of energy, speed, money, and so on.

internal gear   A gear in the form of a ring with its teeth on the inside rather than the outside.

inverter   An electronic device to generate AC electricity from a DC source.

islanding   A state of isolation of part of an AC network from the rest, thus resulting in the possibility of frequency and voltage drift while operating.

ISO   International Standards Organization

isolation transformer   A transformer whose purpose is to electrically isolate (separate) a circuit or device from the main circuit.

K

kilowatt   A thousand watts, a unit for measuring power, mostly used in electric devices and circuits.

kilowatt-hour   The most common unit for measuring electrical energy consumption. It can also be used for measuring work and other types of energy.

kinetic energy   A type of mechanical energy associated with a moving object; its magnitude depends on the mass and speed of the object.

L

lagging   A term concerning the *power factor* in an AC circuit, when the current is behind the voltage (voltage reaches its maximum value in its cycle before the current reaches its maximum).

lanyard   Part of the *personal protective equipment* required to climb a turbine. A lanyard is used to hook a person to a safe point for fall arrest. It contains a spring-like piece that reduces the impact of a sudden stop on a person's body in case of falling.

lattice tower   A tower made from many smaller pieces that are welded or fastened together like a truss. Most of the older turbines have lattice towers.

leading   A term concerning the *power factor* in an AC circuit, when the current is ahead of the voltage; thus, the circuit has a *leading power factor* as opposed to the *lagging power factor* (see also *lagging*).

lift   A force perpendicular to the airflow direction caused by the difference in pressure on the two sides of a thin object (or an airfoil) in a wind stream.

lift coefficient   The ratio of lift force to the aerodynamic force causing it. This depends on the profile of an object and not its size.

lift force   The component of an aerodynamic force perpendicular to the direction of wind.

lift-type turbine   A wind turbine whose power grasp from wind is based on the lift force on the blades, as contrary to a drag-type turbine.

lightning bolt   Very high voltage and bright arc of a lightning.

line current   The current in the supply lines of a three-phase electrical system.

line voltage   The voltage between the supply lines of a three-phase system.

load (electrical)   Inside electrical circuits, any device that works with electricity is a load when consuming electricity.

lockout   The action of locking a dangerous area or device to restrict the access.

lockout and tagout   The action of locking a dangerous area or device to restrict the access and putting a note to warn people and inform them whom to contact.

look-up table   A table of calculated values of a variable that is used to determine a numerical value without doing the calculations. Using such a table can be much faster.

Lorentz force   Force exerted on a wire carrying a current when placed inside a magnetic field.

M

magnetic field   The area between two magnetic poles where a magnetic force exists. A magnetic field has a direction; outside a magnet, the direction of the magnetic field is from the north pole to the south pole.

main shaft   The low-speed shaft in a wind turbine gearbox, which is connected to the rotor.

maintenance   Keeping an operation, a machine, or equipment in operating condition in one of the following forms:

> **corrective maintenance**   The repair work and maintaining the normal condition after a fault has happened, that is, when a part has broken or failed.

> **preventive maintenance**   The systematic repair and maintenance work to prevent a fault or failure before it happens.

> **scheduled maintenance**   Regularly planned repair and maintenance work as part of preventive maintenance.

mass center   The point in an object where the whole mass can be assumed to be concentrated.

megawatt   A unit larger than a *watt* often used to measure electric power, equal to 1 million watts.

megawatt-hour   A unit larger than a *watt-second* used to measure work and energy. A watt-second is one joule, and a megawatt-hour is 3,600,000,000 joules.

met tower   An abbreviation for meteorological tower.

meteorological tower   A tower with instruments to study wind and weather.

misalignment   The state of not being aligned for two matching shafts or gears that must be aligned (i.e., their axes make a straight line).

motor   A machine that converts electrical energy to mechanical energy.

> **rotor**   The rotating part.

> **stator**   The stationary part.

N

nacelle   A room at the top of a turbine tower that houses the gearbox, the generator, and other equipment.

natural frequency   The frequency (or frequencies in a more complex system) inherent to any flexible object, part of a machine, or structure that can vibrate in any way. If the object receives a shock or an impulsive force, it vibrates with a frequency equal to its natural frequency.

net present value   A term used in economics to represent all the revenues minus the money invested and all the costs during the life of a project, converted to today's money. Today's money has more value than future money. The corresponding numbers, thus, are smaller than if expressed in future money.

O

offshore   On a lake, sea, or ocean, away from land.

onshore   On the land.

operating cost   Continuous expenses of running a business, manufacturing a product, or producing goods, such as salaries, rent for property, purchase of raw material, maintenance cost, insurance, and so on. This is as opposed to a one-time cost for purchase of land, machinery, patents, and the like.

operating speed   The speed at which a machine normally works.

OSHA   Occupational Safety and Health Authority

output   The product or result of processing of a machine, system, and so on. The output of a gearbox, for example, is a rotational motion with a required speed and torque.

overspeed   Speeding up and reaching speeds that are beyond the designed capacity.

P

pad mount (transformer)   A transformer suitable to be mounted on a flat slab.

pad transformer   A pad-mount (or pad mounted) transformer

peak hours   The (normally few) hours during a 24-hr period when the electricity consumption is the highest.

performance curve   A curve that exhibits some relationship regarding the performance of a device, a machine, or equipment, in terms of the variation of a major parameter. *Performance curve* and *characteristic curve* are often interchangeable terms.

permanent magnet   A magnet that has continuous and constant magnetism, as opposed to an electromagnet that can be turned on and off.

personal protective equipment (PPE)   Equipment for protection of a person in a hazardous working environment. PPE is not the same for all situations and depends on a job and its environments.

phase angle (in electricity)   The angle between the expressions for voltage waveform and the current waveform in alternating current electricity. Physically, it translates in a time delay between the two waveforms in an electrical circuit.

phase current   The current passing through a load connected between each of two lines of a three-phase system. The relationship between *phase current* and *line current* depends on the type of connection (delta or wye) used.

phase voltage   The voltage across each individual load connected to a three-phase system. The relationship between *phase voltage* and *line voltage* depends on the type of connection (delta or wye) used.

pinion   The smaller gear in a pair of gears or the output gear in a drive train.

pitch   A term used in machinery for various purposes. For application in wind turbines and gearboxes see the following lines.

pitch angle   The angle of a blade from a reference in rotation about its axis. Modifying the pitch angle of a turbine blade alters the angle of attack and, thus, the wind power capability of the blade.

pitch circle   An imaginary circle in gears representing a gear size. If two meshing gears are replaced by two rubber cylinders that roll on each other and transfer the motion with the same gear ratio, the circle representing each cylinder is the *pitch circle* of that gear. For two meshing gears the pitch circles are tangent to each other.

pitch control   The action of controlling the pitch angle of the blades in a wind turbine in order to modify the turbine performance.

pitch diameter   The diameter of the *pitch circle*.

planet gear   Any of the gears in a planetary gear that stand between the sun gear and the ring gear.

planetary gear   A set of gears arranged in a special form, consisting of a sun gear in the middle, a number of (usually three) planet gears engaging with the sun gear and the outer ring gear with internal teeth. The planet gears are mounted on a bracket called an arm that can turn independently. A planetary gear can accept two input speeds. A planetary gear is also called an epicyclic gear.

pole mount (transformer)   A transformer that can be installed at the top of an electricity distribution post.

power   The amount of *work* in 1 sec by a machine, a motor, an engine, and so on. In AC electrical circuits, the following terms correspond to *power*:

active power   The net power that converts to heat or work.

reactive power   The amount of power that is stored in a capacitor or an inductor in each half cycle and then is given back to the circuit in the next half cycle. This power is used for magnetization in motors, for instance.

apparent power   The power that a generator must supply to cover for the consumed active power and reactive power in a circuit; but that is not equal to their algebraic sum, due to the phase difference.

power coefficient   A number always less than 0.59 that determines what percent of the power in the wind stream can be harnessed by a wind turbine. This number corresponds to the aerodynamics of turbine blades and does not include gearbox and generator efficiency.

power curve   A curve for the operation of a wind turbine, indicating the expected power from the turbine versus wind speed.

power factor   The ratio of active power to apparent power. This ratio is smaller than unity.

power factor correction   Improving the power factor of alternating current circuits by introducing a bank of capacitors.

power plant   An industrial unit for generation of electric power.

power quality   The degree of accordance of electric power parameters with an expected requirement; for example, the stability of frequency at 50 or 60 Hz.

PPE   Personal protection equipment.

present value of money   The value of an amount of money, expressed for a future time (say in 10 yr from now), in terms of today's money; i.e., a smaller amount obtained by subtracting the interests for the period.

prevailing wind   A most-often-blowing wind.

preventive maintenance   See *maintenance*.

prime mover   A source of mechanical power that turns an electric generator.

propeller turbine   A turbine with blades similar to a propeller; the most common wind turbine.

R

rectifier   An electronic device that converts AC electricity to DC.

renewable energy (RE)   Energy from natural sources that do not pollute the environment.

reluctance force   A force by an electromagnet trying to shorten the path of a magnetic field.

ring gear   A gear in the form of a ring with inside teeth that engages with planet gears in a planetary gear.

ripple   The (undesirable) rapid fluctuations of voltage about the nominal value in a DC voltage

root of a blade   The end of a blade where it attaches to a hub.

rotating magnetic field   A magnetic field that rotates about an axis. This exists inside the stator of AC motors and constitutes the basis of their operation.

rotor   The rotating part of electric machinery.

Rotor (in a wind turbine)   The rotating part of a turbine comprising the hub and the blades.

S

savonius rotor   A type of drag-type wind turbine that basically consists of two half cylinders put together in the form of a letter S.

scheduled maintenance   See *maintenance*.

scuffing   Transfer of metal particles by tearing from a tooth and adhering to another tooth by welding (due to heat and pressure).

shadow flicker   The moving shadow cast of a rotating blade.

shear stress   A type of stress due to a cutting force. The force from a pair of scissors causes shear stress in the object to be cut.

sine wave   A wave form whose variation is according to a sinusoidal function.

single-phase   The simplest form of AC electricity, transmitted by two wires, as opposed to three-phase, which is transmitted by three or four wires.

sinusoidal wave   Same as *sine wave*.

skew wind   Wind whose direction is very off from horizontal.

slip   The act of not rotating at synchronous speed in induction motors and generators.

slip-ring   A metallic ring on the rotor of certain types of AC electric machines where brushes slip on the slip-rings to transfer electric current.

slip speed   The difference between the actual rotating speed of an induction motor or generator and the corresponding synchronous speed. Slip speed is not constant and can change based on the load on the machine shaft.

smart grid   A grid equipped with added intelligent capabilities (measurement devices, control devices, switching devices, and so on).

solidity   The measure of percentage of solid area in a circle traced by a turbine rotor; for example, a four-blade turbine has more solidity than a three-blade turbine of the same size.

spar platform   A platform in the form of a vertical cylinder at the top of which a wind turbine can be installed. The cylinder is hollow, but partly filled with sand and stone to have a heavy weight. It must stay afloat and remain vertical. Some cables limit its horizontal displacement on the water.

spur gear   A gear in which the teeth are parallel to the axis of rotation.

square-wave   A waveform each cycle of which consists of two equal rectangles, one on the positive side followed by one on the negative side.

squirrel-cage motor   A type of induction motor without windings on the rotor. Instead of windings, the rotor consists of thick copper (or aluminum) bars connected to each other at both ends by two rings, such that these metallic bars form a cage-like structure. The rotor is not connected to electricity.

stall   The effect of a decrease in the lift force with an increase in the angle of attack in airplane wings and wind turbine blades. This happens at some higher values of the angle of attack. Before stall takes place, the lift force increases with an increase in the angle of attack.

stall control   The effect of decrease in the wind power capture at higher wind speeds (due to stall) is a regulating mechanism to prevent them from going into overspeed or overload. A turbine using this feature does not use blade pitch control, and is called a stall-controlled turbine.

star connection   A method of connecting wires in a three-phase load, where the three connected wires form a letter Y, as opposed to forming a Δ.

stator   The stationary windings of an electric machine.

step-down transformer   A transformer used to decrease voltage.

step-up transformer   A transformer used to increase voltage.

substation   An electricity utility consisting of a number of transformers and other equipment to increase or decrease and regulate the voltage of incoming three-phase electricity.

sun gear   The middle gear in a planetary gear around which the planet gears move.

synchronous generator   An AC generator whose rotor carries one or more magnets. Electricity is generated in the stator windings when the rotor rotates. The rotor must spin at a constant speed for the generated electricity to have a fixed frequency. The rotor magnets can be permanent or generated by DC current through rotor windings.

synchronous machine   A synchronous motor or generator.

synchronous motor   An AC motor that works based on one or more magnets on a rotor inside a *rotating magnetic field*. The magnetic forces cause the rotor to follow the rotating field. A synchronous motor runs at a constant speed. It needs additional means for starting.

synchronous speed   The constant speed of a synchronous machine. This speed depends on the frequency of the supply electricity.

T

tagout   The action of putting a note on a device or to the entrance of a dangerous location to warn people.

tension leg platform (TLP)   A floating platform that is kept in place by cables attached to the seabed. The cables are under tension in order to prevent the platform from moving with water.

three-phase system   Three-phase electricity; the sort of AC electricity that requires at least three wires for transmission. The voltage between each pair of wires has a sinusoidal form and has the same frequency and maximum amplitude, but there is a fixed delay between the instants when each of the three voltages reaches its maximum value.

thyristor   A type of transistor, used in controlled rectifiers and inverters.

tip speed   The speed of the tip of a blade in a wind turbine.

tip speed ratio   The ratio of the speed of the tip of a turbine blade to the wind speed.

TLP   Tension leg platform.

torque   The turning effort effect of a force about a point away from the line of action of the force, also called the *moment* of the force about that point.

tower   A necessary part of a propeller-type wind turbine whose function is to hold, and support the forces from, the rotor and nacelle.

lattice tower   See *lattice tower*.

tubular tower   A tower that has a circular structure in the form of a cylinder or a conic shape with a slight slant.

transformer   An electrical device to increase or decrease the voltage in AC electricity. In the simplest form it has one primary winding where the supply electricity is connected, and one secondary winding where the output is taken from.

autotransformer   See *autotransformer*.

ideal transformer   A transformer for which all the losses are assumed to be zero and, thus, the efficiency is 100%.

isolation transformer   A transformer the main purpose of which is not to change voltage, but to electrically isolate (separate) a device from the supply circuit, or isolate part of a circuit from the rest.

pad mount transformer   A transformer that can be installed on a flat base.

pole mount transformer   A transformer, normally with a round structure, more suitable for mounting on a distribution pole.

step-up transformer   A transformer used to increase voltage.

step-down transformer   A transformer used to decrease voltage.

tubular tower   A wind turbine tower in the form of a cylinder or a conic section with a slight slant.

turbine   A converter of energy of a different form to rotational mechanical energy. Examples include a gas turbine, a steam turbine, a hydro turbine, and a wind turbine.

Darreius turbine   A vertical-axis wind turbine in the form of an eggbeater.

propeller turbine   A horizontal-axis wind turbine with (mostly, but not necessarily three) blades in the form of an airplane propeller.

Savonius rotor   A wind turbine made by installing two half-cylindrical sections on a shaft, in reverse to each other to form a letter S, and capture the wind energy.

turns ratio   The ratio between the number of turns in the primary and secondary windings in a transformer.

U

universal motor   A type of motor that can work with both DC electricity and single-phase AC electricity

upwind turbine   A wind turbine installed on its tower in such a way that the blades are in front of the tower and wind reaches the blades before reaching the tower, as opposed to the downwind turbine.

V

variable slip mode   A way of operation control of wind turbines with an induction generator for connecting to a grid. In this mode, a wound-rotor induction generator is to be used and some power is extracted from the rotor winding. The turbine must have an almost constant rpm.

variable speed mode   A way of operation control of a wind turbine for connection to a grid. In this mode, a synchronous generator is used, but it is not necessary to run the turbine with a constant rpm, as opposed to *fixed speed mode*.

VAWT   Vertical-axis wind turbine.

vector   A quantity that has both magnitude and direction, such as a force, as opposed to a scalar value that has only magnitude, such as temperature.

vertical axis   Having a vertical axis of rotation.

vertical-axis wind turbine (VAWT)   A turbine whose axis of rotation is vertical. Not all turbines can have a vertical axis. A vertical-axis turbine is not sensitive to wind direction changes.

vibration   Physical oscillations in a mechanical system. A continuous and cyclic variation around a nominal value in a system.

   amplitude   The amount of displacement in a vibrating component at each instant of time. The quantity that varies between a maximum and a minimum in a vibratory system.

   frequency   The number of oscillations (back and forth cycles) in 1 sec.

   period   The duration of completing one cycle in a vibrating action.

volt   The unit to measure voltage, the electrical potential difference.

volt-ampere (VA)   A unit of measurement of electric power in AC electricity. It is normally used to express *apparent power*. It is particularly used when voltage and current are out of phase.

volt-amperes-reactive (VAR)   A unit for measuring *reactive power* in AC circuits. Reactive power exists in a circuit containing inductors (coils, magnetism, and motors) and capacitors.

voltage   Electrical potential difference or electromotive force (the active level for electric charge, forcing electrons to flow when a path exists).

W

watt   The unit of measurement of power; it equals 1 joule per second.

watt-hour   A unit for measuring energy and work. 1 W-hr is 3600 watt-seconds, or 3600 joules. It is 0.001 of a kW-hr.

wind data   The data containing the statistics of wind (speed, frequency, duration, direction, etc.) in a region.

wind farm   A region where a number of wind turbines are installed for generating electric power. Also called a wind park.

wind harnessing   Capturing of wind energy by wind turbines.

wind speed   The speed at which wind blows at each instant.

wind turbine power curve   A curve specific to a wind turbine that determines the output power versus the wind speed. This curve is used by the wind turbine controller during operation to adjust the blade pitch.

wind vane   An instrument to measure the wind direction.

work   The resultant of spending energy. Work can be mechanical or electrical, or it can be converted to heat. Mechanical work is more tangible; when a weight is lifted, *work* is performed. Lifting a weight of 1 lb by 1 ft results in 1 ft-lb of work.

worm gear   A pair of gears with a high gear ratio, and used only to reduce speed. The axes are at 90°. A worm gear has the property that it is impossible for the driven gear to force the drive gear to rotate; thus, if the drive gear stops it also locks the driven gear.

wound rotor induction machine (WRIM)   A type of induction motor or generator whose rotor contains windings, as opposed to a squirrel-cage machine that has no windings in the rotor.

wound rotor induction motor (WRIM)   An induction motor (and not generator) with a wound rotor.

WRIM   This abbreviation is used for both a *wound rotor induction machine* and a *wound rotor induction motor*.

wye connection   Also called *star connection*. A way of connecting the terminals of a three-phase load to the supply wires. A three-phase load has six terminals. If three of them are connected together and the other three are connected to the supply voltage, this forms a Y connection. The order of which terminals to choose is important.

Y

yaw   The action of orienting a wind turbine to the direction of the wind.

yaw gear   The gear system to rotate the turbine (the nacelle and rotor) for a yaw motion.

yaw system   The entire gears and motors involved in the yaw motion of a wind turbine.

Index